A Thermodynamic Introduction to Transport Phenomena

Henning Struchtrup

A Thermodynamic Introduction to Transport Phenomena

Henning Struchtrup
Faculty of Engineering
University of Victoria
Victoria, BC, Canada

ISBN 978-3-031-61870-3 ISBN 978-3-031-61868-0 (eBook)
https://doi.org/10.1007/978-3-031-61868-0

This Springer imprint is published by the registered company Springer Nature Switzerland AG
The registered company address is: Gewerbestrasse 11, 6330 Cham, Switzerland

Preface

In engineering programs, courses on *Thermodynamics*, *Fluid Mechanics*, and *Heat and Mass Transfer* are the pillars of *Thermofluids* education. Typically these courses are taught side by side, and at best it is made vaguely clear that the pillars not only stand on the same ground, but are as well connected in many ways.

Studying transport processes in *Thermofluids* from a unified viewpoint, with a clear mathematical language, enhances each sub-field, by deepening insight, and opening possibilities for exploration.

This book provides material for a one semester course on *Transport Phenomena* in which the common ground and the connections are thoroughly explored. The target audience is senior undergraduate and incoming graduate students in engineering and applied sciences, and, of course, their instructors.

The emphasis of the presentation lies on principles, methods, and techniques, and not on the deep exploration of particular transport processes or systems. The development of the transport equations—global and local balance laws for mass, momentum, energy and entropy—is universally valid. However, the study of transport processes through solutions of the equations considers mostly simple materials, i.e., ideal gases and incompressible liquids, in simple geometries to allow for (mostly) analytical solutions.

This approach emphasizes the general understanding of *Transport Phenomena*, visualizes the interplay between the different branches of *Thermofluids*, and thus will enhance the understanding of each field, as well as their interconnections. More technical detail such as advanced computational methods and extraction of property data from tables or databases are left for further studies in specialized courses or books.

In other words, this book is intended, indeed, as an *Introduction*, serving as an access ramp into this exciting and wide field.

The line of arguments requires some previous knowledge and understanding of *Thermofluids*, in particular of *Thermodynamics*, which appears in the title for good reason. Originally, the material was developed as additional chapters for my textbook on *Thermodynamics and Energy Conversion* (2nd

Edition, Springer 2024). In order to provide a self-consistent presentation of *Transport Phenomena*, some relevant material from that book is repeated.

The study of *Thermodynamic Transport Phenomena* is a technical subject, focussing on transport equations and their solutions. About 140 end-of-chapter problems of varied length and difficulty are included to teach the required technical skills while giving further insight into the multitude of *Transport Phenomena*. Engagement with the problems is an integral and important part of learning. For a successful learning experience, the problems contain hints and intermediate results to guide towards complete solutions.

The material was developed and tuned over several years of teaching *Transport Phenomena* to groups of undergraduate and graduate students at the University of Victoria. I am thankful for their feedback, corrections, and constructive criticism, which led to the balance of theoretical background and interesting applications presented.

My personal path in *Thermodynamics* and *Thermofluids* started at the Technical University of Berlin, Germany, as student of Prof. Ingo Müller who taught me maybe not all, but most of what I needed to know for my career as scientist and teacher. Ever since, the material in this book, here presented for beginners, was relevant for my research, which informs at least some of the choice of subjects included.

It is my sincere hope all readers will find this book accessible, illuminating, and useful.

Victoria, BC
Spring 2024

Henning Struchtrup
(struchtr@uvic.ca).

Contents

Chapter 1
Introduction and Background

A complete understanding of the behavior of thermofluid systems requires full resolution of the local state. The development and understanding of the equations governing the space-time evolution of processes is the main aim of this book. Classical engineering thermodynamics is mostly concerned with the study of global thermodynamic systems which exchange mass, heat and work with their surroundings, while the interior state of the systems studied is not resolved. This introductory chapter highlights the differences between local and global descriptions. The transport equations for systems and thermodynamic property relations are recalled—they form the relevant background for the proceedings.

1.1 Introduction

Transport Phenomena provides the common foundation for the fields known as *Thermodynamics*, *Fluid Mechanics*, and *Heat and Mass Transfer*, sometimes summarized as *Thermofluids*.

The fields of *Thermofluids* can be differentiated as follows:

Most of *Thermodynamics* concentrates on system behavior, while the locally resolved description provided by the Navier-Stokes equations of *Fluid Mechanics* is mentioned only in passing.

Fluid Mechanics explores flow behavior in great detail for cases where temperature effects play no, or only a marginal, role. Nevertheless, both fields overlap, using different language to describe the same processes or systems. For example, pumps and turbines are evaluated in *Fluid Mechanics* and *Thermodynamics*, where one discusses device performance either through "hydraulic head loss", or "entropy generation rate" which leads to the concept of "isentropic efficiencies".

Heat and Mass Transfer relies on *Fluid Mechanics* in particular to describe convective transport, but, despite the importance of heat transfer in

H. Struchtrup, *A Thermodynamic Introduction to Transport Phenomena*,
https://doi.org/10.1007/978-3-031-61868-0_1

thermodynamics systems, typically does not study irreversibility and entropy generation, which is all important in *Thermodynamics.*

The approach to *Transport Phenomena* presented in this text is from the viewpoint of *Thermodynamics.*

A thermodynamic system is characterized by its global properties, such as total mass, energy, entropy within the system boundaries. Processes in the system occur due to interaction between the system and its surroundings through transfer of mass ($\Rightarrow$ mass flow) and energy ($\Rightarrow$ heat and work) across system boundaries, and the generation of entropy through irreversible processes within the system.

The conservation laws for mass and energy, and the balance law for entropy—with non-negative production—provide the mathematical description. Apart from the study of, e.g., aircraft propulsion, the balance of momentum typically is not considered, it would yield the forces required to keep the system in place. At the global level of description, details of the state inside the system cannot be determined, unless the system state is homogeneous, which is the case for closed systems undergoing reversible processes, where the systems are in equilibrium states at all times.

Only the look inside, at the detailed state of a system in each point in space, at all times, allows us to get a complete picture of the local and global system behavior. Just as for the system level, this requires a proper set of equations for the mathematical description on the local level. Typically, these are partial differential equations—local balance laws—which describe the change of local thermodynamic properties due to exchange of mass, momentum and energy with the neighboring points.

On the level of classical *Fluid Mechanics* and *Heat and Mass Transfer*, these equations are continuity and mass diffusion equations, Navier-Stokes equations for momentum, and the heat equation for energy.

This book aims at providing a sound foundation for all relevant transport equations in the broader area of *Thermofluids.* In particular, we shall study

- transport equations in local and global (= system) description,
- interface conditions for, e.g., wall-liquid interaction, phase boundaries, shocks in gases, flames, semi-permeable membranes.

The required equations depend on the material studied, as well as on the desired level of description. Accordingly, we will aim for a general framework, where all required assumptions, and hence limitations of applicability, of a set of transport equations are clearly stated.

In particular, we highlight the importance of the second law of thermodynamics for the development of successful stable equations for thermofluidic systems. *Transport Phenomena* is the study of equilibration processes: A system that is not in thermodynamic equilibrium will undergo changes—transport processes—to reach a suitable equilibrium state. The second law of thermodynamics is the mathematical formulation of this behavior.

Thermodynamic equilibrium has many facets:

In thermal equilibrium temperature is uniform. A deviation from thermal equilibrium—i.e., non-uniform temperature—causes heat transfer from warm to cold until equilibrium is reached.

In mechanical equilibrium forces are balanced. Deviation from mechanical equilibrium, such as non-uniform pressures in the atmosphere or in pipelines, induces mass transfer by winds or fluid flow, that aim to level the deviations.

Chemical reactions and diffusion processes occur to bring a mixture towards chemical equilibrium, in which a mixture will not further change its composition.

Local transport equations describe these equilibration processes, e.g., the law of Navier-Stokes for viscous stresses, Fourier's law for heat conduction, and Fick's law for diffusion. Their transport coefficients for mass, momentum, and energy—diffusion coefficients, viscosity, and heat conductivity—describe irreversible processes, and are linked to entropy generation.

The systematic development of the transport equations based on the second law will highlight their common ground.

Apart from helping with developing models, the second law is important in itself. In engineering systems entropy generation implies work loss to irreversibilities. On the system level, the prediction of these losses is often not possible, and one has to use, e.g., isentropic efficiencies, to model losses in a steam or gas turbine. The local description provides a detailed expression for entropy generation through local properties such as temperature, mass fraction, and velocity, and their gradients. With this, entropy generation and work losses can be fully quantified and predicted.

The material presented here embeds the equations of *Fluid Mechanics* and *Heat and Mass Transfer* into the realm of *Thermodynamics*, with emphasis on the general framework of transport phenomena. Only few simple but insightful solutions of the transport equations are discussed.

Topic courses on *Fluid Mechanics* and *Heat and Mass Transfer* consider applications of the material in great detail, but often ignore the deep interconnection between their respective field and *Thermodynamics*, in particular the discussion of entropy generation and the corresponding work losses.

All through the following, it is assumed that the reader has sufficient familiarity with thermodynamics of systems, which provides background and context for the study of transport phenomena. The required knowledge on thermodynamics will be presented as facts, for detailed derivations and discussions the reader is referred to author's textbook[1] *Thermodynamics and Energy Conversion* (2nd Edition, Springer 2024), or any other presentation of thermodynamics.

As becomes clear from the above, *Transport Phenomena* are described and studied through equations. To proceed, we require clear mathematical notation to account for vectors and tensors, the operations between them, and their derivatives in space and time.

[1] We shall refer to chapters or sections of this book, as e.g., [TEC, Chap. 4].

Appendix A offers a compact summary of symbolic and index notation for tensors and their operations; it should be studied before continuing with the main text.

1.2 Chapter Summaries

The line of arguments and the topics studied in the book is best visible from the chapter summaries. All chapters and the appendix are accompanied by end-of-chapter problems—about 140 in total—of varied difficulty and length.

Chapter 1: Background

The transport equations for systems and thermodynamic property relations are recalled—they form the relevant background for the proceedings.

Chapter 2: An Inert Mixture in Equilibrium

The study of transport phenomena focusses on nonequilibrium states and processes. A clear idea of nonequilibrium requires a proper understanding of what states a system will assume in equilibrium. In this chapter we study equilibrium states for a non-reacting mixture in the gravitational field, and the requirements for its stability. The second law is evaluated by maximizing entropy, and using methods of variational calculus. The importance of the chemical potential in the thermodynamic description of processes involving mixtures is highlighted.

Chapter 3: Balance Laws

Balance laws for mass, momentum and energy form the base of all descriptions of transport phenomena in the realm of thermofluids theory and applications. Depending on the desired level of description, they assume different forms. In this chapter, we first formulate the general balance laws for arbitrary volumes, and the local balances for regular points (i.e., excluding discontinuities), and singular surfaces (i.e., across discontinuities). The general balance laws are then specified for total and partial mass, momentum, and energy. Along the way we discuss convective and non-convective fluxes, as well as supply and production terms.

Chapter 4: Linear Irreversible Thermodynamics

The conservation laws for (partial) mass, momentum and (total, internal, kinetic) energy contain non-convective fluxes as well as chemical reaction rates. These quantities depend on the material and the nonequilibrium flow state of the system through constitutive equations that must be specified based on rational arguments. In this chapter we employ the assumption of Local Thermodynamic Equilibrium (LTE) to construct the second law of thermodynamics—the balance law for entropy—accompanying the conservation laws. The constitutive equations are obtained from the requirement that the local entropy generation rate must be non-negative, and vanishes in equilibrium.

Chapter 5: Navier-Stokes-Fourier Models

For simple substances, i.e., substances where no change of composition occurs, the transport equations of the previous chapter reduce to the Navier-Stokes-Fourier (NSF) equations of classical hydrodynamics. This chapter presents the NSF equations for compressible flows and their reduced form for incompressible flows. The Boussinesq approximation for quasi-incompressible liquids with thermal expansion is discussed in some detail. The equations are solved for Couette, Poiseuille and Boussinesq flows. The chapter closes with a discussion of co- and counterflow heat exchangers.

Chapter 6: Linear Problems and Waves

Solutions of transport equations in simple settings are invaluable for the deeper understanding of the various transport phenomena. In this chapter we linearize the transport equations for small deviations from an equilibrium ground state, and study one-dimensional transport behavior. We focus on wave solutions, and study dispersion and damping for wave and diffusion equations. The Cattaneo equation is derived and discussed as an example of non-local transport, where the system is not in local equilibrium states.

Chapter 7: Boundaries and Interfaces

So far we have studied the partial differential equations that describe the behavior of fluids in bulk regions without discontinuities. These bulk regions are connected by sharp transitions—described as singular surfaces—such as wall-fluid boundaries, or phase interfaces. In this chapter we develop boundary and interface conditions, where once more the second law serves as the guiding principle. Topics discussed include velocity slip and temperature jump at wall-fluid boundaries, as well as resistivities in evaporation and condensation processes.

Chapter 8: Shocks and Flames

Shocks and flames are irreversible changes of flow properties in inert or reacting gas flows over short distance, which are described as singular surfaces. In this chapter, after reviewing sub- and supersonic nozzle flows, we study both phenomena for the accessible case of ideal gases with constant specific heats, and a simplified reaction model. Shock and flame solutions arise from the conservation of mass, momentum, and energy across the shock, with the second law determining which solutions are possible. Specifically, we study normal and oblique shocks as well as Chapman-Jouguet flames.

Chapter 9: Mixtures in Nonequilibrium

Due to cross-coupling and, possibly, a large number of components, transport processes in mixtures are rather rich. This chapter aims at giving some insight by studying binary mixtures in simple settings. Driving forces for diffusion are discussed, including separation of gas mixtures in centrifuges. Combined convective and diffusive transport across semi-permeable membranes for desalination and pressure retarded osmosis is used as an instructive example. The chapter closes with a motivation of the Arrhenius law for reaction rates by extension of the principles of Linear Irreversible Thermodynamics towards nonlinear expressions.

Appendix: Scalars, Vectors, Tensors
Throughout the main text, transport equations and constitutive relations are written either in direct tensor notation, or in tensor index notation, where the latter is preferred. This appendix offers a compact summary of symbolic and index notation for tensors and their operations.

1.3 Further Reading

One aim of this book is to provide a compact and accessible *Thermodynamic Introduction to Transport Phenomena*, so that successful students of the material will find easy access to the more specialized and thorough literature.

The moniker *Transport Phenomena* goes back to the now classical textbook by Bird, Stewart & Lightfoot. As suggestion for further studies, the list below collects some of the many excellent books that look at *Transport Phenomena* and *Nonequilibrium Thermodynamics* from different angles.

- A. Bejan: *Advanced Engineering Thermodynamics*, Wiley 2016
- R.B. Bird, W.E. Stewart, E.N. Lightfoot: *Transport Phenomena, Revised 2nd Ed.*, Wiley 2007
- H.B. Callen: *Thermodynamics and an Introduction to Thermostatistics, 2nd Ed.*, Wiley 1991
- S.R. De Groot, P. Mazur: *Non-Equilibrium Thermodynamics*, Dover 2011
- E.B. Gyftopoulos, G.P. Beretta: *Thermodynamics: Foundations and Applications*, Dover 2005
- S. Kjelstrup, D. Bedeaux, E. Johannsen, J. Gross: *Non-Equilibrium Thermodynamics for Engineers: Second Edition 2nd Ed.*, World Scientific 2017
- H.W. Liepmann and A. Roshko: Elements of Gasdynamics, Wiley, New York 1957
- I. Müller: *Thermodynamics*, Pitman, 1985
- I. Müller, T. Ruggeri: *Rational Extended Thermodynamics (Springer Tracts in Natural Philosophy)*, Springer 1998
- I. Müller, W. Weiss: *Entropy and Energy: A Universal Competition*, Springer 2005
- I. Müller: *A History of Thermodynamics–The Doctrine of Energy and Entropy*, Springer 2007
- H.C. Öttinger: *Beyond Equilibrium Thermodynamics*, Wiley 2005
- A. Sommerfeld: *Lectures on Theoretical Physics: Thermodynamics and Statistical Mechanics*, Academic Press 1955
- H. Struchtrup: *Macroscopic Transport Equations for Rarefied Gas Flows—Approximation Methods in Kinetic Theory*, Springer 2005
- H. Struchtrup: *Thermodynamics and Energy Conversion, 2nd Edition*, Springer 2024
- D.C. Venerus, H.C. Öttinger: *A Modern Course in Transport Phenomena*, Cambridge University Press 2018

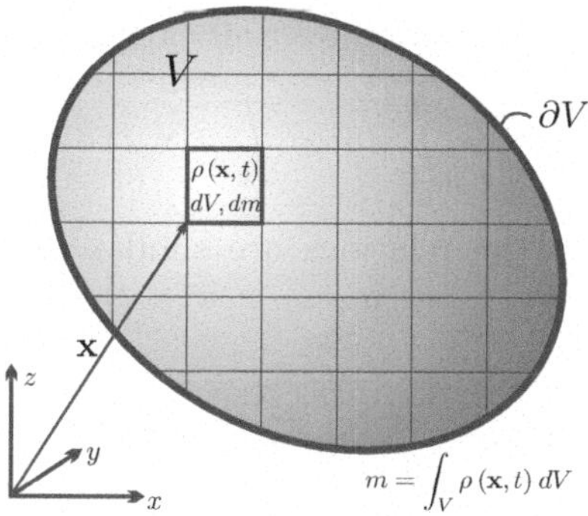

Fig. 1.1 Inhomogeneous distribution of mass density $\rho(\mathbf{x})$ in a system of volume V.

1.4 Local and Global Description

We aim to describe irreversible processes, in which the system under considerations will be found in arbitrary inhomogeneous states. Hence, it does not suffice to know global system properties, such as total mass and energy, but it is necessary to fully resolve the state in space and time. In the classical description of continuum thermodynamics, the state of a fluid (liquid or gas) mixture with ν components $\alpha = 1, 2, \dots, \nu$ is described through the fields[2]

$$\begin{array}{ll} \text{mass density} & \rho(\mathbf{x},t) \\ \text{composition} & \mathsf{c}_\alpha(\mathbf{x},t) = \frac{\rho_\alpha(\mathbf{x},t)}{\rho(\mathbf{x},t)} \\ \text{velocity} & \mathbf{v}(\mathbf{x},t) \\ \text{temperature} & T(\mathbf{x},t) \end{array}$$

in all points $\mathbf{x}$ of a body, at all times t. As defined, c_α is the (local) mass fraction of component α, ρ_α is the partial mass density of the component, and $\rho = \sum_{\beta=1}^{\nu} \rho_\beta$ is the total mass density of the mixture.

The underlying idea for this description is to divide the physical space into volume elements dV which are infinitesimally small for our senses, but nevertheless contain a large number of interacting atoms or molecules. Indeed, with our senses we cannot resolve the volume element, and perceive the material as a continuum, not as a group of particles.

Figure 1.1 shows a sketch of a system of volume $V = \int dV$ and its volume elements dV. The mass within the volume element dV located at $\mathbf{x}$ is denoted as dm, such that the local mass density is[3] $\rho(\mathbf{x},t) = \frac{dm}{dV}$. The total mass of

[2] In the present section we use symbolic tensor notation, where bold letters indicate vectors, e.g., location vector $\mathbf{x}$, flow velocity $\mathbf{v}$, see Appendix A. For velocity we use the symbols $\mathbf{v}$, v_i. To avoid confusion, specific volume will be denoted by v, but will hardly be used.

[3] Of course, $\frac{dm}{dV}$ is not a derivative, but the ratio of two infinitesimal quantities, the volume element dV and the mass dm in the volume element.

the system results from summation over all elements, as

$$m(t) = \int_V dm = \int_V \rho(\mathbf{x}, t)\, dV \ . \tag{1.1}$$

For other properties, the relations are similar. When the local specific internal energy (i.e., energy per mass) is $u(\mathbf{x}, t)$, the total internal energy within the volume V is

$$U(t) = \int_V u(\mathbf{x}, t)\, dm = \int_V \rho(\mathbf{x}, t)\, u(\mathbf{x}, t)\, dV = \int_V \rho u dV \ . \tag{1.2}$$

For more compact notation, we often will not make the space-time arguments $(\mathbf{x}, t)$ explicit, such as in the last relation above, where ρu denotes the local density of internal energy (energy per volume).

The total momentum vector $\mathbf{M}$ results from summing the momentum of the volume elements as

$$\mathbf{M}(t) = \int_V \mathbf{v}(\mathbf{x}, t)\, dm = \int_V \rho(\mathbf{x}, t)\, \mathbf{v}(\mathbf{x}, t)\, dV = \int_V \rho \mathbf{v} dV \ . \tag{1.3}$$

Kinetic energy in the system is the sum of the kinetic energies of the volume elements,

$$E_{\mathrm{kin}}(t) = \int_V \frac{1}{2} v^2(\mathbf{x}, t)\, dm = \int_V \frac{1}{2} \rho(\mathbf{x}, t)\, v^2(\mathbf{x}, t)\, dV = \int_V \frac{\rho}{2} v^2 dV \ , \tag{1.4}$$

where $v^2 = \mathbf{v} \cdot \mathbf{v}$ is the square of the absolute local velocity.

The potential energy of the system results from summation over the potential energies (mass times gravitational acceleration) of the cells,

$$E_{\mathrm{pot}} = \int \rho(\mathbf{x}, t)\, \mathrm{g} z dV = \int \rho \mathrm{g} z dV \ , \tag{1.5}$$

where $\mathrm{g} = 9.81 \frac{\mathrm{m}}{\mathrm{s}^2}$ denotes gravitational acceleration on Earth.

Finally, the total entropy of the system results from summation over the local specific entropy,

$$S(t) = \int_V s(\mathbf{x}, t)\, dm = \int_V \rho(\mathbf{x}, t)\, s(\mathbf{x}, t)\, dV = \int_V \rho s dV \ . \tag{1.6}$$

In thermodynamic language, total mass, momentum, energy and entropy are extensive properties, they depend on system size.

Intensive properties are independent of system size, such as pressure $p(\mathbf{x}, t)$, temperature $T(\mathbf{x}, t)$, as well as the specific properties (u, s, etc.) and densities.

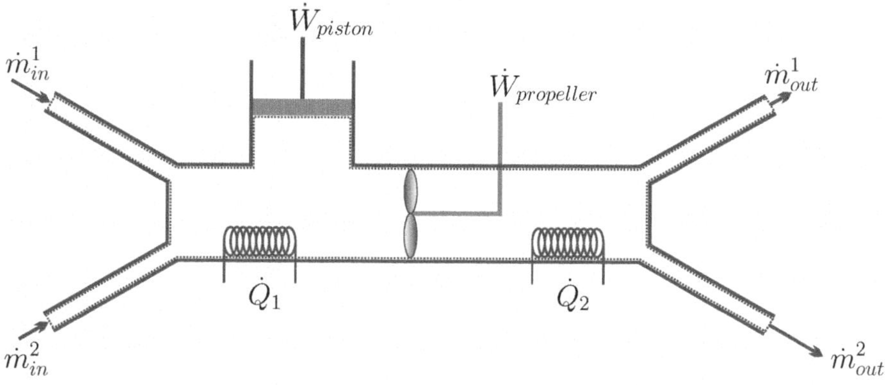

Fig. 1.2 Open system with two inflows, two outflows and two heat sources. The dotted line indicates the system boundary.

1.5 Global Balance Laws

We recall the balance laws for systems as used in engineering thermodynamics, for general open systems as shown in Fig. 1.2.

The conservation law for mass states that the total mass m within the system boundaries changes due to inflow and outflow,

$$\frac{dm}{dt} = \sum_{\text{in}} \dot{m}_i - \sum_{\text{out}} \dot{m}_e \,. \tag{1.7}$$

Here, $\dot{m}_i$ denotes incoming mass flows and $\dot{m}_e$ denotes exiting mass flows ($\dot{m} = \rho v A$, where A denotes cross section), as indicated in Fig. 1.2.

The first law of thermodynamics, i.e., the conservation law for the total energy $E = U + E_{kin} + E_{pot}$ of the system, reads

$$\frac{dE}{dt} = \sum_{\text{in}} \dot{m}_i \left(h + \frac{1}{2}v^2 + \mathrm{g}z\right)_i - \sum_{\text{out}} \dot{m}_e \left(h + \frac{1}{2}v^2 + \mathrm{g}z\right)_e + \dot{Q} - \dot{W} \,. \tag{1.8}$$

The indices (i-inlet, e-exit) indicate the values of the properties at the location where the respective flows cross the system boundary, that is their average values at the inlets and outlets, respectively.

This equation states that the total energy E within the system changes due to convective inflow and outflow ($h = u + \frac{p}{\rho}$ denotes enthalpy), as well as due to heat transfer and work, where $\dot{Q}$ is heating rate and $\dot{W}$ is power. Note that the flow energy includes the work required to move the mass across the boundaries (flow work), so that $\dot{W}$ accounts only for piston and propeller work. There can be several contributions to work and heat transfer, overall power and heating rate result from summation,

$$\dot{W} = \sum_j \dot{W}_j \quad \text{and} \quad \dot{Q} = \sum_k \dot{Q}_k \; . \tag{1.9}$$

The second law of thermodynamics—the balance of entropy—for open systems reads

$$\frac{dS}{dt} = \sum_{\text{in}} \dot{m}_i s_i - \sum_{\text{out}} \dot{m}_e s_e + \sum_k \frac{\dot{Q}_k}{T_k} + \dot{S}_{gen} \quad \text{with} \quad \dot{S}_{gen} \geq 0 \; , \tag{1.10}$$

where $\dot{Q}_k$ is the heat crossing the system boundary at local temperature T_k.

The entropy S of the system changes due to entropy flux with heat, $\frac{\dot{Q}_k}{T_k}$, convective in- and outflow $\dot{m}s$, and entropy generation $\dot{S}_{gen}$. The entropy generation rate $\dot{S}_{gen}$ is positive in nonequilibrium, and vanishes in equilibrium states.

The second law describes the tendency of an isolated system to approach global equilibrium states. Manipulation of the system, e.g., by transfer of mass or energy, and inhomogeneous boundary conditions, typically induces nonequilibrium states and processes. Reversible processes are an idealization, where the entropy generation vanishes.

1.6 States and State Properties

1.6.1 Equilibrium States

Evaluation of equilibrium states shows that these are homogeneous such that in thermodynamic equilibrium all elements of a system have the same temperature.

Furthermore the Gibbs free energy (for simple substances) or the chemical potential (for mixtures with or without reactions) play an important role in determining equilibrium states. For a simple substance in equilibrium, the Gibbs free energy is homogeneous in phase equilibrium. For a mixture divided by a semipermeable membrane, the chemical potential of components that can pass the membrane is homogeneous. When gravity effects play a role, the sum of chemical potential and potential energy is homogeneous, which yields hydrostatic (for fluids) and barometric (for ideal gases) pressure laws.

In thermodynamics of systems, equilibrium conditions are obtained from maximizing entropy, see Chap. 2. In the realm of *Transport Phenomena* we will find the same relations from equilibrium solutions of the resolved transport equations for mass, momentum and energy, which will be constructed such that they and their solutions stand in agreement with the second law at all times.

1.6.2 Properties and their Relations

The basic laws of thermodynamics contain properties that can be measured, such as pressure p, mass density ρ, specific volume $\mathsf{v} = 1/\rho$, temperature T, mole fractions X_α, mass fractions c_α, velocity $\mathbf{v}$. Other properties are not directly measurable, most importantly internal energy u, enthalpy $h = u + \frac{p}{\rho}$, entropy s, and chemical potential μ_α.

Experience shows that thermodynamic properties are related by *property relations*, or constitutive relations, that give properties in terms of measurable ones [TEC, Chap. 6]. These constitutive relations depend on the material and its state. Typically, property relations are determined for equilibrium states, important examples are the thermal equation of state $p(\rho, T)$, the caloric equation of state $u(\rho, T)$, or the relation $s(\rho, T)$ for entropy. Property relations for equilibrium states are found from measurements and application of the Gibbs equation [TEC, Chap. 16]. Often the relations are available in the form of *property tables*, not as equations.

As important examples that will be used throughout the text, we list property relations for ideal gases and incompressible liquids.

An ideal gas with constant specific heats has the explicit property relations

$$\begin{aligned} p(\rho, T) &= \rho R T \ , \\ u(T) &= c_{\mathsf{v}}(T - T_0) + u_0 \ , \\ h(T) &= c_p(T - T_0) + h_0 \ , \\ s(\rho, T) &= c_{\mathsf{v}} \ln \frac{T}{T_0} - R \ln \frac{\rho}{\rho_0} + s_0 = c_p \ln \frac{T}{T_0} - R \ln \frac{p}{p_0} + s_0 \ , \end{aligned} \tag{1.11}$$

where $R = \frac{\bar{R}}{M}$ is the gas constant, $\bar{R} = 8.314 \frac{\mathrm{kJ}}{\mathrm{kg\,K}}$ is the universal gas constant, M is the molar mass of the substance, and c_{v} and $c_p = c_{\mathsf{v}} + R$ are specific heats at constant volume and constant pressure, respectively. Moreover, T_0, p_0 are temperature and pressure of a reference state, $\rho_0 = \frac{p_0}{R T_0}$, and $u_0 = u(T_0)$, $h_0 = h(T_0) = u_0 + R T_0$, $s_0 = s(T_0, p_0)$ are the assigned values of the properties at the reference state. For ideal gases, internal energy and enthalpy are functions only of temperature, but entropy depends on temperature and mass density or pressure.

For ideal incompressible liquids and solids the property relations read

$$\begin{aligned} \rho &= \rho_0 = const. \ , \\ u(T) &= c(T - T_0) + u_0 \ , \\ h(T) &= c(T - T_0) + \frac{p - p_0}{\rho_0} + h_0 \ , \\ s(T) &= c \ln \frac{T}{T_0} + s_0 \end{aligned} \tag{1.12}$$

where c is the constant specific heat of the material, and $h_0 = u_0 + \frac{p_0}{\rho_0}$.

Property data for ideal gases with temperature dependent specific heats $c_v(T)$, $c_p(T)$, non-ideal gases, vapors, compressible liquids etc. is widely available, e.g., see [TEC, Chap. 6 and Appendix].

1.6.3 Local Thermodynamic Equilibrium

For many nonequilibrium states the assumption of *Local Thermodynamic Equilibrium* (LTE) is valid, which states that the above equilibrium property relations hold locally even when the system as a whole is not in equilibrium, so that, e.g.,

$$\begin{aligned} p(\mathbf{x},t) &= p(\rho(\mathbf{x},t),T(\mathbf{x},t)) = p(\rho,T) \ , \\ u(\mathbf{x},t) &= u(\rho(\mathbf{x},t),T(\mathbf{x},t)) = u(\rho,T) \ , \\ s(\mathbf{x},t) &= s(\rho(\mathbf{x},t),T(\mathbf{x},t)) = s(\rho,T) \ , \end{aligned} \tag{1.13}$$

where the functions on the right hand side are the established equilibrium relations.

LTE breaks down under extreme conditions, such as gas rarefaction, but holds for most technical processes involving gases and liquids. For an intuitive understanding, recall that within the continuum viewpoint the volume elements contain a significant number of particles, that interact at very high frequencies; e.g., a particle in air at standard conditions undergoes about 5×10^9 collisions per second. The approach to equilibrium occurs through exchange of energy and momentum between the particles. As long as the volume element contains a sufficiently large number of particles, and as long as these interact with sufficiently high frequency within the volume element, local equilibrium is maintained.

Kinetic Theory of Gases provides a description on the particle level, where the continuum equations can be obtained by averaging and limiting procedures.[4] Hence, the framework allows to formalize the conditions under which LTE is valid. An important measure for the behavior of gases is the *mean free path*, which is the average distance a particle travels between two collisions. LTE holds, when the dimension of the volume element, e.g., the side length of a cube, is considerably larger than the mean free path so that most particle collisions are within the volume element, establishing local equilibrium.

Exchange of momentum and energy between volume elements is effected by particles leaving one element and colliding with particles in another. Under LTE, this exchange occurs between adjacent elements, so that transport is local. This locality of transport will be reflected in first order gradients in the conductive fluxes of mass, momentum and energy.

---

[4] H. Struchtrup, *Macroscopic Transport Equations for Rarefied Gas Flows,* Springer, Heidelberg 2005, doi:10.1007/3-540-32386-4

If the desired resolution is of the order of the mean free path, gas particles do not collide locally but travel larger distances passing through several volume elements without collisions, and transport of momentum and energy becomes non-local.

The size of the volume element depends on the desired resolution of the description, which also depends on the level of accuracy of measurements. The mean free path in air at standard conditions is of the order of 7×10^{-8} m, hence for resolutions of 10^{-4} m and higher the assumption is very well fulfilled. Moreover, time variation must be slow relative to the the collision time scale, which at standard conditions is the case for frequencies below 10^6 Hz.

For vacuum flows, that is at very low pressures, the mean free path becomes relatively large and the assumption of LTE breaks down at macroscopic dimensions. The best example for this is re-entry of space craft into the thin upper atmosphere, where the mean free path is as large as the size of the space craft, or even larger. On the other hand, for a gas with small mean free path undergoing processes in very small devices the desired resolution scale must be as small as the mean free path, and again, the LTE assumption breaks down.

While in a gas the particles move over the distance of the mean free path without interaction, in liquids and solids the atoms are densely packed, and are in constant contact. Hence, the particles interact all the time, making the time and space scales for validity of LTE considerably smaller than for a gas. Unless one is interested in resolving the molecular scale, liquids and solids are well described under the LTE assumption.

1.6.4 Gibbs Equation

As we proceed, we will make frequent use of the Gibbs equation in its many forms. The Gibbs equation is a differential relation between properties, which is universally valid for equilibrium states, and plays an important role for finding property relations [TEC, Chap. 16].

For a closed equilibrium system, the most familiar form of the Gibbs equation reads

$$TdS = dU + pdV \tag{1.14}$$

Division by the (constant) mass m in the system gives the corresponding form for mass specific properties,

$$Tds = du + pd\mathsf{v} \ . \tag{1.15}$$

where $\mathsf{v} = \frac{V}{m} = \frac{1}{\rho}$ is specific volume. A Legendre transform, $pd\mathsf{v} = d\left(p\mathsf{v}\right) - \mathsf{v}dp$, and the definition of enthalpy $h = u + p\mathsf{v}$, give the alternative form

$$Tds = dh - \mathsf{v}dp \ . \tag{1.16}$$

The above expressions (1.11, 1.12) for entropy for ideal gases and incompressible fluids are obtained from integration of the Gibbs equation in one of these forms.

In what follows we shall not use specific volume as a variable, but use mass density instead, for which the standard Gibbs equation reads[5]

$$Tds = du - \frac{p}{\rho^2} d\rho \,. \tag{1.17}$$

Alternatively, the Gibbs equation can be written for densities of energy and entropy (i.e., property per volume). Multiplying the above with ρ, and pulling the density into the differential, yields[6]

$$Td(\rho s) = d(\rho u) - gd\rho \,, \tag{1.18}$$

where

$$g = u + \frac{p}{\rho} - Ts = h - Ts \tag{1.19}$$

denotes Gibbs free energy.

1.6.5 Thermodynamic Potentials and Maxwell Relations

The Gibbs equation yields a large number of relations between thermodynamic properties, which we briefly summarize here.

Writing the Gibbs equation as

$$du = Tds + \frac{p}{\rho^2} d\rho \tag{1.20}$$

suggests considering internal energy as function of entropy and mass density, $u(s, \rho)$, with the partial derivatives

$$\left(\frac{\partial u}{\partial s}\right)_\rho = T \quad , \quad \left(\frac{\partial u}{\partial \rho}\right)_s = \frac{p}{\rho^2} \,. \tag{1.21}$$

The order of mixed second derivatives can be exchanged, $\left(\frac{\partial^2 u}{\partial \rho \partial s}\right) = \left(\frac{\partial^2 u}{\partial s \partial \rho}\right)$ which yields the accompanying Maxwell relation

$$\left(\frac{\partial T}{\partial \rho}\right)_s = \frac{1}{\rho^2}\left(\frac{\partial p}{\partial s}\right)_\rho \,. \tag{1.22}$$

Correspondingly, the Gibbs equation with enthalpy, in the form

[5] With the chain rule of calculus: $dv = d\left(\frac{1}{\rho}\right) = -\frac{1}{\rho^2} d\rho$.

[6] In more detail: $\rho Tds = \rho du - \frac{p}{\rho} d\rho \Rightarrow T\left[d(\rho s) - sd\rho\right] = \left[d(\rho u) - ud\rho\right] - \frac{p}{\rho} d\rho$

$$dh = Tds - \frac{1}{\rho}dp \ , \tag{1.23}$$

suggests considering enthalpy as $h\,(s,p)$ which yields the relations

$$\left(\frac{\partial h}{\partial s}\right)_p = T \quad , \quad \left(\frac{\partial h}{\partial p}\right)_s = \frac{1}{\rho} \quad , \quad \left(\frac{\partial T}{\partial p}\right)_s = -\frac{1}{\rho^2}\left(\frac{\partial \rho}{\partial s}\right)_p \ . \tag{1.24}$$

Introducing the Helmholtz free energy

$$f = u - Ts \tag{1.25}$$

gives another form of the Gibbs equation,

$$df = -sdT + \frac{p}{\rho^2}d\rho \ , \tag{1.26}$$

and considering Helmholtz free energy as $f\,(T,\rho)$ yields

$$\left(\frac{\partial f}{\partial T}\right)_\rho = -s \quad , \quad \left(\frac{\partial f}{\partial \rho}\right)_T = \frac{p}{\rho^2} \quad , \quad -\left(\frac{\partial s}{\partial \rho}\right)_T = \frac{1}{\rho^2}\left(\frac{\partial p}{\partial T}\right)_\rho \ . \tag{1.27}$$

Finally, with the Gibbs free energy the Gibbs equation reads

$$dg = -sdT + \frac{1}{\rho}dp \tag{1.28}$$

and considering Gibbs free energy as $g\,(T,p)$ yields

$$\left(\frac{\partial g}{\partial T}\right)_p = -s \quad , \quad \left(\frac{\partial g}{\partial p}\right)_T = \frac{1}{\rho} \quad , \quad -\left(\frac{\partial s}{\partial p}\right)_T = -\frac{1}{\rho^2}\left(\frac{\partial \rho}{\partial T}\right)_p \ . \tag{1.29}$$

The functions $u\,(s,\rho)$, $h\,(s,p)$, $f\,(T,\rho)$ and $g\,(T,p)$ are known as thermodynamic potentials, or fundamental relations. All thermodynamic state properties can be determined from one of these functions and its derivatives. For example considering the Gibbs free energy $g\,(T,p)$ we have

$$\rho\,(T,p) = \frac{1}{\left(\frac{\partial g}{\partial p}\right)_T} \tag{1.30}$$

$$s\,(T,p) = -\left(\frac{\partial g}{\partial T}\right)_p \tag{1.31}$$

$$h\,(T,p) = g + Ts = g - T\left(\frac{\partial g}{\partial T}\right)_p = -T^2\left(\frac{\partial \frac{g}{T}}{\partial T}\right)_p \tag{1.32}$$

$$u\,(T,p) = h - \frac{p}{\rho} = g - T\left(\frac{\partial g}{\partial T}\right)_p - p\left(\frac{\partial g}{\partial p}\right)_T \tag{1.33}$$

1.7 Mixtures

1.7.1 Property Relations

In mixtures with ν components, the property relations also depend on composition, expressed either through partial mass densities $\rho_\alpha = \frac{dm_\alpha}{dV}$ or local mass fractions $\mathsf{c}_\alpha = \frac{dm_\alpha}{dm} = \frac{\rho_\alpha}{\rho}$, where dm_α is the mass of component α in the volume element dV, $dm = \sum_{\alpha=1}^{\nu} dm_\alpha$ is the total mass in the element, and $\rho = \frac{dm}{dV} = \sum_{\alpha=1}^{\nu} \rho_\alpha$ is the total mass density of the mixture [TEC, Chap. 18]. This definition implies $\sum_{\alpha=1}^{\nu} \mathsf{c}_\alpha = 1$, hence only $\nu - 1$ mass fractions c_α are independent.

Accordingly, the property relations for a component in the mixture are of the form, e.g., for enthalpy and entropy,

$$h_\alpha\left(\rho, T, \mathsf{c}_{\beta\ (\beta=1,\ldots,\nu-1)}\right) \quad , \quad s_\alpha\left(\rho, T, \mathsf{c}_{\beta\ (\beta=1,\ldots,\nu-1)}\right) , \tag{1.34}$$

respectively. As always in thermodynamics, one can chose other variables, for instance pressure instead of total mass density,

$$h_\alpha\left(p, T, \mathsf{c}_{\beta\ (\beta=1,\ldots,\nu-1)}\right) \quad , \quad s_\alpha\left(p, T, \mathsf{c}_{\beta\ (\beta=1,\ldots,\nu-1)}\right) , \tag{1.35}$$

or the ν partial mass densities and temperature, so that[7]

$$h_\alpha\left(T, \rho_{\beta\ (\beta=1,\ldots,\nu)}\right) \quad , \quad s_\alpha\left(T, \rho_{\beta\ (\beta=1,\ldots,\nu)}\right) . \tag{1.36}$$

1.7.2 Chemical Potential

When the composition of a mixture changes, so that the masses m_α of components change by dm_α, or the mass fractions change by $d\mathsf{c}_\alpha$, other properties will be affected and change their values. To account for these changes, a term is added to the Gibbs equation for mixtures—inert or reacting—as

$$TdS = dU + pdV - \sum_{\alpha=1}^{\nu} \mu_\alpha dm_\alpha , \tag{1.37}$$

where μ_α is the mass-based chemical potential of component α.

The chemical potential plays an important role in the thermodynamics of mixtures [TEC, Chaps. 20-26], and is formally defined through its introduction in the Gibbs equation as above. As with other thermodynamic properties, one has to find property relations that relate the chemical potential to measurable properties such as density, mass fraction, temperature, etc.

---

[7] These sets of variables all have $(\nu + 1)$ independent variables.

To obtain the corresponding Gibbs equation for specific properties, we insert $S = ms$, $U = mu$, $V = \frac{m}{\rho}$, $m_\alpha = m\mathsf{c}_\alpha$ and use the product rule of calculus as, e.g., $d(ms) = mds + sdm$, to find, after some reordering,

$$Tds = du - \frac{p}{\rho^2}d\rho - \sum_{\alpha=1}^{\nu} \mu_\alpha d\mathsf{c}_\alpha + \left[u + \frac{p}{\rho} - Ts - \sum_{\alpha=1}^{\nu} \mathsf{c}_\alpha \mu_\alpha\right] \frac{dm}{m} \tag{1.38}$$

A change dm of overall mass of the system does not affect its specific properties. For instance, when removing a portion of an equilibrium mixture, only the size of the system changes, but there will be no change of the specific properties, i.e., $ds = du = d\rho = d\mathsf{c}_\alpha = 0$. Accordingly, the expression in the square bracket must vanish

$$g = u + \frac{p}{\rho} - Ts = h - Ts = \sum_{\alpha=1}^{\nu} \mathsf{c}_\alpha \mu_\alpha \tag{1.39}$$

This equation relates the specific Gibbs free energy of the mixture, g, to the chemical potentials μ_α of the components.

The total Gibbs free energy, enthalpy, and entropy of a mixture are given by

$$G = mg = \sum_{\alpha=1}^{\nu} m_\alpha \mu_\alpha, \quad H = mh = \sum_{\alpha=1}^{\nu} m_\alpha h_\alpha, \quad S = ms = \sum_{\alpha=1}^{\nu} m_\alpha s_\alpha. \tag{1.40}$$

Thus, the chemical potential is the Gibbs free energy of a component in the mixture.[8] With $G = H - TS$, we identify

$$\sum_{\alpha=1}^{\nu} m_\alpha \mu_\alpha = G = H - TS = \sum_{\alpha=1}^{\nu} m_\alpha (h_\alpha - Ts_\alpha) \tag{1.41}$$

so that

$$\mu_\alpha = h_\alpha - Ts_\alpha \ . \tag{1.42}$$

Here, all properties, μ_α, h_α, s_α refer to component α in the mixture—due to the influence of the other components there will be mixing contributions to enthalpy and entropy!

1.7.3 Gibbs Equation

The Gibbs equation for the mixture (1.38) becomes

[8] In the discussion of mixtures in [TEC], we distinguish between the Gibbs free energy g_α of the component alone (not in the mixture), and the chemical potential μ_α (in the mixture).

$$Tds = du - \frac{p}{\rho^2} d\rho - \sum_{\alpha=1}^{\nu} \mu_\alpha d\mathsf{c}_\alpha \, . \tag{1.43}$$

Due to $\sum_{\alpha=1}^{\nu} \mathsf{c}_\alpha = 1$ only $(\nu - 1)$ of the ν mass fractions are independent. Later we will need the equation in a form that has only differentials of independent properties; with $\mathsf{c}_\nu = 1 - \sum_{\alpha=1}^{\nu-1} \mathsf{c}_\alpha$

$$Tds = du - \frac{p}{\rho^2} d\rho - \sum_{\alpha=1}^{\nu-1} (\mu_\alpha - \mu_\nu) \, d\mathsf{c}_\alpha \, . \tag{1.44}$$

Multiplying (1.43) with total density, and pulling density into the differential leads to[9]

$$Td(\rho s) = d(\rho u) - \sum_{\alpha=1}^{\nu} \mu_\alpha d\rho_\alpha \, . \tag{1.45}$$

Another manipulation of the Gibbs equation relies on Legendre transforms. We start with

$$du - \frac{p}{\rho^2} d\rho - Tds = \sum_{\alpha=1}^{\nu} \mu_\alpha d\mathsf{c}_\alpha \, . \tag{1.46}$$

Using the product rule backwards in two terms, this can be rewritten as

$$d\left(u + \frac{p}{\rho} - Ts\right) - \frac{dp}{\rho} + sdT = d\left(\sum_{\alpha=1}^{\nu} \mu_\alpha \mathsf{c}_\alpha\right) - \sum_{\alpha=1}^{\nu} \mathsf{c}_\alpha d\mu_\alpha \, . \tag{1.47}$$

With (1.39), the first terms left and right cancel, and we find the Gibbs-Duhem relation

$$0 = \frac{dp}{\rho} - sdT - \sum_{\alpha=1}^{\nu} \mathsf{c}_\alpha d\mu_\alpha \, . \tag{1.48}$$

Under LTE conditions, all of the above equations for specific properties or densities are valid locally.

1.7.4 Mixing Volume, Heat and Entropy of Mixing

In general, the chemical potential and other properties in a mixture are functions of mixture temperature, pressure, and composition (T, p, c_β).

As components are mixed at constant temperature and pressure the extensive properties might change. To properly account for the change, we consider ν components in an initial unmixed state (I) where each component is at the

[9] Not all calculations can be shown—it is recommended to use pencil and paper to confirm anything that is stated here and elsewhere without much detail!

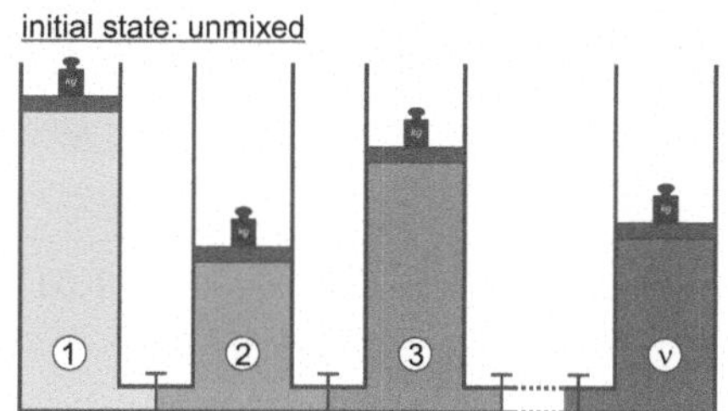

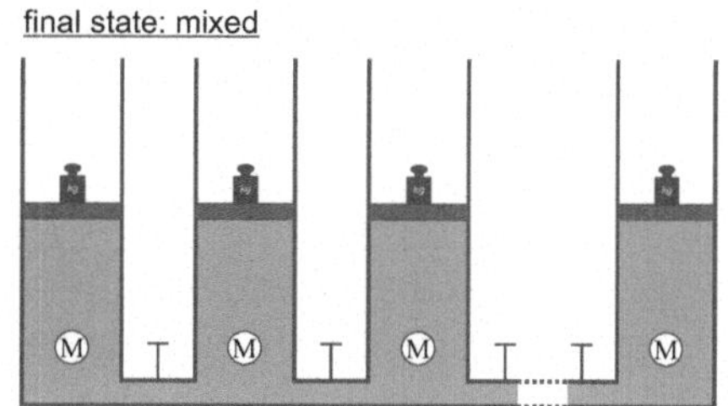

Fig. 1.3 Mixing of components at constant T and p.

same temperature, T, and pressure, p. The components are mixed while keeping temperature and pressure constant, the final state (E) is a homogeneous mixture, see Fig. 1.3.

The volume change between initial and final state is determined as

$$V_{\text{mix}} = V_E - V_I = m_\gamma \mathsf{v}_\gamma \left(T, p, \mathsf{c}_\beta\right) - \sum_{\alpha=1}^{\nu} m_\alpha \mathsf{v}_\alpha \left(T, p\right) \ , \tag{1.49}$$

where $\mathsf{v}_\alpha \left(T, p\right)$ denotes the specific volume of component α alone at (T, p) and $\mathsf{v}_\gamma \left(T, p, \mathsf{c}_\beta\right)$ denotes the specific volume of *any* component γ in the mixture of composition c_β $(\beta = 1, \ldots, \nu - 1)$ at (T, p). Note that in the mixed state all components are distributed over the volume of the mixture, V_E.

The change of volume is due to spatial hindrances or advantages on the molecular scale. For instance a mixture of 1 litre of water with 1 litre of ethanol (C_2H_5OH) yields a mixing volume of 1.93 litres.

Isothermal mixing of components might release or require heat, which must be transferred. The first law applied to the isothermal and isobaric mixing process yields

$$H_{\text{mix}} = H_E - H_I = \sum_{\alpha=1}^{\nu} m_\alpha \left[h_\alpha \left(T, p, \mathsf{c}_\beta\right) - h_\alpha \left(T, p\right)\right] \ , \tag{1.50}$$

where H_{mix} is the heat that must be exchanged in order to keep the temperature T constant for the isobaric mixing process. Here, $h_\alpha \left(T, p\right)$ is the specific enthalpy of component α alone at (T, p) and $h_\alpha \left(T, p, \mathsf{c}_\beta\right)$ is the specific enthalpy of the component α in the mixture of composition c_β $(\beta = 1, \ldots, \nu - 1)$ at (T, p).

Enthalpy and internal energy are influenced by the interaction potential between molecules. In a pure substance, particles of type α interact only with particles of the same type. In a mixture, however, particles of type α are surrounded by different types of particles β $(\beta = 1, \ldots, \nu)$, which leads to different molecular interaction potentials, and thus a change in internal energy u_α and enthalpy h_α for the particles of type α as compared to the pure substance.

The entropy of mixing is determined in the same way, as

$$S_{\text{mix}} = S_E - S_I = \sum_{\alpha=1}^{\nu} m_\alpha \left[s_\alpha \left(T, p, \mathsf{c}_\beta \right) - s_\alpha \left(T, p \right) \right] . \tag{1.51}$$

1.7.5 Ideal Mixtures

Ideal mixtures have neither heat nor volume of mixing, that is when components are mixed at the same temperature and pressure, temperature does not change, and the volume of the mixture is the sum of the volumes before mixing.

For ideal gas mixtures with constant specific heats, the property relations for components are particularly simple, e.g., with variables (ρ_α, T) or $(\rho, T, \mathsf{c}_\alpha)$

$$\begin{aligned}
p_\alpha &= \rho_\alpha R_\alpha T = \mathsf{c}_\alpha \rho R_\alpha T \; , \\
u_\alpha &= c_{\mathsf{v}}^\alpha \left(T - T_0 \right) + u_{\alpha,0} \; , \\
h_\alpha &= c_p^\alpha \left(T - T_0 \right) + h_{\alpha,0} \; , \\
s_\alpha &= c_{\mathsf{v}}^\alpha \ln \frac{T}{T_0} - R_\alpha \ln \frac{\rho_\alpha}{\rho_{\alpha,0}} + s_{\alpha,0} \; . \\
g_\alpha &= c_p^\alpha \left(T - T_0 \right) + h_{\alpha,0} - c_{\mathsf{v}}^\alpha T \ln \frac{T}{T_0} + R_\alpha T \ln \frac{\rho_\alpha}{\rho_0} - T s_{\alpha,0}
\end{aligned} \tag{1.52}$$

where $R_\alpha = \frac{\bar{R}}{M_\alpha}$, $\rho_{\alpha,0} = \frac{p_0}{R_\alpha T_0}$. Here, p_α is the partial pressure of the component in the mixture, with total pressure

$$p = \sum_{\alpha=1}^{\nu} p_\alpha = \sum_{\alpha=1}^{\nu} \rho_\alpha R_\alpha T \; . \tag{1.53}$$

Considering variables (T, p_α), component entropies and Gibbs free energies are

$$s_\alpha \left(T, p_\alpha \right) = c_p^\alpha \ln \frac{T}{T_0} - R_\alpha \ln \frac{p_\alpha}{p_0} + s_{\alpha,0} \; , \tag{1.54}$$

$$g_\alpha \left(T, p_\alpha \right) = c_p^\alpha \left(T - T_0 \right) + h_{\alpha,0} - c_p^\alpha T \ln \frac{T}{T_0} + R_\alpha T \ln \frac{p_\alpha}{p_0} - T s_{\alpha,0} \tag{1.55}$$

With the above property relations, mixing volume and enthalpy of mixing vanish, $V_{\text{mix}} = H_{\text{mix}} = 0$ (Prob. 1.8). The total entropies before and after mixing at (T, p) are

$$S_I = \sum_{\alpha=1}^{\nu} m_\alpha \left[c_p^\alpha \ln \frac{T}{T_0} - R_\alpha \ln \frac{p}{p_0} \right] , \tag{1.56}$$

$$S_E = \sum_{\alpha=1}^{\nu} m_\alpha \left[c_p^\alpha \ln \frac{T}{T_0} - R_\alpha \ln \frac{p_\alpha}{p_0} \right] = S_I - \sum_{\alpha=1}^{\nu} m_\alpha R_\alpha \ln \frac{p_\alpha}{p} . \tag{1.57}$$

hence the entropy of mixing is

$$S_{\text{mix}} = S_E - S_I = - \sum_{\alpha=1}^{\nu} m_\alpha R_\alpha \ln \frac{p_\alpha}{p} = - \sum_{\alpha=1}^{\nu} m_\alpha R_\alpha \ln X_\alpha > 0 , \tag{1.58}$$

where

$$X_\alpha = \frac{p_\alpha}{p} = \frac{\frac{\rho_\alpha}{M_\alpha}}{\sum_{\beta=1}^{\nu} \frac{\rho_\beta}{M_\beta}} = \frac{\frac{\mathsf{c}_\alpha}{M_\alpha}}{\sum_{\beta=1}^{\nu} \frac{\mathsf{c}_\beta}{M_\beta}} = \frac{\bar{\rho}_\alpha}{\bar{\rho}} \tag{1.59}$$

is the mole fraction in the mixture, with $\bar{\rho}_\alpha = \frac{\rho_\alpha}{M_\alpha}$ and $\bar{\rho} = \sum_{\beta=1}^{\nu} \bar{\rho}_\beta$ denoting the partial and total molar densities.

Thus, the chemical potential in an ideal gas mixture is the sum of the Gibbs free energy of the component alone at mixture conditions (T, p) plus the contribution from the entropy of mixing,

$$\mu_\alpha (T, p, \mathsf{c}_\beta) = g_\alpha (T, p) + R_\alpha T \ln X_\alpha (\mathsf{c}_\beta) . \tag{1.60}$$

Note that here the dependence on all mass fractions is reduced to dependence on only the mole fraction X_α of the component α itself.

Ideal mixtures have the same mixing properties as ideal gas mixtures, that is vanishing mixing volume and heat of mixing, and an entropy of mixing just as that of the ideal gas,[10] i.e.,

$$V_{\text{mix}} = H_{\text{mix}} = 0 \quad , \quad S_{\text{mix}} = - \sum_\alpha n_\alpha \bar{R} \ln X_\alpha . \tag{1.61}$$

Ideal gas mixtures are a special case of ideal mixtures. In particular, the theory of ideal mixtures can be applied to dilute liquid solutions, e.g., salt in water.

The description of non-ideal mixtures relies on activity and fugacity coefficients to describe deviation from ideal mixtures [TEC, Chap. 22].

---

[10] The entropy of mixing is best motivated through Boltzmann's microscopic interpretation of entropy, $S = k_B \ln \Omega$, where $k_B = \bar{R}/N_A = 1.3806 \times 10^{-23} \frac{\text{J}}{\text{K}}$ is Boltzmann's constant, $N_A = 6.022 \times 10^{23} \frac{1}{\text{mol}}$ is Avogadro number, and Ω is the number of microscopic realizations of a given macroscopic state, see [TEC, Sec. 18.12]

Problems

1.1. Bathroom Mold

In badly ventilated shower rooms, mold appears on the ceiling, in particular in corners. What does that have to do with thermodynamics and transport theory?

1.2. Propane Tank

Barbecue season is back! Most gas barbecues are powered by propane from standard liquid propane tanks. After the barbecue runs for a while, water droplets begin to form on the tank. Use thermodynamic language to explain this observation.

1.3. Heating of a Room

A room of volume $75\,\mathrm{m}^3$ contains air, initially at $T_0 = 273\,\mathrm{K}$. A heater supplies heat at a rate of $2000\,\mathrm{W}$. Since air can leave or enter the room through small gaps in windows and doors, the pressure in the room is equal to the outside pressure of $1\,\mathrm{bar}$ at all times. Assume that no air *enters* the room. Consider air as an ideal gas with constant specific heats, assume that all properties inside the room are homogeneous, and use the global balance laws.

1. Assuming that there are no heat losses to the environment, determine mass and temperature of the air in the room as a function of time and plot the result.
2. Consider the same problem with heat losses. For this, assume that the heat loss is given by Newton's law of cooling, which states the loss is proportional to the difference between the temperatures inside and outside (assume $T_{outside} = T_0 = 273\,\mathrm{K}$), that is $\dot{Q}_{loss} = \alpha A (T_0 - T)$. Here, $\alpha = 90\frac{\mathrm{kJ}}{\mathrm{m}^2\,\mathrm{h}\,\mathrm{K}}$, is the heat transfer coefficient, and $A = 60\,\mathrm{m}^2$ is the wall surface. Plot the resulting curves.

1.4. Friction Loss in Pipe Flow

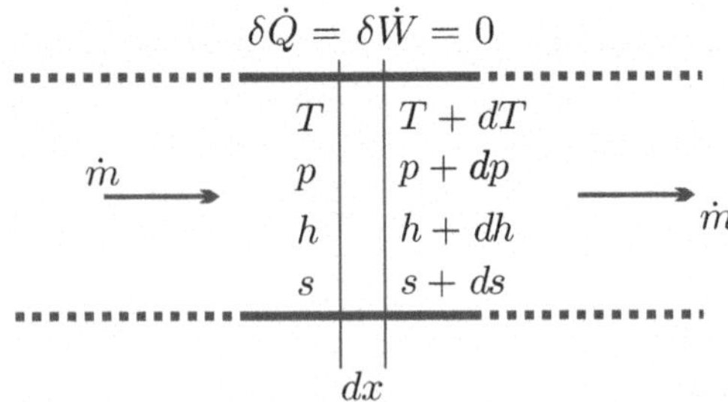

Consider a mass flow $\dot{m}$ through an adiabatic pipe of length L at steady state with pressure loss. The fluid enters the pipe at temperature T_1 and pressure p_1. Due to friction, the pressure drops as $\delta p = -\beta dx$ along the distance dx.

To solve the following, consider first and second law for open systems in steady state for the infinitesimal system in the figure, and integrate. Ignore potential and kinetic energies, and assume that system boundaries are adiabatic and no work is done.

1. Show that the first law of thermodynamics states that enthalpy stays constant, $dh = 0$.
2. Assume that the fluid is an ideal gas with constant specific heats (that is $du = c_v dT$, and $dh = c_p dT$ with $c_p = c_v + R$) and determine the temperature $T(x)$.
3. For the ideal gas, use the Gibbs equation $Tds = dh - \frac{dp}{\rho}$ and the ideal gas law to determine the entropy change ds. Integrate over a control volume just around the fluid in the pipe, and determine entropy generation and work loss in that volume.
4. Assume that the fluid is incompressible ($\rho = const$) and has constant specific heat c (that is $du = cdT$, and $dh = du + \frac{dp}{\rho}$). Determine the temperature of the fluid, $T(x)$.
5. For the incompressible liquid, use the Gibbs equation $Tds = dh - \frac{dp}{\rho}$ to determine the entropy change ds. Integrate over a control volume just around the fluid in the pipe, and determine entropy generation and work loss in that volume.
6. Discuss the differences between gas and liquid.

1.5. Heat Transfer Loss in Isobaric Pipe Flow

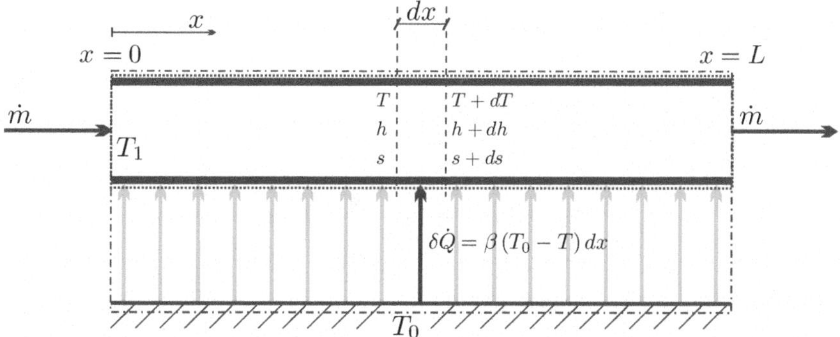

Consider a mass flow $\dot{m}$ through a pipe of length L at steady state without pressure loss. The fluid enters the pipe at temperature $T(x = 0) = T_1$. The heat exchange of the pipe section of length dx with the environment at T_0 can be described by Newton's law of cooling as $\delta\dot{Q} = \alpha(T_0 - T)\,dx$. For the following, assume that variation of temperature within a cross section at x can be ignored, the fluid has constant specific heat c_p and ignore kinetic and potential energies.

1. Consider the first law for the infinitesimal interval dx and determine the temperature of the fluid, $T(x)$, by integration.

2. Consider a control volume just around the fluid in the pipe (dotted line), and determine entropy generation rate in that volume. Discuss.
3. Consider a control volume around the pipe whose boundaries are at the environmental state (dash-dotted line) and determine entropy generation rate in that volume.
4. Show that the entropy generation is positive (for any choice of temperatures T_0, T_1).

1.6. Gibbs Equation for Simple Substance

The Gibbs equation for a simple substance in a closed system reads $TdS = dU + pdV$. If the mass m of substance in the system is allowed to change, a term must be added to account for this, i.e.,

$$TdS = dU + pdV + \Gamma dm$$

Determine the factor Γ in terms of well-known thermodynamic properties, and the Gibbs equation for specific properties, $Tds = \cdots$.

1.7. Ideal Gas with Constant Specific Heats

For an ideal gas at constant specific heats, the thermal and caloric equations of state read $p(\rho, T) = \rho RT$, $u(\rho, T) = c_v(T - T_0)$ with gas constant R, specific heat at constant volume c_v. Determine enthalpy $h(\rho, T)$, entropy $s(\rho, T)$ and Gibbs free energy $g(\rho, T)$. Use definition of properties and Gibbs equation $Tds = du - \frac{p}{\rho^2} d\rho$.

1.8. Ideal Gas Mixture

Show that ideal gas mixtures have neither mixing volume nor enthalpy of mixing, that is $V_{mix} = H_{mix} = 0$.

1.9. Gibbs Equation for Mixtures

We found the Gibbs equation for mixtures as

$$Tds = du - \frac{p}{\rho^2} d\rho - \sum_{\alpha=1}^{\nu} \mu_\alpha d\mathsf{c}_\alpha$$

1. Introduce the Gibbs free energy $g = u + \frac{p}{\rho} - Ts = h - Ts = \sum_\alpha \mathsf{c}_\alpha \mu_\alpha$ and show that the Gibbs equation can be written in the alternative form

$$dg = -sdT + \frac{1}{\rho} dp + \sum_{\alpha=1}^{\nu} \mu_\alpha d\mathsf{c}_\alpha$$

2. Because of $\sum_{\alpha=1}^{\nu} \mathsf{c}_\alpha = 1$, the c_α are not independent. Show that therefore the Gibbs equation can be re-written as

$$dg = -sdT + \frac{1}{\rho} dp + \sum_{\alpha=1}^{\nu-1} (\mu_\alpha - \mu_\nu) d\mathsf{c}_\alpha$$

3. Show that this implies

$$\mu_\beta - \mu_\nu = \left(\frac{\partial g}{\partial \mathsf{c}_\beta}\right)_{T,p,\mathsf{c}_\alpha\ (\alpha\neq\beta)} \quad \text{and hence} \quad \frac{\partial\left(\mu_\beta - \mu_\nu\right)}{\partial \mathsf{c}_\alpha} = \frac{\partial\left(\mu_\alpha - \mu_\nu\right)}{\partial \mathsf{c}_\beta}$$

4. Multiply the original Gibbs equation with total mass density ρ and show that

$$Td\left(\rho s\right) = d\left(\rho u\right) - \sum_{\alpha=1}^{\nu} \mu_\alpha d\rho_\alpha$$

5. From this form of the Gibbs equation, show that

$$\mu_\beta = \left(\frac{\partial\left(\rho u - T\rho s\right)}{\partial \rho_\beta}\right)_{T,\rho_\alpha\ (\alpha\neq\beta)} \quad \text{and hence} \quad \frac{\partial \mu_\beta}{\partial \rho_\alpha} = \frac{\partial \mu_\alpha}{\partial \rho_\beta}$$

1.10. Gibbs-Duhem Equation

1. Using $g = \sum_\alpha \mu_\alpha \mathsf{c}_\alpha$ and the Gibbs equation, derive the Gibbs-Duhem equation

$$0 = -sdT + vdp - \sum_\alpha \mathsf{c}_\alpha d\mu_\alpha$$

2. Show that the Gibbs-Duhem equation implies

$$\sum_\alpha \mathsf{c}_\alpha \left(\frac{\partial \mu_\alpha}{\partial \mathsf{c}_\beta}\right)_{T,p,\mathsf{c}_\gamma(\gamma\neq\beta)} = \sum_\alpha \mathsf{c}_\alpha \left(\frac{\partial \mu_\beta}{\partial \mathsf{c}_\alpha}\right)_{T,p,\mathsf{c}_\gamma(\gamma\neq\beta)} = 0$$

3. Write the differential of the chemical potential as

$$\begin{aligned} d\mu_\alpha &= \left(\frac{\partial \mu_\alpha}{\partial T}\right)_{p,\mathsf{c}_\beta} dT + \left(\frac{\partial \mu_\alpha}{\partial p}\right)_{T,\mathsf{c}_\beta} dp + \sum_\gamma \left(\frac{\partial \mu_\alpha}{\partial \mathsf{c}_\gamma}\right)_{T,p,\mathsf{c}_\beta} d\mathsf{c}_\gamma \\ &= -s_\alpha\left(T,p,\mathsf{c}_\beta\right) dT + \mathsf{v}_\alpha\left(T,p,\mathsf{c}_\beta\right) dp + \sum_\gamma \left(\frac{\partial \mu_\alpha}{\partial \mathsf{c}_\gamma}\right)_{T,p,\mathsf{c}_\beta} d\mathsf{c}_\gamma \end{aligned}$$

and show that this is sound, if

$$s = \sum_\alpha \mathsf{c}_\alpha s_\alpha\left(T,p,\mathsf{c}_\beta\right) \quad \text{and} \quad \mathsf{v} = \sum_\alpha \mathsf{c}_\alpha \mathsf{v}_\alpha\left(T,p,\mathsf{c}_\beta\right) .$$

1.11. Chemical Potential of Ideal Mixture

Consider an ideal mixture with $\mu_a = g_\alpha\left(T,p\right) + R_\alpha T \ln X_\alpha$, where $g_\alpha\left(T,p\right)$ is Gibbs free energy of α alone at temperature and pressure of the mixture, and X_α is the mole fraction.

1. Relate entropy $s_\alpha\left(T,p,\mathsf{c}_\beta\right)$ and specific volume $\mathsf{v}_\alpha\left(T,p,\mathsf{c}_\beta\right)$ of the component in the mixture to entropy and volume of the component alone, which are $s_\alpha\left(T,p\right) = -\left(\frac{\partial g_\alpha}{\partial T}\right)_p$ and $\mathsf{v}_\alpha\left(T,p\right) = \left(\frac{\partial g_\alpha}{\partial p}\right)_T$.

2. Determine the matrix $\left(\frac{\partial(\mu_\alpha-\mu_\nu)}{\partial \mathsf{c}_\gamma}\right)_{T,p,\mathsf{c}_\beta}$ and show that it is symmetric. Note: this is non-trivial, since μ_a is given in terms of X_α, but derivative is with respect to c_α.
3. Confirm by explicit calculations that $\sum_\alpha \mathsf{c}_\gamma \left(\frac{\partial \mu_\alpha}{\partial \mathsf{c}_\gamma}\right)_{T,p,\mathsf{c}_\beta} = 0$.

1.12. Mixing of Argon and Helium

An adiabatic cylinder is closed by a moveable piston. The cylinder contains 4 litres of argon at 150 kPa and a rubber balloon which contains 1 litre of helium at 3 bar. Both gases are initially at a temperature of 25 °C, and the piston rests due to its own weight. Assume that the pressure-volume characteristic of the balloon is of the form $\Delta p = a\,(V_B - V_0)^2$ where Δp is the pressure difference between inside and outside, V_B is the actual volume, the reference volume V_0 is 0.25 litres, and a is a material constant. This implies that the balloon shell stores some energy when stretched. As soon as the balloon has reached the volume V_0, there are no further stresses in the balloon, and the balloon shell will just collapse. For simplicity, ignore the thermal mass of the balloon.

A small hole opens in the balloon through which all helium escapes; in the final equilibrium state the gases are mixed.

1. Determine pressure, temperature and cylinder volume in the final equilibrium state.
2. Determine the entropy changes for both gases, and the total entropy generated in the process. Compare to the entropy of mixing, S_{mix}, and discuss the difference.

1.13. Taylor Series

Throughout the next chapters we will make frequent use of Taylor expansions. Refresh your knowledge on Taylor series and expand the functions $\frac{1}{1+x}$, $\ln(1+x)$, $\cos x$ into series around $x = 0$ and $x = 1$ accounting for n terms in the series. Plot original function and series for $n = 1, \ldots, 10$ (or so). Chose the plot range carefully, and discuss quality and convergence of the series.

Chapter 2
An Inert Mixture in Equilibrium

The study of transport phenomena focusses on nonequilibrium states and processes. A clear idea of nonequilibrium requires a proper understanding of what states a system will assume in equilibrium. In this chapter we study equilibrium states for a non-reacting mixture in the gravitational field, and the requirements for its stability. The second law is evaluated by maximizing entropy, and using methods of variational calculus. The importance of the chemical potential in the thermodynamic description of processes involving mixtures is highlighted.

2.1 Equilibrium States

Expressed in words, the second law of thermodynamics states that an isolated system left to itself will, after some time, assume a stable global equilibrium state in which it will remain until it is manipulated by some action [TEC, Chap. 4]. When a manipulation occurs, for instance a change of the boundary conditions, the system is brought out of the equilibrium state, and processes occur within the system, which lead to change of the local state (temperature, pressure, etc.). When the system is isolated after the manipulation stops, processes will continue to occur in the system, which move it towards a new equilibrium state. That is, transport processes such as mass and heat transfer, or macroscopic motion occur due to deviation from equilibrium states.

The sound knowledge of equilibrium states forms the base for deeper understanding of transport behavior, which will be seen as driven by processes that move the system towards equilibrium states. Below, we determine equilibrium states in mixtures of liquids or gases in the gravitational field, which result from maximizing entropy in equilibrium and use of variational calculus. Stability conditions follow from evaluation of the second variation.

The material presented below is the extension to mixtures of the discussion in [TEC, Chap. 17], here it serves as a review of important elements of

H. Struchtrup, *A Thermodynamic Introduction to Transport Phenomena*,
https://doi.org/10.1007/978-3-031-61868-0_2

equilibrium thermodynamics. In a first reading of this book, or in a course focussing on transport, one might skip this chapter without loss of context.

2.2 Problem Statement

We consider a mixture of ν components in an isolated container in the gravitational field g, and ask for the equilibrium values of the partial mass densities $\rho_\alpha(\mathbf{x})$, partial velocities $\mathbf{v}_\alpha(\mathbf{x})$, and system temperature $T(\mathbf{x})$ as functions in space.

Since the container is isolated, and no reactions occur, all partial masses, total momentum, and total energy remain fixed at all times, so that

$$m_\alpha = \int_V \rho_\alpha dV \tag{2.1}$$

$$\mathbf{M} = \sum_\alpha \int_V \rho_\alpha \mathbf{v}_\alpha dV \tag{2.2}$$

$$E = U + E_{\text{kin}} + E_{\text{pot}} = \sum_\alpha \int_V \rho_\alpha \left(u_\alpha + \frac{1}{2} v_\alpha^2 + \mathrm{g}z \right) dV \tag{2.3}$$

remain constant. The total mass in the system is $m = \sum_\alpha m_\alpha$.

For theis isolated system the second law of thermodynamics (1.10) reduces to

$$\frac{dS}{dt} = \dot{S}_{gen} \geq 0 \ , \tag{2.4}$$

with entropy generation positive in non-equilibrium, and vanishing in equilibrium. Accordingly, the total entropy of the system will grow ($\dot{S}_{gen} > 0$) until it reaches a maximum in the final equilibrium state ($\dot{S}_{gen} = 0$), i.e.,

$$S = \sum_\alpha \int_V \rho_\alpha s_\alpha dV \Rightarrow \max \ . \tag{2.5}$$

This behavior is just how entropy and the second law are introduced: Experience shows that an isolated system will evolve toward a final equilibrium state, and remain in this state as long as the system remains isolated.

To find the detailed equilibrium state, that is the location dependent fields $\rho_\alpha(\mathbf{x})$, $\mathbf{v}_\alpha(\mathbf{x})$, $T(\mathbf{x})$, we maximize entropy under constraints of given values for partial masses m_α, momentum $\mathbf{M}$, and energy E.

2.3 Evaluation with Variational Calculus

We formulate the maximization problem using Lagrange multipliers Λ_α, $\mathbf{\Lambda}_M$, Λ_E that account for the constraints.[1] The Lagrange multipliers are constants for the maximization problem, and will appear in the solutions for the final equilibrium states. Their values can be found from the fixed values for m_α, $\mathbf{M}$, E and the equilibrium fields that result from the maximization procedure.

Taking care of the constraints through the Lagrange multipliers, we need to maximize, without constraints, the function

$$\begin{aligned}\Phi = &\sum_\alpha \int_V \rho_\alpha s_\alpha dV \\ &- \sum_\alpha \Lambda_\alpha \left[\int_V \rho_\alpha dV - m_\alpha \right] \\ &- \mathbf{\Lambda}_M \cdot \left[\sum_\alpha \int_V \rho_\alpha \mathbf{v}_\alpha dV - \mathbf{M} \right] \\ &- \Lambda_E \left[\sum_\alpha \int_V \rho_\alpha \left(u_\alpha + \frac{1}{2} v_\alpha^2 + \mathrm{g}z \right) dV - E \right] \end{aligned} \tag{2.6}$$

Note that Φ is a function of the location dependent fields $\rho_\alpha(\mathbf{x})$, $\mathbf{v}_\alpha(\mathbf{x})$, $T(\mathbf{x})$, which also appear in the property relations for energy and entropy,

$$u_\alpha(\mathbf{x}) = u_\alpha(T(\mathbf{x}), \rho_\beta(\mathbf{x})) \quad , \quad s_\alpha(\mathbf{x}) = s_\alpha(T(\mathbf{x}), \rho_\beta(\mathbf{x})) \quad . \tag{2.7}$$

The problem formulation is valid for arbitrary constituents and phases, hence there is no need to specify the property relations for energies and entropies of the constituents. We emphasize, however, that in a mixture energy and entropy of any constituent (α) might depend on the mass densities of *all other* constituents, ρ_β ($\beta = 1, \ldots, \nu$), as well as on temperature, i.e., $u_\alpha\left(T, \rho_{\beta|(\beta=1,\ldots,\nu)}\right)$, $s_\alpha\left(T, \rho_{\beta|(\beta=1,\ldots,\nu)}\right)$. The above notation accounts for this in a compact way.[2]

For conveniently compact notation, we combine the variables for this problem in the vector

$$\mathbf{Y}(\mathbf{x}) = \{\rho_\alpha(\mathbf{x}), \mathbf{v}_\alpha(\mathbf{x}), T(\mathbf{x})\} \quad , \tag{2.8}$$

---

[1] The method of Lagrange multipliers is briefly explained at the end of this chapter, in Sec. 2.6.

[2] The notation is such that if the index is the same on the property and its arguments, e.g., $f_\alpha(\rho_\alpha)$, then dependence is on just the one ρ_α with the same index as f_α. When the indices of property and arguments differ, e.g. $f_\alpha(\rho_\beta)$, then the dependence on *all* ρ_β ($\beta = 1, \ldots, \nu$) is implied. Note that indices α, β are place holders, hence any other letters can be used, e.g., $f_\gamma(\rho_\gamma)$, $f_\gamma(\rho_\delta)$, etc.

and introduce the functions ($\alpha = 1, \ldots, \nu$)

$$X_\alpha(\mathbf{Y}) = \rho_\alpha \left[s_\alpha - \Lambda_\alpha - \mathbf{\Lambda}_M \cdot \mathbf{v}_\alpha - \Lambda_E \left(u_\alpha + \frac{1}{2} v_\alpha^2 + \mathrm{g} z \right) \right] , \qquad (2.9)$$

so that the maximization problem (2.6) can be written as

$$\hat{\Phi} = \int \sum_\alpha X_\alpha(\mathbf{Y})\, dV \Rightarrow \max \ . \qquad (2.10)$$

The new function $\hat{\Phi} = \Phi - \left(\sum_\alpha \Lambda_\alpha m_\alpha + \mathbf{\Lambda}_M \cdot \mathbf{M} + \Lambda_E E\right)$ differs from the function Φ only by subtracted constants.

We are interested in the equilibrium fields $\mathbf{Y}_E(\mathbf{x})$ for which $\hat{\Phi}$ assumes the maximum value. The maximization problem is solved using the methods of variational calculus.

For this we consider arbitrary small variations $\delta\mathbf{Y}(\mathbf{x})$ of the equilibrium solution $\mathbf{Y}_E(\mathbf{x})$, so that nonequilibrium states close to equilibrium are given by

$$\mathbf{Y}(\mathbf{x}) = \mathbf{Y}_E(\mathbf{x}) + \delta\mathbf{Y}(\mathbf{x}) \ . \qquad (2.11)$$

Next, we use a Taylor series to evaluate

$$X_\alpha(\mathbf{Y}) = X_\alpha(\mathbf{Y}_E + \delta\mathbf{Y}) = X_\alpha(\mathbf{Y}_E) + \frac{\partial X_\alpha}{\partial \mathbf{Y}}_{|E} \cdot \delta\mathbf{Y} + \frac{1}{2}\delta\mathbf{Y} \cdot \frac{\partial^2 X_\alpha}{\partial \mathbf{Y} \partial \mathbf{Y}}_{|E} \cdot \delta\mathbf{Y}, \qquad (2.12)$$

where the index $|E$ indicates evaluation with $\mathbf{Y}_E(\mathbf{x})$. One must be careful with the rather compact notation: $\delta\mathbf{Y}$ is a vector with $(2\nu + 1)$ elements, $\frac{\partial X_\alpha}{\partial \mathbf{Y}}$ are vectors (for each α) and $\frac{\partial^2 X_\alpha}{\partial \mathbf{Y}\partial \mathbf{Y}}$ are matrices (for each α).

The function to be maximized appears as the sum of three contributions,

$$\hat{\Phi} = \int \sum_\alpha X_\alpha(\mathbf{Y}_E)\, dV + \int \sum_\alpha \frac{\partial X_\alpha}{\partial \mathbf{Y}}_{|E} \cdot \delta\mathbf{Y} dV + \int \sum_\alpha \frac{1}{2}\delta\mathbf{Y} \cdot \frac{\partial^2 X_\alpha}{\partial \mathbf{Y}\partial \mathbf{Y}}_{|E} \cdot \delta\mathbf{Y} dV \qquad (2.13)$$

Since $\mathbf{Y}_E$ is the equilibrium solution, and in equilibrium the function $\hat{\Phi}$ assumes a maximum $\hat{\Phi}_E$, we have

$$\hat{\Phi}_E = \int \sum_\alpha X_\alpha(\mathbf{Y}_E)\, dV \geq \hat{\Phi} \quad \text{for all} \quad \delta\mathbf{Y}(\mathbf{x}) \ . \qquad (2.14)$$

Accordingly, both the second and third terms in (2.13) must be independently non-positive (i.e., negative or zero) for *all* possible deviations $\delta\mathbf{Y}(\mathbf{x})$,

$$\int \sum_\alpha \frac{\partial X_\alpha}{\partial \mathbf{Y}}_{|E} \cdot \delta\mathbf{Y} dV \leq 0 \ , \qquad (2.15)$$

$$\int \sum_{\alpha} \frac{1}{2} \delta \mathbf{Y} \cdot \frac{\partial^2 X_\alpha}{\partial \mathbf{Y} \partial \mathbf{Y}}_{|E} \cdot \delta \mathbf{Y} dV \leq 0 \quad \text{for all} \quad \delta \mathbf{Y}(\mathbf{x}) \ . \tag{2.16}$$

The variation $\delta\mathbf{Y}(\mathbf{x})$ is an arbitrary function in space, hence we can always consider a variation $\delta\mathbf{Y}(\mathbf{x})$ that is zero everywhere in the volume V apart from just one location.[3] With that, we have the pointwise condition that $\sum_\alpha \frac{\partial X_\alpha}{\partial \mathbf{Y}}_{|E} \cdot \delta\mathbf{Y}$ must be negative for *any* value of $\delta\mathbf{Y}$. This is only possible for the equilibrium condition

$$\sum_{\alpha} \frac{\partial X_\alpha}{\partial \mathbf{Y}}_{|E} = 0 \ . \tag{2.17}$$

Indeed, if this was non-zero, we just have to chose a vector $\delta\mathbf{Y}$ pointing in the direction of $\sum_\alpha \frac{\partial X_\alpha}{\partial \mathbf{Y}}_{|E}$ to violate the statement that $\hat{\Phi}_E$ is maximum.

For the third term it suffices that

$$\frac{1}{2} \sum_{\alpha} \frac{\partial^2 X_\alpha}{\partial \mathbf{Y} \partial \mathbf{Y}}_{|E} \quad - \quad \text{negative definite} \tag{2.18}$$

this condition guarantees stability of the equilibrium solution.[4]

Relations (2.17, 2.18) are the multidimensional generalizations of the requirements $f'(x_0) = 0$ and $f''(x_0) < 0$ for the relative maximum of a function $f(x)$ at x_0. We proceed with evaluating both conditions.

2.4 Thermodynamic Equilibrium States

We need to evaluate the stability requirement $\sum_\alpha \frac{\partial X_\alpha}{\partial \mathbf{Y}}_{|E} = 0$ for the variables $\mathbf{Y} = \{\rho_\beta, \mathbf{v}_\beta, T\}$. The indices used are carefully chosen: α and β are both place holders, and α is used for the summation. When we write $X_\alpha(\rho_\beta, \mathbf{v}_\beta, T)$ it is implied that X_α (for any α) depends on densities and velocities of *all* components ($\beta = 1, \ldots, \nu$).

Making the derivatives explicit, we find the individual requirements

$$\sum_{\alpha} \frac{\partial X_\alpha}{\partial \rho_\beta}_{|E} = 0 \quad , \quad \beta = 1, \ldots, \nu \tag{2.19}$$

$$\sum_{\alpha} \frac{\partial X_\alpha}{\partial \mathbf{v}_\beta}_{|E} = 0 \quad , \quad \beta = 1, \ldots, \nu \tag{2.20}$$

---

[3] For instance $\delta\mathbf{Y}(\mathbf{x})$ can be chosen as a Dirac delta function centered at arbitrary points in the system.

[4] A matrix $\mathbf{A}$ is negative definite, if $\mathbf{a} \cdot \mathbf{A} \cdot \mathbf{a} < 0$ for any vector $\mathbf{a}$. This is the case if all eigenvalues of $\mathbf{A}$ are negative.

$$\sum_\alpha \frac{\partial X_\alpha}{\partial T}_{|E} = 0 , \tag{2.21}$$

where X_α is given by (2.9). For the next steps, recall that the Lagrange multipliers are constants, i.e., they do neither depend on $\mathbf{Y}$ nor $\mathbf{x}$.

2.4.1 Mechanical Equilibrium

For convenience we first evaluate the condition for mechanical equilibrium (2.20), which reduces to

$$0 = \sum_\alpha \frac{\partial X_\alpha}{\partial \mathbf{v}_\beta}_{|E} = \sum_\alpha \rho_\alpha \left[-\mathbf{\Lambda}_M \cdot \frac{\partial \mathbf{v}_\alpha}{\partial \mathbf{v}_\beta} - \frac{1}{2} \Lambda_E \frac{\partial v_\alpha^2}{\partial \mathbf{v}_\beta} \right]_{|E} . \tag{2.22}$$

To further simplify, we use that

$$\frac{\partial \mathbf{v}_\alpha}{\partial \mathbf{v}_\beta} = \delta_{\alpha\beta} = \begin{cases} 1 \text{ if } \alpha = \beta \\ 0 \text{ if } \alpha \neq \beta \end{cases} , \tag{2.23}$$

$$\frac{1}{2} \frac{\partial v_\alpha^2}{\partial \mathbf{v}_\beta} = \frac{1}{2} \frac{\partial \mathbf{v}_\alpha \cdot \mathbf{v}_\alpha}{\partial \mathbf{v}_\beta} = \mathbf{v}_\alpha \cdot \frac{\partial \mathbf{v}_\alpha}{\partial \mathbf{v}_\beta} = \mathbf{v}_\alpha \delta_{\alpha\beta} , \tag{2.24}$$

so that for all β

$$\sum_\alpha \rho_\alpha \left[-\mathbf{\Lambda}_M \delta_{\alpha\beta} - \frac{1}{2} \Lambda_E \mathbf{v}_\alpha \delta_{\alpha\beta} \right]_{|E} = \rho_\beta \left[-\mathbf{\Lambda}_M - \Lambda_E \mathbf{v}_\beta \right]_{|E} = 0 . \tag{2.25}$$

We finally conclude that

$$\mathbf{v}_{\beta|E} = -\frac{\mathbf{\Lambda}_M}{\Lambda_E} , \quad \beta = 1, \ldots, \nu . \tag{2.26}$$

With this the equilibrium velocities of *all* components at *all* points are given by the ratio of the Lagrange multipliers. For the total momentum this implies

$$\mathbf{M} = \sum_\alpha \int_V \left[\rho_\alpha \mathbf{v}_\alpha \right]_{|E} dV = -\frac{\mathbf{\Lambda}_M}{\Lambda_E} \sum_\alpha \int_V \rho_{\alpha|E} dV = -\frac{\mathbf{\Lambda}_M}{\Lambda_E} \sum_\alpha m_\alpha = -\frac{\mathbf{\Lambda}_M}{\Lambda_E} m \tag{2.27}$$

The Lagrange multipliers $\mathbf{\Lambda}_M$, Λ_E are related to the fixed total momentum $\mathbf{M}$ of the system. For an observer, momentum depends on their velocity relative to the system. An observer resting with the center of mass of the system measures the momentum $\mathbf{M} = 0$, which implies $\mathbf{\Lambda}_M = \mathbf{v}_{\alpha|E} = 0$. Hence for such an observer, in equilibrium the mixture itself, and each of its components, are at rest. We evaluate the other conditions in the system rest frame.

2.4.2 Thermal Equilibrium

We proceed with the condition for thermal equilibrium (2.21), which becomes (with $\mathbf{\Lambda}_M = \mathbf{v}_{\alpha|E} = 0$)

$$0 = \sum_\alpha \left[\frac{\partial \rho_\alpha s_\alpha}{\partial T} - \Lambda_E \frac{\partial \rho_\alpha u_\alpha}{\partial T} \right]_{|E} . \tag{2.28}$$

Since entropy and energy density of the mixture are given by

$$\rho s = \sum_\alpha \rho_\alpha s_\alpha \quad , \quad \rho u = \sum_\alpha \rho_\alpha u_\alpha \ , \tag{2.29}$$

this reduces further to

$$0 = \left[\left(\frac{\partial \rho s}{\partial T} \right)_{\rho_\beta} - \lambda_E \left(\frac{\partial \rho u}{\partial T} \right)_{\rho_\beta} \right]_{|E} , \tag{2.30}$$

where the index ρ_β indicates the other variables of entropy and energy, in the usual notation of thermodynamics.[5] With the Gibbs equation in the form (1.45), we find

$$T \left(\frac{\partial \rho s}{\partial T} \right)_{\rho_\beta} = \left(\frac{\partial \rho u}{\partial T} \right)_{\rho_\beta} \tag{2.31}$$

and finally arrive at

$$0 = \left[\left(\frac{1}{T} - \Lambda_E \right) \left(\frac{\partial \rho u}{\partial T} \right)_{\rho_\beta} \right]_{|E} . \tag{2.32}$$

We conclude that in equilibrium the temperature equals the inverse Lagrange multiplier,

$$T(\mathbf{x})_{|E} = T_{|E} = \frac{1}{\Lambda_E} . \tag{2.33}$$

Since the Lagrange multiplier is a constant for the system, it follows that in equilibrium temperature is homogeneous. Just as with vanishing component velocities, this result is not surprising, but satisfying. Indeed, temperature is introduced as that measurable property which is homogeneous in equilibrium states [TEC, Sec. 2.11].

The value of the equilibrium temperature, and hence the Lagrange multiplier, can be determined from the total energy E in the system.

---

[5] See, e.g., the definitions of specific heats at constant volume and constant pressure, $c_{\mathsf{v}} = \left(\frac{\partial u}{\partial T}\right)_{\mathsf{v}}$ and $c_p = \left(\frac{\partial h}{\partial T}\right)_p$, respectively.

2.4.3 Chemical Equilibrium

The conditions for chemical equilibrium (2.19) give a less intuitive result; they read

$$0=\sum_{\alpha}\frac{\partial X_{\alpha}}{\partial\rho_{\beta}}_{|E}=\sum_{\alpha}\left[\frac{\partial\rho_{\alpha}s_{\alpha}}{\partial\rho_{\beta}}-\Lambda_{\alpha}\frac{\partial\rho_{\alpha}}{\partial\rho_{\beta}}-\Lambda_{E}\frac{\partial\rho_{\alpha}u_{\alpha}}{\partial\rho_{\beta}}-\Lambda_{E}\mathrm{g}z\frac{\partial\rho_{\alpha}}{\partial\rho_{\beta}}\right]_{|E}. \tag{2.34}$$

To evaluate, we use that the partial densities are independent, $\frac{\partial\rho_{\alpha}}{\partial\rho_{\beta}}=\delta_{\alpha\beta}$, and introduce energy and entropy densities as above, so that for each $\beta=1,\ldots,\nu$

$$0=\left[\left(\frac{\partial\rho s}{\partial\rho_{\beta}}\right)-\Lambda_{\beta}-\Lambda_{E}\left(\frac{\partial\rho u}{\partial\rho_{\beta}}\right)-\lambda_{E}\mathrm{g}z\right]_{|E}. \tag{2.35}$$

With the Gibbs equation in the form (1.45) we find

$$T\left(\frac{\partial\rho s}{\partial\rho_{\beta}}\right)=\left(\frac{\partial\rho u}{\partial\rho_{\beta}}\right)-\sum_{\alpha=1}^{\nu}\mu_{\alpha}\frac{\partial\rho_{\alpha}}{\partial\rho_{\beta}}=\left(\frac{\partial\rho u}{\partial\rho_{\beta}}\right)-\mu_{\beta}. \tag{2.36}$$

With this, and $T=T_{|E}=1/\Lambda_{E}$ in equilibrium, the equilibrium condition reduces to

$$0=\left[-\frac{\mu_{\beta}}{T}-\Lambda_{\beta}-\frac{1}{T}\mathrm{g}z\right]_{|E} \tag{2.37}$$

The result becomes a bit more instructive when we write

$$\mu_{\beta|E}+\mathrm{g}z=-T_{|E}\Lambda_{\beta}=-\frac{\Lambda_{\beta}}{\Lambda_{E}}. \tag{2.38}$$

Since the Lagrange multipliers Λ_{E}, Λ_{β} are space independent, in equilibrium the sum of chemical potential $\mu_{\beta|E}=\mu_{\beta}\left(\rho_{\gamma|E},T|_{E}\right)$ and potential energy $\mathrm{g}z$ is a constant; we identify $-\frac{\Lambda_{\beta}}{\Lambda_{E}}=\mu^{0}_{\beta|E}$ as the chemical potential of component β at height $z=0$.

The equilibrium values for the partial mass densities $\rho_{\beta|E}(\mathbf{x})$ follow from evaluation of these conditions and depend on the property relations for the components.

For single substances the chemical potential equals the Gibbs free energy, hence (2.38) reduces to

$$g\left(T,p\left(z\right)\right)+\mathrm{g}z=g\left(T,p^{0}\right), \tag{2.39}$$

where p^{0} is the pressure at reference height $z=0$ and $p(z)$ is pressure at z.

For an ideal gas $g(T,p)=h(T)-T\left(s^{0}(T)-R\ln\frac{p}{p_{0}}\right)$, where p_{0} is the standard reference pressure, hence pressure obeys the barometric formula

$$p(z) = p^0 \exp\left[-\frac{gz}{RT}\right] . \tag{2.40}$$

For incompressible fluids, $\rho = \rho_0 = const.$, and internal energy and entropy depend only on temperature, so that $g(T,p) = u(T) + \frac{p}{\rho_0} - Ts(T)$, hence pressure obeys the hydrostatic pressure law, which is often written for depth $h = -z$,

$$p = p^0 - \rho_0 gz = p^0 + \rho_0 gh . \tag{2.41}$$

Evaluation of (2.38) for mixtures yields extensions of these relations. In ideal gas mixtures one finds the barometric formula for each component (Prob. 2.3), while for liquids mixtures one finds a variation of the hydrostatic pressure equation (Prob. 2.4).

In systems of moderate size, the influence of gravity is small, and can be neglected. Then, in equilibrium the chemical potentials of all components are homogenous, $\mu_\beta = -\Lambda_\beta/\Lambda_E$. If there is a single phase, the partial mass densities ρ_β and partial pressures p_β are homogeneous as well, while in multiphase equilibrium systems, the partial mass densities will be homogeneous within a phase, but will differ between phases.

In summary, we found that in equilibrium, temperature T, component velocities $\mathbf{v}_\alpha$, and chemical potentials μ_α are all homogeneous, and all component velocities agree. The body of chemical equilibrium thermodynamics relies on further evaluation of these requirements [TEC, Chaps. 20-26].

We note that this analysis can only determine the final equilibrium state, but gives no insight on how the state is approached over time by irreversible processes. Full resolution of the *process towards equilibrium* requires transport laws as partial differential equations, which will be discussed in the subsequent chapters.

2.5 Thermodynamic Stability

The condition of vanishing first derivatives does not suffice as criterion for maximum entropy, we also need to consider second derivatives, for which variational calculus gives the stability condition (2.18)

$$\mathbf{A} = \frac{1}{2}\sum_\alpha \frac{\partial^2 X_\alpha}{\partial \mathbf{Y} \partial \mathbf{Y}}_{|E} \quad - \text{ negative definite} . \tag{2.42}$$

We proceed with the evaluation of this stability requirement. Note that we must *first* take all derivatives, and *then* insert equilibrium values from above.

To clarify the derivatives with respect to the variables summarized in $\mathbf{Y}$, we write the (symmetric) matrix explicitly

$$\mathbf{A} = \frac{1}{2}\begin{bmatrix} \sum_\alpha \frac{\partial^2 X_\alpha}{\partial\rho_\beta\partial\rho_\gamma} & \sum_\alpha \frac{\partial^2 X_\alpha}{\partial\rho_\beta\partial\mathbf{v}_\gamma} & \sum_\alpha \frac{\partial^2 X_\alpha}{\partial\rho_\beta\partial T} \\ \sum_\alpha \frac{\partial^2 X_\alpha}{\partial\mathbf{v}_\beta\partial\rho_\gamma} & \sum_\alpha \frac{\partial^2 X_\alpha}{\partial\mathbf{v}_\beta\partial\mathbf{v}_\gamma} & \sum_\alpha \frac{\partial^2 X_\alpha}{\partial\mathbf{v}_\beta\partial T} \\ \sum_\alpha \frac{\partial^2 X_\alpha}{\partial T\partial\rho_\gamma} & \sum_\alpha \frac{\partial^2 X_\alpha}{\partial T\partial\mathbf{v}_\gamma} & \sum_\alpha \frac{\partial^2 X_\alpha}{\partial T\partial T} \end{bmatrix}_{|E} \tag{2.43}$$

Here we use a rather compact notation, in which the elements of this matrix are matrices themselves. We show some, but not all details of the ensuing evaluation.

For the elements in the first row, we find from (2.9)

$$\sum_\alpha \frac{\partial^2 X_\alpha}{\partial\rho_\beta\partial\rho_\gamma} = \sum_\alpha \left[\frac{\partial^2 \rho_\alpha s_\alpha}{\partial\rho_\beta\partial\rho_\gamma} - \Lambda_E \frac{\partial^2 \rho_\alpha u_\alpha}{\partial\rho_\beta\partial\rho_\gamma}\right] \tag{2.44}$$

$$\sum_\alpha \frac{\partial^2 X_\alpha}{\partial\rho_\beta\partial\mathbf{v}_\gamma} = -\sum_\alpha \delta_{\alpha\beta}\left[\mathbf{\Lambda}_M\delta_{\alpha\gamma} + \Lambda_E\mathbf{v}_\alpha\delta_{\gamma\alpha}\right] \tag{2.45}$$

$$\sum_\alpha \frac{\partial^2 X_\alpha}{\partial\rho_\beta\partial T} = \sum_\alpha \left[\frac{\partial^2 \rho_\alpha s_\alpha}{\partial\rho_\beta\partial T} - \Lambda_E \frac{\partial^2 \rho_\alpha u_\alpha}{\partial\rho_\beta\partial T}\right] \tag{2.46}$$

Inserting equilibrium, with $\mathbf{\Lambda}_M = \mathbf{v}_\alpha = 0$, $\Lambda_E = \frac{1}{T_{|E}}$, and use of the Gibbs equation (1.45) gives

$$\left[\sum_\alpha \frac{\partial^2 X_\alpha}{\partial\rho_\beta\partial\rho_\gamma}\right]_{|E} = \frac{1}{T}\left[\frac{\partial}{\partial\rho_\gamma}\left(T\frac{\partial\rho s}{\partial\rho_\beta} - \frac{\partial\rho u}{\partial\rho_\beta}\right)\right]_{|E} = -\frac{1}{T}\left(\frac{\partial\mu_\beta}{\partial\rho_\gamma}\right)_T \tag{2.47}$$

$$\left[\sum_\alpha \frac{\partial^2 X_\alpha}{\partial\rho_\beta\partial\mathbf{v}_\gamma}\right]_{|E} = 0 \tag{2.48}$$

$$\left[\sum_\alpha \frac{\partial^2 X_\alpha}{\partial\rho_\beta\partial T}\right]_{|E} = \left[\frac{\partial}{\partial\rho_\beta}\left(\frac{\partial\rho s}{\partial T} - \frac{1}{T}\frac{\partial\rho u}{\partial T}\right)\right]_{|E} = \frac{\partial}{\partial\rho_\beta}(0) = 0 \tag{2.49}$$

The first entry above is symmetric, which implies that $\left(\frac{\partial\mu_\beta}{\partial\rho_\gamma}\right)_T = \left(\frac{\partial\mu_\gamma}{\partial\rho_\beta}\right)_T$ is symmetric—this is just another of the many bits of information one finds from the Gibbs equation.

The relevant elements of the second row give,

$$\sum_\alpha \frac{\partial^2 X_\alpha}{\partial\mathbf{v}_\beta\partial\mathbf{v}_\gamma} = \sum_\alpha \rho_\alpha\left[-\Lambda_E\delta_{\alpha\gamma}\delta_{\alpha\beta}\right] = -\Lambda_E\rho_\beta\delta_{\beta\gamma} \tag{2.50}$$

$$\sum_\alpha \frac{\partial^2 X_\alpha}{\partial\mathbf{v}_\beta\partial T} = 0 \tag{2.51}$$

and in equilibrium

$$\left[\sum_\alpha \frac{\partial^2 X_\alpha}{\partial \mathbf{v}_\beta \partial \mathbf{v}_\gamma}\right]_{|E} = -\frac{\rho_\beta}{T}\delta_{\beta\gamma} \tag{2.52}$$

$$\left[\sum_\alpha \frac{\partial^2 X_\alpha}{\partial \mathbf{v}_\beta \partial T}\right]_{|E} = 0 \tag{2.53}$$

Due to symmetry, we must look only at the last element from the third row,

$$\sum_\alpha \frac{\partial^2 X_\alpha}{\partial T \partial T} = \sum_\alpha \frac{\partial^2 \left[\rho_\alpha s_\alpha - \Lambda_E \rho_\alpha u_\alpha\right]}{\partial T \partial T} = \frac{\partial^2 \left[\rho s - \Lambda_E \rho u\right]}{\partial T \partial T} \tag{2.54}$$

Again, we must take the derivatives *before* evaluating in equilibrium. With the Gibbs equation (1.45), we find

$$\sum_\alpha \frac{\partial^2 X_\alpha}{\partial T \partial T} = \frac{\partial}{\partial T}\left[\left(\frac{1}{T} - \Lambda_E\right)\frac{\partial \rho u}{\partial T}\right] = -\frac{1}{T^2}\rho\frac{\partial u}{\partial T} + \left(\frac{1}{T} - \Lambda_E\right)\rho\frac{\partial^2 u}{\partial T^2}\,. \tag{2.55}$$

Here we have used that $\rho = \sum_\alpha \rho_\alpha$ and T are independent variables. Finally, inserting the equilibrium result $\Lambda_E = 1/T$ we find

$$\left[\sum_\alpha \frac{\partial^2 X_\alpha}{\partial T \partial T}\right]_{|E} = -\frac{1}{T^2}\rho\left(\frac{\partial u}{\partial T}\right)_\rho = -\frac{\rho}{T^2}c_v\,, \tag{2.56}$$

where $c_v = \left(\frac{\partial u}{\partial T}\right)_\rho$ is the specific heat of the mixture at constant volume (or, at constant density).

Finally, we summarize the complete matrix as

$$\mathbf{A} = \frac{1}{2}\begin{bmatrix} -\frac{1}{T}\left(\frac{\partial \mu_\beta}{\partial \rho_\gamma}\right)_T & 0 & 0 \\ 0 & -\frac{\rho_\beta}{T}\delta_{\beta\gamma} & 0 \\ 0 & 0 & -\frac{\rho}{T^2}c_v \end{bmatrix} \tag{2.57}$$

Stability requires this to be negative definite. Since density is always positive, this is the case when

$$\left(\frac{\partial \mu_\beta}{\partial \rho_\gamma}\right)_T \; - \text{ positive def.} \quad , \quad T > 0 \quad , \quad c_v > 0\,. \tag{2.58}$$

In order, these are the conditions for chemical stability (2nd variation of densities), mechanical stability (2nd variation of velocities), and thermal stability (2nd variation of temperature).

Specifically, positive thermodynamic temperature guarantees that it is impossible to obtain work from a single reservoir in equilibrium. Positive specific heat guarantees that heating (i.e., the increase of internal energy) leads to an increase in temperature. As always, the second law coincides with our experience—after all, it was constructed to do just that!

The condition for chemical equilibrium is less intuitive. To get an idea of its meaning, we evaluate for a single substance, for which it reduces to $\left(\frac{\partial g}{\partial \rho}\right)_T > 0$. To proceed, we consider the Gibbs equation in the form (1.28),

$$dg = -sdT + \frac{1}{\rho}dp \;, \tag{2.59}$$

so that the derivative is

$$\left(\frac{\partial g}{\partial \rho}\right)_T = \frac{1}{\rho}\left(\frac{\partial p}{\partial \rho}\right)_T > 0 \;. \tag{2.60}$$

Accordingly, in a stable system an increase of pressure must result in an increase of density, i.e., a decrease in volume.

2.6 Lagrange Multipliers

We briefly review the method of Lagrange multipliers.

The goal is to find the extremum of a function $f(x_i)$ of the variables x_i $(i = 1, \cdots, n)$ under a constraint $g(x_i) = 0$. At the extremum the variation of f vanishes,

$$df = \sum_{i=1}^{n} \frac{\partial f}{\partial x_i} dx_i = 0 \;. \tag{2.61}$$

Without the constraint, the variables x_i are independent, and one could arbitrarily chose $dx_i = 0$ for $i \neq k$ and $dx_k \neq 0$ so that $\partial f / \partial x_k = 0$ for all k. However, because of the constraint, the variables x_i are not independent, and their variations dx_i are related through

$$dg = \sum_{i=1}^{n} \frac{\partial g}{\partial x_i} dx_i = 0 \;. \tag{2.62}$$

This implies that the variations dx_i cannot be chosen independently. To solve the problem, the last equation is multiplied by a Lagrange multiplier Λ and added to the other, so that

$$df - \Lambda dg = \sum_{i=1}^{n} \left(\frac{\partial f}{\partial x_i} - \Lambda \frac{\partial g}{\partial x_i}\right) dx_i = 0 \;. \tag{2.63}$$

Still, the variations cannot be chosen independently, but now the Lagrange multiplier can be used to eliminate one term from the sum, by choosing it such that

$$\frac{\partial f}{\partial x_n} - \Lambda \frac{\partial g}{\partial x_n} = 0 \,. \tag{2.64}$$

The remaining condition reads

$$\sum_{i=1}^{n-1} \left(\frac{\partial f}{\partial x_i} - \Lambda \frac{\partial g}{\partial x_i} \right) dx_i = 0 \,, \tag{2.65}$$

where only $(n-1)$ variations dx_i $(i = 1, \cdots, n-1)$ appear, and these can be chosen independently, so that

$$\frac{\partial f}{\partial x_i} - \Lambda \frac{\partial g}{\partial x_i} = 0 \text{ for } i = 1, \cdots, n-1 \,. \tag{2.66}$$

With this, we have $\frac{\partial}{\partial x_i}(f - \Lambda g) = 0$ for all variables $(i = 1, \cdots, n)$, which is just the requirement for an extremum of $f - \Lambda g$.

That is instead of explicitly trying to find the extremum of $f(x_i)$ with constraint $g(x_i) = 0$, the method of Lagrange Multipliers offers as an elegant alternative the maximization of $(f - \Lambda g)$ without constraint. The Lagrange multiplier Λ is a constant for the problem at hand, and can be determined at the end of the calculations.

When several constraints must be considered, $g_\alpha(x_i) = 0$, one needs to use more Lagrange multipliers, that is one has to find the extremum of $f - \sum_\alpha \Lambda_\alpha g_\alpha$.

Problems

2.1. Equilibrium State

N blocks of different metals with masses m_α, specific heats c_α and initial temperatures T_α^{i} are enclosed in an adiabatic rigid chamber. All blocks are brought into thermal contact and equilibrate.

Show that all blocks will assume the same final temperature $T_\alpha^{\mathrm{f}} = T$, and determine this temperature.

Hint: In equilibrium total entropy $S = \sum_\alpha S_\alpha$ must be a maximum under constraint of given total energy $U = \sum_\alpha U_\alpha$. Energy and entropy of an incompressible solid with constant specific heat c_α are given by $U_\alpha = m_\alpha c_\alpha (T_\alpha - T_0)$, $S_\alpha = m_\alpha c_\alpha \ln \frac{T_\alpha}{T_0}$, where T_0 is a suitable reference temperature. You can either use the method of Lagrange multipliers to take care of the constraint, or consider it explicitly.

2.2. Equilibrium State

An insulated space contains air (mass m_a, specific heat c_a) and a rigid shelf on which rests a metal ball (mass m_b, specific heat c_b). The initial temperature of the system is T_0, and the shelf is at height H above the floor.

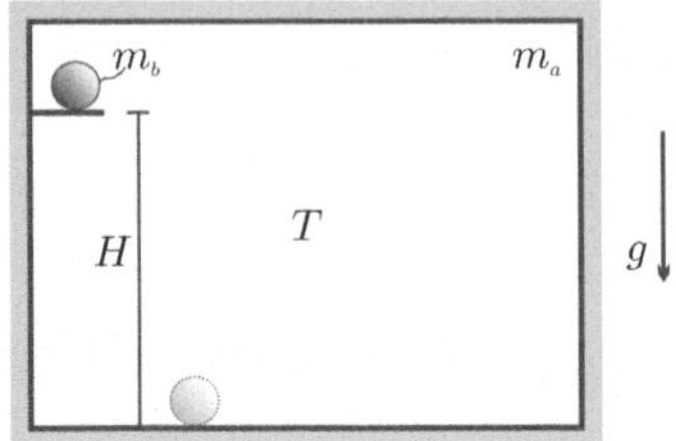

1. Use first and second law to decide which of the two states (ball on shelf, or on floor) is the thermodynamic equilibrium state.
2. Why are we not surprised if we find this system in the thermodynamically less favored configuration?

2.3. Chemical Potential of an Ideal Gas and Barometric Formula
Recall that for a component α of an ideal gas mixture, the partial pressure is $p_\alpha = \rho_\alpha R_\alpha T$, the total pressure of the mixture is $p = \sum_\alpha p_\alpha$, and energy, enthalpy and entropy of the component are given by $u_\alpha = u_\alpha(T)$, $h_\alpha = h_\alpha(T) = u_\alpha(T) + R_\alpha T$, $s_\alpha = s_\alpha^0(T) - R_\alpha \ln \frac{p_\alpha}{p_0}$. Note that $u_\alpha(T)$, $h_\alpha(T)$, $s_\alpha^0(T)$ are independent of composition.

The chemical potential of a component in the mixture is $\mu_\alpha = h_\alpha - T s_\alpha$, where h_a and s_α are enthalpy and entropy of the component in the mixture.

1. Use the above, to write the chemical potential of a component as explicit function of temperature and partial densities, $\mu_\alpha(\rho_\beta, T)$, and as explicit function of temperature, total mass density $\rho = \sum \rho_\alpha$, and mass fractions $\mathsf{c}_\alpha = \frac{\rho_\alpha}{\rho}$. Write the resulting expression for $\mu_\alpha(\rho, T, \mathsf{c}_\beta)$ as compact as possible
2. In equilibrium under a gravitational field with acceleration g, the chemical potential for a component obeys $\mu_\alpha + \mathsf{g}z = \mu_\alpha^0$, where z is height and μ_α^0 are the chemical potentials at $z = 0$. Show that this implies that each gas component obeys its own barometric formula.
3. Consider a mixture of oxygen, carbon dioxide, and helium in equilibrium, where the partial pressures at height $z = 0$ are $p^0_{O_2} = 0.4\,\text{bar}$, $p^0_{CO_2} = 0.2\,\text{bar}$, $p^0_{He} = 0.1\,\text{bar}$. For $\mathsf{g} = 32\frac{\text{m}}{\text{s}^2}$ and $T = 260\,\text{K}$ plot: total and partial pressures, $p(z)$ and $p_\alpha(z)$; total and partial mass densities, $\rho(z)$ and $\rho_\alpha(z)$, and the mass fractions $\mathsf{c}_\alpha(z)$ for heights up to 30km
4. Discuss how changes in temperature or gravitational acceleration affect these curves.

2.4. Binary Liquid Mixture in Gravitational Field
The chemical potential of an ideal mixture is given by the relation $\mu_\alpha = g_\alpha(T, p) + R_\alpha T \ln X_\alpha$, where $g_\alpha(T, p)$ is the Gibbs free energy of α alone at temperature T and pressure p of the mixture, and X_α is the mole fraction. Moreover, for an incompressible liquid with constant specific mass density ρ_α, one finds $g_\alpha(T, p) = g_\alpha(T, p_0) + (p - p_0)/\rho_\alpha$.

We consider an ideal liquid mixture of incompressible components $\alpha = 1, 2$ in the gravitational field, and ask for composition, pressure and mass density of the mixture as function of height. As in the previous problem, this requires evaluation of the equilibrium condition $\mu_\alpha + \text{g}z = \mu_\alpha^0$ for $\alpha = 1, 2$. The problem is non-trivial, the following steps lead you through:

1. Explicitly write the two equilibrium conditions; introduce the mole fraction $X = X_1 = 1 - X_2$. This gives two equations connecting $\{p, X, z\}$ with parameters g, R, M_1, M_2, ρ_1, ρ_2, T, and, of course, the reference values μ_α^0.
2. At $z = 0$ the mixture has the pressure $p\,(z = 0) = p_0$ and the mole fraction $X\,(z = 0) = X_0$. Use this to eliminate the μ_α^0.
3. The mole fraction X cannot be eliminated explicitly from the resulting two equations, but the following works:

    a. Solve the two equations to determine pressure and height as functions of the mole fraction, $p\,(X)$, $z\,(X)$.
    b. Use X as a parameter for plotting, i.e., to plot $p\,(z)$, plot the points $\{z\,(X)\,, p\,(X)\}$, and for $X\,(z)$ plot the points $\{z\,(X)\,, X\}$.
    c. Also plot the total mass density ρ as function of height. Chose a wide range to clearly see the non-linearity.

Use the following data: $\bar{R} = 8.314 \frac{\text{kJ}}{\text{kmol K}}$, $\text{g} = 9.81 \frac{\text{m}}{\text{s}^2}$, $M_1 = 12 \frac{\text{kg}}{\text{kmol}}$, $\rho_1 = 500 \frac{\text{kg}}{\text{m}^3}$, $M_2 = 80 \frac{\text{kg}}{\text{kmol}}$, $\rho_2 = 3000 \frac{\text{kg}}{\text{m}^3}$, $T = 298\,\text{K}$, $X_0 = 0.01$.

2.5. Extremum under Several Constraints

Extend the argument from Sec. 2.6 to finding the extremum of $f\,(x_i)$ under several constraints $g_\alpha\,(x_i) = 0$, with $\alpha = 1, \cdots, \nu$; $\nu < n$.

Chapter 3
Balance Laws

Balance laws for mass, momentum and energy form the base of all descriptions of transport phenomena in the realm of thermofluids theory and applications. Depending on the desired level of description, they assume different forms. In this chapter, we first formulate the general balance laws for arbitrary volumes, and the local balances for regular points (i.e., excluding discontinuities), and singular surfaces (i.e., across discontinuities). The general balance laws are then specified for total and partial mass, momentum, and energy. Along the way we discuss convective and non-convective fluxes, as well as supply and production terms.

3.1 Thermodynamic Systems

A thermodynamic system is a well defined subset of the universe.

Specifically, we consider a system described by matter inside a volume V, with closed boundary surface ∂V that moves with velocity[1] $w_i(\mathbf{x},t)$. Due to the motion of the boundary, the system volume changes over time, hence we write $V(t)$. Such system is sketched in Fig. 3.1, where n_i indicates the outer normal vector on a surface element dA of the system boundary ∂V moving with w_i.

The velocity $w_i(\mathbf{x},t)$ of the system boundary must be distinguished from the velocity $v_i(\mathbf{x},t)$ of the matter within the system. The choice of $w_i(\mathbf{x},t)$ determines the system and its temporal evolution. Most often one will consider one of the following choices:

For $w_i = v_i$ the system boundary moves with the matter at the boundary. Accordingly, no matter crosses the system boundary and thus the system is

[1] See Appendix A for detailed discussion of notation for vectors and tensors in symbolic and index notation, tensor operations etc.

H. Struchtrup, *A Thermodynamic Introduction to Transport Phenomena*,
https://doi.org/10.1007/978-3-031-61868-0_3

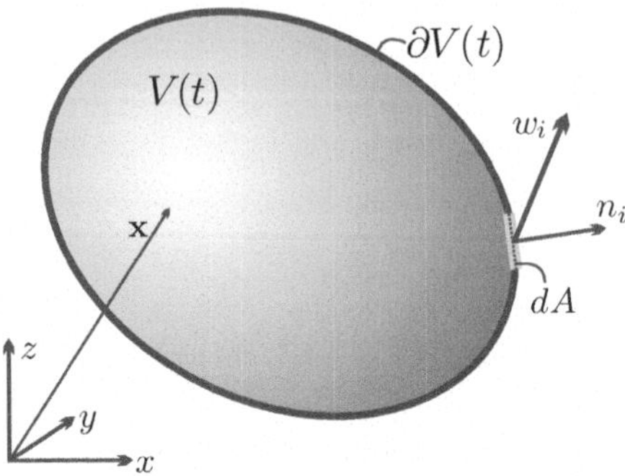

Fig. 3.1 General system of volume $V(t)$ with boundary $\partial V(t)$ in space; dA denotes a surface element with outer normal n_i moving with velocity w_i.

always comprised of the same material particles. We speak of a *closed system*, or, alternatively, a *material volume*, denoted as $V_m(t)$.

For $w_i = 0$, the system boundary is at rest relative to the observer frame. Matter crosses the system boundary with the local velocity $v_i(\mathbf{x}, t)$, that is at each moment in time the system consists of a different set of particles. We speak of a *stationary open system*, or, alternatively, of a *control volume*, denoted as V_0.

We will develop balance laws that describe the change of the system properties, in particular the momentary amounts of mass, momentum, energy (internal and kinetic), and entropy in the system, which depend on the chosen system volume $V(t)$ and the thermodynamic state within the system:

$$m(t) = \int_{V(t)} \rho dV \tag{3.1}$$

$$M_i(t) = \int_{V(t)} \rho v_i dV \tag{3.2}$$

$$E(t) = \int_{V(t)} \rho \left(u + \frac{1}{2} v^2 \right) dV \tag{3.3}$$

$$S(t) = \int_{V(t)} \rho s dV \tag{3.4}$$

Other properties, e.g., partial mass of components, will be considered as well. Note that potential energy is not included in this list, it will appear in the equations through the supply terms.

For the general discussion of balance laws, we use placeholders Ψ, ψ for the properties, and write

$$\Psi(t) = \int_{V(t)} \rho \psi dV \tag{3.5}$$

where $\psi = \left\{1, v_i, u + \frac{1}{2}v^2, s, \ldots\right\}$ and $\Psi = \{m, M_i, E, S, \ldots\}$.

3.2 General Balance

The change of a system property Ψ within a well-defined system $V(t)$ with boundary $\partial V(t)$ can be effected by a number of mechanisms. Put into words, the most general balance law reads

$$\begin{bmatrix}\text{rate of change of}\\ \text{amount of } \Psi\\ \text{in the system } V(t)\end{bmatrix} = \begin{bmatrix}\text{rate of amount of } \Psi \text{ transferred}\\ \text{across } \partial V \text{ with mass (convective)}\end{bmatrix}$$

$$+ \begin{bmatrix}\text{rate of amount of } \Psi \text{ transferred}\\ \text{across } \partial V \text{ without mass (conductive)}\end{bmatrix}$$

$$+ \begin{bmatrix}\text{rate of production of } \Psi\\ \text{inside } V\end{bmatrix}$$

$$+ \begin{bmatrix}\text{rate of supply of } \Psi\\ \text{into / out of } V\end{bmatrix}$$

Convective transfer takes into account that all mass entering and leaving the system carries thermodynamic properties, e.g., momentum or energy, hence if mass crosses the system boundary, its properties are entering or leaving the system as well.

Conductive transfer is effected without the transport of mass, for instance the exchange of energy and momentum between neighboring particles. Conductive heat transfer from hot to cold (e.g., if one touches a hot plate), is the typical example.

A *production* cannot be controlled from the outside, it occurs due to processes within the system. Examples are the generation of entropy due to irreversible processes, such as friction, heat transfer, and mixing, as well as production or consumption of species in chemical reactions. In general a production can be positive or negative, but entropy production cannot be negative.

In contrast to a production, a *supply* can be controlled from the outside. For instance, in a microwave oven radiation is generated and then travels across boundaries into the system where it is absorbed somewhere within the volume. Also the gravitational field is considered as a supply—it acts on all elements inside the volume, and is a result of the presence of matter outside the system, which could, at least in principle, be removed. A supply can be positive or negative, e.g., radiative loss of energy can be described as a negative supply.

As an equation, we write the general balance as

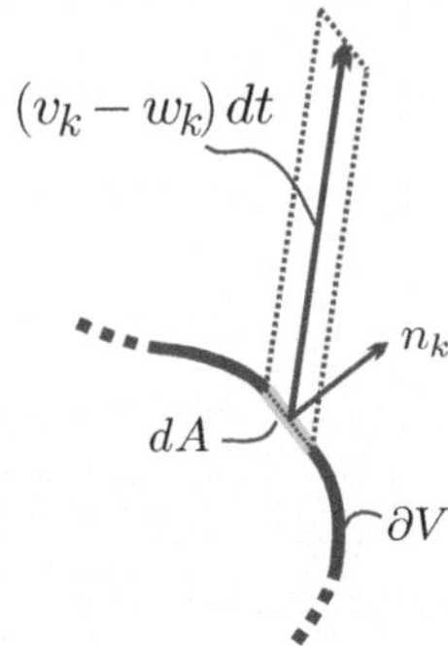

Fig. 3.2 Flow across the surface element dA with ourward normal vector n_k of the system boundary ∂V.

$$\frac{d}{dt}\int_{V(t)}\rho\psi dV = -\oint_{\partial V(t)}\rho\psi\left(v_k - w_k\right)n_k dA - \oint_{\partial V(t)}\phi_k n_k dA + \int_{V(t)}\rho\pi dV + \int_{V(t)}\rho\zeta dV \tag{3.6}$$

where we introduced the symbols $\phi_k(\mathbf{x},t)$ for the non-convective (or conductive) flux density vector, $\pi(\mathbf{x},t)$ for the mass specific production rate, $\zeta(\mathbf{x},t)$ for the mass specific supply rate.

The convective flux term, the first term on the right hand side of the above equation, results from the following discussion. Figure 3.2 shows a surface element dA with outer normal vector n_k that is passed by a flow with velocity $(v_k - w_k)$ relative to the surface element, which itself moves with velocity w_k. The figure indicates a cylinder with base area dA and directed length $(v_k - w_k)\,dt$ with the volume $[dA\,(v_k - w_k)\,n_k dt]$. All mass in this cylinder has passed the element dA in the time interval dt, i.e., the mass that has passed during dt is $[\rho dA\,(v_k - w_k)\,n_k dt]$ and the rate of mass transfer through dA—i.e., the mass flux—results from division by the time interval as $[\rho\,(v_k - w_k)\,n_k dA]$. Finally, multiplication with the mass specific property ψ gives the convective rate of transfer of Ψ across dA as $[\rho\psi\,(v_k - w_k)\,n_k dA]$, and integration over the complete system boundary ∂V gives the convective transfer rate as shown in the equation. The sign is set such that Ψ decreases when net transfer velocity $(v_k - w_k)$ and outward normal vector n_k point in the same direction.

The non-convective flux vector ϕ_k can be decomposed into a tangential and a normal contribution with respect to the surface element—only the normal contribution $\phi_k n_k$ will lead to a transfer across the surface. The conductive transfer contribution must be positive when the flux ϕ_k points into the system, opposite to the outward normal vector n_k, the minus sign in front of the integral accounts for this.

Production and supply for the system result from integration of local production and supply densities $\rho\pi$ and $\rho\zeta$ over the system volume.

3.3 Local Balance in Regular Points

The general balance law (3.6) is valid for arbitrary volumes $V(t)$. We shall now reduce the global law to the local balance law, which allows for the pointwise resolution of processes. For this, we consider a system at rest relative to the observer, so that $w_i = 0$, and $V = V_0 = const.$

Since the boundary is fixed, we can draw the time derivative into the integral, where it turns into a partial derivative, i.e.,[2]

$$\frac{d}{dt}\int_{V_0} \rho\psi dV = \int_{V_0} \frac{\partial \rho\psi}{\partial t} dV \,. \tag{3.7}$$

Note, that in general $\rho(\mathbf{x},t)$ and $\psi(\mathbf{x},t)$ are functions of space and time, while the integral depends on time only. With this, (3.6) assumes the form

$$\int_{V_0} \frac{\partial \rho\psi}{\partial t} dV = -\oint_{\partial V_0} [\rho\psi v_k + \phi_k]\, n_k dA + \int_{V_0} \rho(\pi + \zeta)\, dV \,. \tag{3.8}$$

The surface integral on the right describes the transfer of ψ across the system boundary ∂V_0 by convection and conduction. Assuming that all fields within the volume are smooth, so that their gradients exist, we use the Gauss theorem[3] to convert the surface integral into a volume integral. Then, all terms in the equation can be combined under a volume integral,

$$\int_{V_0} \left\{ \frac{\partial \rho\psi}{\partial t} + \frac{\partial}{\partial x_k}[\rho\psi v_k + \phi_k] - \rho(\pi + \zeta) \right\} dV = 0 \,. \tag{3.9}$$

This equation must hold for *arbitrary* volumes V_0, hence the integrand must vanish, which yields the general *local balance law in regular points* as

[2] Notation is such that all functions on the line are affected by the derivative, $\frac{\partial \rho\psi}{\partial t} = \frac{\partial(\rho\psi)}{\partial t}$.

[3] The Gauss theorem, or divergence theorem, states that for a vector $\xi_k(\mathbf{x})$ the closed surface integral is equal to a volume integral as $\oint_{\partial V} \xi_k n_k dA = \int_V \frac{\partial \xi_k}{\partial x_k} dV$, provided the derivative of ξ_k exists everywhere in V, which is the case if $\xi_k(\mathbf{x})$ has no discontinuities.

$$\frac{\partial \rho \psi}{\partial t}+\frac{\partial}{\partial x_k}\left[\rho \psi v_k+\phi_k\right]=\rho\left(\pi+\zeta\right) \tag{3.10}$$

Here, the restriction to *regular points* ensures that the space derivatives exist, that is discontinuities in any of the properties are excluded.

3.4 Balance Law at Singular Surfaces

Discontinuities occur, e.g., as density jumps in heterogeneous systems. Typical examples are phase interfaces, wall-liquid interfaces, as well as shock waves and flames. Then regions that are described by the balance in regular points (3.10) are separated by discontinuities which appear as singular surfaces.

Balance laws for the singular surface are required to describe the transport across these surfaces, and also these follow from the general balance (3.6).

For resolution on microscopic scales apparent singularities such as phase interfaces and shock waves exhibit, in fact, smooth but steep changes of properties. For instance, the resolved liquid-vapor interface has a thickness of about ten atomic diameters, while resolved shock waves in gases have a thickness of a few mean free paths. Any production (e.g., of entropy) and supply in the interface are concentrated in the extremely small volume in which the steep variation of properties occurs. From the macroscopic perspective it suffices to not resolve the interface structure, and treat these interfaces as jumps, with surface production and supply.

Here, we study only the simplest case, where the singular surface itself does not have properties such as mass, momentum, energy, entropy, hence we exclude processes such as the accumulation of surfactants. With this, no transport occurs within the surface, and all transport is between the bulk material at both sides of the surface.

With a singular surface in the system, the global balance in the form (3.8) remains valid, but the Gauss theorem cannot be applied due to the discontinuities. To extract the balance laws across such a surface, we consider a singular surface $\mathcal{S}$, where the jump just lies in the surface, and apply the global balance to a "pill-box" shaped volume element $dV = hdA$ of thickness h and base surface dA with same orientation as the singular surface, see Fig. 3.3.

In the limit that the pill-box thickness h approaches zero, the volume integral of the finite property $\rho\psi$ vanishes,

$$\lim_{h\to 0}\int_{V=hdA}\rho\psi dV=0\ . \tag{3.11}$$

The volume densities of production and supply, however, due to lack of resolution might appear infinite at the singular surface, in the sense that

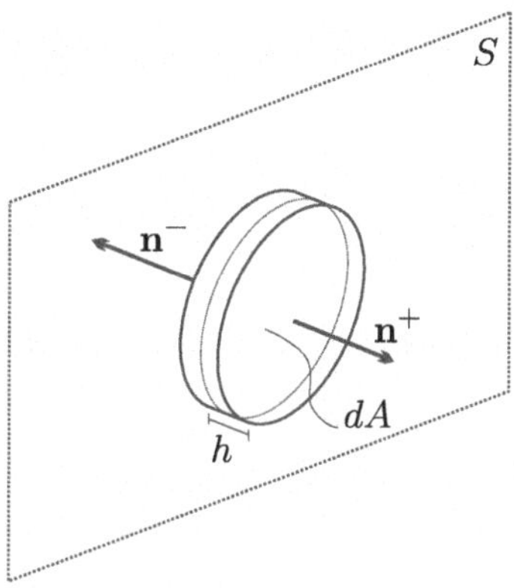

Fig. 3.3 The "pill-box" argument.

$$\lim_{h\to 0}\int_{V=hdA}\rho\left(\pi+\zeta\right)dV=\left(\pi_{\mathcal{S}}+\zeta_{\mathcal{S}}\right)dA\ . \tag{3.12}$$

Here, we have introduced the surface densities of production, $\pi_{\mathcal{S}}$, and supply, $\zeta_{\mathcal{S}}$, which are the limits of usual volume densities in the limit of vanishing thickness (or, rather, vanishing resolution). Examples for these are the generation of entropy at an phase interface, a wall-fluid interface, or in a shock, and the absorption or emission of radiation at a wall.

As $h\to 0$, the rim of the pill-box vanishes, hence there will be no surface flux across the rim. Accordingly, the surface integral $-\oint_{\partial V_0}\left[\rho\psi v_k+\phi_k\right]n_k dA$ reduces to contributions from both sides of $\mathcal{S}$, indicated with "+" and "-". After division by dA, we find the balance for the pill-box shaped volume in the limit $h\to 0$ as

$$\left[\rho\psi\left(v_k-w_k\right)+\phi_k\right]^-\ n_k^-+\left[\rho\psi\left(v_k-w_k\right)+\phi_k\right]^+\ n_k^+=\pi_{\mathcal{S}}+\zeta_{\mathcal{S}}\ . \tag{3.13}$$

As indicated in Fig. 3.3, normal vectors n_i^+, n_i^- point in opposite directions, and with $n_k=n_k^+=-n_k^-$, and the definition of the *jump* of a function f as

$$[[f]]=f^+-f^-\ , \tag{3.14}$$

we finally write the general balance law for singular surfaces in the compact form

$$\left[\left[\rho\psi\left(v_k-w_k\right)+\phi_k\right]\right]n_k=\pi_{\mathcal{S}}+\zeta_{\mathcal{S}}\ . \tag{3.15}$$

Here, w_k is the local velocity of the singularity, and v_k is the velocity of matter at the singularity, which might be different on both sides.

3.5 Material—or Convective—Derivative

Let us consider an arbitrary function $\psi(\mathbf{x}, t)$ that evolves in space-time. The observed rate of change of the function depends on the motion of the observer. Typically one distinguishes between the Eulerian and Lagrangian descriptions.

In the *Eulerian description*, the observer is resting in an inertial frame. We denote with $\frac{\partial \psi}{\partial t}$ the rate of change of ψ for this observer, that is the rate of change at a fixed location $\mathbf{x}$.

In the *Lagrangian description,* the observer is moving with the flow, that is with the local velocity v_i. We denote with $\frac{D\psi}{Dt}$ the rate of change of ψ for this observer, that is the rate of change of ψ for a material fluid element.

To establish the relation between both descriptions, we use the following line of arguments:

The trajectory of a particle[4] X in space, i.e., its location in space at any time t is given by $x_i(X, t)$. We identify the particle by its location at a reference time t_0, when it is at $x_i^0 = x_i(X, t_0)$. Since only one particle can be at a location at any given time, we can eliminate the particle name (X) and write $x_i\left(x_i^0, t_0; t\right)$ as the location at time t of the particle that at time t_0 was at location x_i^0. By inversion, $x_i^0(x_i, t; t_0)$ is the starting position at time t_0 of the particle that now, at time t, is at location x_i.

Any thermodynamic property ψ is a property of matter, i.e., of a continuum particle. In the Euler description we write $\psi(x_i, t)$ for that particle that at time t is at the considered location x_i. In the Lagrange description we write $\psi\left(x_i^0, t_0; t\right)$ for that particle that started at $\left(x_i^0, t_0\right)$, or simply $\psi(X, t)$. Note that in the Lagrange description the property ψ depends only on one variable, the time t.

Velocity is the change of position of a material particle, hence

$$v_i = \frac{Dx_i(X, t)}{Dt}, \tag{3.16}$$

where the notation $\frac{D}{Dt}$ indicates the material time derivative.

The function $\psi(x_i, t)$ describes the space-time evolution of the material property ψ. In the Euler description, its differential is

$$d\psi = \frac{\partial \psi}{\partial t} dt + \frac{\partial \psi}{\partial x_i} dx_i . \tag{3.17}$$

During the time Dt a material particle moves the distance Dx_i. Hence, the variation $D\psi$ for a material particle is obtained by setting $dx_i \to Dx_i$, $dt \to Dt$, so that for a material particle

[4] Recall that we describe the material as a continuum, where resolution is such that the continuum volume element contains a larger number of atoms or molecules. In this context, *particle* refers to the amount of continuum at a location, i.e., in a local element dV.

$$D\psi = \frac{\partial \psi}{\partial t} Dt + \frac{\partial \psi}{\partial x_i} Dx_i \ . \tag{3.18}$$

After division by the time interval, and with (3.16), we find the relation between material and partial time derivative as

$$\frac{D\psi}{Dt} = \frac{\partial \psi}{\partial t} + v_i \frac{\partial \psi}{\partial x_i} \ . \tag{3.19}$$

Often, local balance laws are written with the material derivative for more compact expressions.

3.6 Reynolds Transport Theorem

Above, when we reduced the integral balance to the local balance, we took advantage of using a fixed volume V_0, which allowed us to draw the time derivative into the integral. That is, we relied on the Eulerian description.

For completeness, we look at the rate of change of a system property in the Lagrangian description. That is, we consider the time derivative of the integral of a function $\theta(\mathbf{x}, t)$ over a material volume $V_m(t)$, with the difference quotient

$$\frac{d}{dt} \int_{V_m(t)} \theta(\mathbf{x}, t)\, dV = \lim_{\Delta t \to 0} \frac{1}{\Delta t} \left[\int_{V_m(t+\Delta t)} \theta(\mathbf{x}, t + \Delta t)\, dV - \int_{V_m(t)} \theta(\mathbf{x}, t)\, dV \right] \ . \tag{3.20}$$

Taylor expansion in time gives $\theta(\mathbf{x}, t + \Delta t) = \theta(\mathbf{x}, t) + \frac{\partial \theta}{\partial t} \Delta t$, and thus

$$\frac{d}{dt} \int_{V_m(t)} \theta(\mathbf{x}, t)\, dV = \lim_{\Delta t \to 0} \frac{1}{\Delta t} \left[\int_{V_m(t+\Delta t)} \frac{\partial \theta}{\partial t} \Delta t dV \right] + \lim_{\Delta t \to 0} \frac{1}{\Delta t} \left[\int_{V_m(t+\Delta t)} \theta(\mathbf{x}, t)\, dV - \int_{V_m(t)} \theta(\mathbf{x}, t)\, dV \right] \tag{3.21}$$

or

$$\frac{d}{dt} \int_{V_m(t)} \theta(\mathbf{x}, t)\, dV = \int_{V_m(t)} \frac{\partial \theta}{\partial t} dV + \lim_{\Delta t \to 0} \frac{1}{\Delta t} \int_{\Delta V} \theta(\mathbf{x}, t)\, dV \tag{3.22}$$

where $\Delta V = V_m(t + \Delta t) - V_m(t)$ is the change of volume due to the motion of the boundary, which moves with the material velocity v_i. This is, in fact, the volume swept out by the motion of all boundary elements dA with velocity v_i during Δt, hence $dV = v_i n_i \Delta t dA$ and

$$\frac{d}{dt}\int_{V_m(t)} \theta dV = \int_{V_m(t)} \frac{\partial \theta}{\partial t} dV + \lim_{\Delta t\to 0} \frac{1}{\Delta t}\oint_{\partial V_m(t)} \theta v_i n_i dA \Delta t$$
$$= \int_{V_m(t)} \frac{\partial \theta}{\partial t} dV + \oint_{\partial V_m(t)} \theta v_i n_i dA \tag{3.23}$$

Using the divergence theorem once more, and setting $\theta = \rho\psi$, we find

$$\frac{d}{dt}\int_{V_m(t)} \rho\psi dV = \int_{V_m(t)} \left(\frac{\partial \rho\psi}{\partial t} + \frac{\partial \rho\psi v_i}{\partial x_i}\right) dV \ . \tag{3.24}$$

This relation is known as Reynolds transport theorem. It can be used to find the local balance in regular points from the global balance for a material volume $V_m(t)$; this is left to the interested reader.

3.7 Mass Balances

3.7.1 Total Mass Balance

Total mass $m = \int \rho dV$ can only be transferred by convective flux, and it can neither be produced nor destroyed, hence the required entries for the balance laws are $\psi = 1$, and $\phi_i = \pi = \zeta = \pi_s = \zeta_s = 0$. Global, local and surface balance laws read

$$\frac{d}{dt}\int_{V(t)} \rho dV + \oint_{\partial V(t)} \rho (v_k - u_k) n_k dA = 0 \tag{3.25}$$

$$\frac{\partial \rho}{\partial t} + \frac{\partial \rho v_k}{\partial x_k} = 0 \tag{3.26}$$

$$[[\rho (v_k - w_k)]] n_k = 0 \tag{3.27}$$

With the convective derivative (3.19), the local balance is written as

$$\frac{D\rho}{Dt} + \rho \frac{\partial v_k}{\partial x_k} = 0 \ . \tag{3.28}$$

The local mass balance, in either form, is also know as *continuity equation.*

With the mass balance and (3.19), the general local balance in regular points (3.10) assumes the compact form,

$$\rho \frac{D\psi}{Dt} + \frac{\partial \phi_k}{\partial x_k} = \rho (\pi + \zeta) \ . \tag{3.29}$$

3.7.2 Partial Mass Balances

While total mass is conserved, the mass of a component might change locally due to diffusion or chemical reactions. For a component α of a mixture, the partial mass balance in regular points reads

$$\frac{\partial \rho_\alpha}{\partial t} + \frac{\partial \rho_\alpha v_k^\alpha}{\partial x_k} = \tau_\alpha \; , \tag{3.30}$$

where v_i^α is the macroscopic velocity of the component, and τ_α is its production rate due to chemical reactions, which can be positive, negative, or zero.

The total mass density of the mixture is

$$\rho = \sum_{\alpha=1}^{\nu} \rho_\alpha \tag{3.31}$$

and summation of the partial mass balances gives

$$\frac{\partial \rho}{\partial t} + \frac{\partial}{\partial x_k}\left(\sum_{\alpha=1}^{\nu} \rho_\alpha v_k^\alpha\right) = \sum_{\alpha=1}^{\nu} \tau_\alpha \; . \tag{3.32}$$

This must agree with the mass balance (3.26), hence we identify

$$\sum_{\alpha=1}^{\nu} \tau_\alpha = 0 \quad , \quad \sum_{\alpha=1}^{\nu} \rho_\alpha v_k^\alpha = \rho v_k \; . \tag{3.33}$$

The first relation guarantees conservation of total mass, and the second identifies the relation between the center of mass velocities of components and the mixture.

Often, instead of partial densities, one considers the mass fractions $\mathsf{c}_\alpha = \frac{\rho_\alpha}{\rho}$, with $\sum_{\alpha=1}^{\nu} \mathsf{c}_\alpha = 1$. With $\rho_\alpha = \rho \mathsf{c}_\alpha$ and the overall mass balance we find

$$\begin{aligned}
\frac{\partial \rho_\alpha}{\partial t} + \frac{\partial \rho_\alpha v_k^\alpha}{\partial x_k} &= \frac{\partial \rho \mathsf{c}_\alpha}{\partial t} + \frac{\partial \rho \mathsf{c}_\alpha v_k^\alpha}{\partial x_k} \\
&= \rho \frac{\partial \mathsf{c}_\alpha}{\partial t} + \mathsf{c}_\alpha \left(-\frac{\partial \rho v_k}{\partial x_k}\right) + \frac{\partial \rho_\alpha v_k^\alpha}{\partial x_k} \\
&= \rho \frac{\partial \mathsf{c}_\alpha}{\partial t} + \rho v_k \frac{\partial \mathsf{c}_\alpha}{\partial x_k} + \frac{\partial \rho_\alpha \left(v_k^\alpha - v_k\right)}{\partial x_k}
\end{aligned} \tag{3.34}$$

Hence, the partial mass balance assumes the compact form

$$\rho \frac{D \mathsf{c}_\alpha}{Dt} + \frac{\partial J_k^\alpha}{\partial x_k} = \tau_\alpha \; . \tag{3.35}$$

Here, we have introduced the diffusion fluxes

$$J_k^\alpha = \rho_\alpha \left(v_k^\alpha - v_k \right) \tag{3.36}$$

which describe mass transport of component α relative to the mixture flowing at v_i. Note that the sum of diffusion fluxes vanishes, since

$$\sum_{\alpha=1}^{\nu} J_k^\alpha = \sum_{\alpha=1}^{\nu} \rho_\alpha \left(v_k^\alpha - v_k \right) = \sum_{\alpha=1}^{\nu} \rho_\alpha v_k^\alpha - \left(\sum_{\alpha=1}^{\nu} \rho_\alpha \right) v_k = 0 \ . \tag{3.37}$$

The production rates τ_α describe the rate of change of the amount of component α due to chemical reactions. We consider n different reactions with reaction rate densities λ^a, $a = 1, \ldots, n$, and stoichiometric coefficients γ_α^a, defined such that γ_α^a gives the number of molecules of type α that are produced (if $\gamma_\alpha^a > 0$) or consumed (if $\gamma_\alpha^a < 0$) in a single reaction[5] of type a. The mass production rate of a component due to all reactions is

$$\tau_\alpha = \sum_{a=1}^{n} \gamma_\alpha^a M_\alpha \lambda^a \ ; \tag{3.38}$$

M_α is the molar mass of component α. For clarity, we note the typical units of these quantities as $[\gamma_\alpha^a] = 1$, $[M_\alpha] = \frac{\text{kg}}{\text{kmol}}$, $[\tau_\alpha] = \frac{\text{kg}}{\text{m}^3\,\text{s}}$, $[\lambda^a] = \frac{\text{kmol}}{\text{m}^3\,\text{s}}$. Mass conservation in an individual reaction yields $\sum_\alpha \gamma_\alpha^a M_\alpha = 0$, which implies $\sum_\alpha \tau_\alpha = 0$ as required.

The reaction rates λ^a will be determined as constitutive equations.

When written with component densities $\rho_\alpha = \rho c_\alpha$, the various forms of the mass balance (global, local, singular surfaces) read

$$\frac{d}{dt} \int_{V(t)} \rho_\alpha dV + \oint_{\partial V(t)} \left[\rho_\alpha \left(v_k - w_k \right) + J_k^\alpha \right] n_k dA = \tau_\alpha \ , \tag{3.39}$$

$$\frac{\partial \rho_\alpha}{\partial t} + \frac{\partial}{\partial x_k} \left[\rho_\alpha v_k + J_k^\alpha \right] = \tau_\alpha \ , \tag{3.40}$$

$$\left[\left[\rho_\alpha \left(v_k - w_k \right) + J_k^\alpha \right]\right] n_k = \tau_S^\alpha \ . \tag{3.41}$$

Verification of these is left to the reader. We identify $\psi = c_\alpha$, $\phi_k = J_k^\alpha$, $\pi = \tau_\alpha$, and $\pi_S = \tau_S^\alpha$. Typical examples for surface chemical production τ_S^α are catalytic surfaces and unresolved reaction zones, such as flames.

The diffusion fluxes J_i^α and the reaction rates λ^a depend on the material properties of the mixture and the flow conditions—one has to find constitutive relations for these, and all other non-convective fluxes and production terms. For this, we will employ the second law of thermodynamics in Chap 4.

---

[5] As an example, for the reaction $H_2 + \frac{1}{2}O_2 \rightleftharpoons H_2O$, the stoichiometric coefficents are $\gamma_{H_2} = -1$, $\gamma_{O_2} = -\frac{1}{2}$, $\gamma_{H_2O} = 1$.

3.8 Momentum Balance

3.8.1 Newton's Second Law

Newton's second law of motion states that the rate of change of momentum of a body is equal to the sum of all forces acting on the body. Momentum is conserved, hence there is no production, $\pi = \pi_S = 0$.

For a system with momentum $M_i = \int_{V(t)} \rho v_i dV$, we obtain the general form of the momentum balance by setting $\psi = v_i$, and introducing the stress tensor t_{ij} for the non-convective flux, and the body force vector f_i as the supply, so that

$$\frac{d}{dt}\int_{V(t)} \rho v_i dV = -\oint_{\partial V(t)} \rho v_i \left(v_k - w_k\right) n_k dA + \oint_{\partial V(t)} t_{ik} n_k dA + \int_{V(t)} \rho f_i dV \,. \tag{3.42}$$

This equation states that the rate of change of momentum in the volume $V(t)$ is equal to the net amount of momentum transferred by convection across the system boundary, and the forces acting on the surface—as described by the stress tensor t_{ik}—and the volume forces—as described by the body forces f_i—acting on all points in the system.

The typical body force is gravity, where $f_i = \{0, 0, -\mathrm{g}\}_i$ with, on Earth, $\mathrm{g} = 9.81\frac{\mathrm{m}}{\mathrm{s}^2}$. Also electromagnetic forces acting within the volume are considered as body forces. Note that body forces are controlled from the outside, at least in principle, hence these appear as supply terms.

The stress tensor is introduced formally as the non-convective momentum flux, which for a vector equation must be a 2-tensor. The physical interpretation of the stress tensor is as follows: The scalar product of the stress tensor t_{ij} with the surface normal n_i yields the stress vector (i.e., force per area) on the surface element, that is $\sigma_i = t_{ik} n_k$ is the stress on dA, and $\sigma_i dA = t_{ik} n_k dA$ is the force on the surface element. The integral over the closed surface $\partial V(t)$ is the total force on that surface. The stress tensor depends on material and flow conditions through constitutive relations, details will be discussed in the next chapter.

From the above, the local balance in regular points is identified as

$$\frac{\partial \rho v_i}{\partial t} + \frac{\partial}{\partial x_k}\left[\rho v_i v_k - t_{ik}\right] = \rho f_i \tag{3.43}$$

or, with help of the mass balance,

$$\rho\frac{Dv_i}{Dt} - \frac{\partial t_{ik}}{\partial x_k} - \rho f_i \,. \tag{3.44}$$

On a singular surface, the momentum balance reduces to

$$[[\rho v_i (v_k - w_k) - t_{ik}]] \, n_k = \zeta_i^S \tag{3.45}$$

where the surface supply ζ_i^S accounts for possible surface forces, such as membrane stress or surface tension.

3.8.2 Angular Momentum

Angular momentum, or moment of momentum, is defined with respect to the origin of the observer frame as the cross product between location and momentum,

$$\mathbf{H} = \rho \, (\mathbf{x} \times \mathbf{v}) \ , \tag{3.46}$$

where $\mathbf{x}$ is the vector from origin to the location considered and $\mathbf{v}\,(\mathbf{x}, t)$ is the local flow velocity.

In index notation we have

$$H_i = \rho \varepsilon_{ijk} x_j v_k \tag{3.47}$$

where ε_{ijk} is the fully antisymmetric tensor of third order (see Appendix A). Multiplication of the local momentum balance[6] (3.44) with $\varepsilon_{ijk} x_j$ gives, after pulling x_j under derivatives, and with $\frac{Dx_j}{Dt} = v_j$, $\frac{\partial x_j}{\partial x_l} = \delta_{jl}$,

$$\rho \frac{D\varepsilon_{ijk} x_j v_k}{Dt} - \rho \varepsilon_{ijk} v_j v_k - \frac{\partial \varepsilon_{ijk} x_j t_{kl}}{\partial x_l} + \varepsilon_{ijk} t_{kl} \delta_{jl} = \rho \varepsilon_{ijk} x_j f_k \ . \tag{3.48}$$

Since $\varepsilon_{ijk} v_j v_k = 0$, and $\varepsilon_{ilk} = -\varepsilon_{ikl}$, we obtain

$$\rho \frac{D\varepsilon_{ijk} x_j v_k}{Dt} - \frac{\partial \varepsilon_{ijk} x_j t_{kl}}{\partial x_l} = \rho \varepsilon_{ijk} x_j f_k + \varepsilon_{ikl} t_{kl} \ . \tag{3.49}$$

Here we have constructed the local balance law for angular momentum with the non-convective flux $\varepsilon_{ijk} x_j t_{kl}$, the supply $\rho \varepsilon_{ijk} x_j f_k$, and the production $\varepsilon_{ikl} t_{kl}$.

For non-polar mixtures, which have no spin, angular momentum is conserved, that is the production must vanish, which is the case for symmetric stress tensor,

$$\varepsilon_{ikl} t_{kl} = 0 \quad \Rightarrow \quad t_{kl} = t_{lk} \ . \tag{3.50}$$

We will study only non-polar materials, with symmetric stress tensors. For these, the balance of angular momentum follows directly from the balance of momentum, and does not give any additional information. Therefore, angular momentum will not be considered further.

---

[6] In (3.44) rename free index $i \to k$, summation index $j \to l$.

3.8.3 Pressure

In a fluid at rest the force on a surface element is perpendicular to the surface, independent of orientation, that is the stress tensor is isotropic and determined by the pressure p acting on the surface opposite to the direction of the normal vector,

$$\sigma_i = t_{ik} n_k = -p n_i \quad \Rightarrow \quad t_{ik} = -p\delta_{ik} \ . \tag{3.51}$$

We consider a fluid at rest, so that $v_i = 0$ and $t_{ij} = -p\delta_{ij}$, to determine barometric and hydrostatic pressure laws. The momentum balance (3.44) reduces to

$$\frac{\partial p}{\partial x_i} = \rho f_i \ . \tag{3.52}$$

In the gravitational field, the body force is $f_i = \{0, 0, -\mathrm{g}\}$, so that the balance of momentum reduces to the three equations

$$\frac{\partial p}{\partial x_1} = 0 \quad , \quad \frac{\partial p}{\partial x_2} = 0 \quad , \quad \frac{\partial p}{\partial x_3} = -\rho \mathrm{g} \ . \tag{3.53}$$

From the horizontal components we learn that there is no horizontal variation of pressure, which therefore depends only on the vertical coordinate $x_3 = z$. The vertical variation of pressure depends on the density, and hence on the material.

For an ideal gas at constant temperature T_0, the ideal gas law gives $\rho = \frac{p}{RT_0}$, and integration yields the barometric formula

$$p = p_0 e^{-\frac{\mathrm{g}z}{RT_0}} \ , \tag{3.54}$$

where again z measures height, and p_0 is the pressure at ground level $z = 0$.

For an incompressible substance ($\rho = \rho_0 = const$), we find upon integration the hydrostatic pressure as

$$p = p_0 - \rho_0 \mathrm{g} z = p_0 + \rho_0 \mathrm{g} h \tag{3.55}$$

where z measures upwards (height), and $h = -z$ measures downwards (depth); p_0 is the pressure at $z = h = 0$.

3.8.4 Buoyancy

We consider a body B of mass density ρ_B and volume V_B immersed in a liquid or gas in the gravitational field $f_i = \{0, 0, -\mathrm{g}\}_i$, as shown in Fig. 3.4.

The forces acting on the body can be read from (3.42) as the gravitational body force

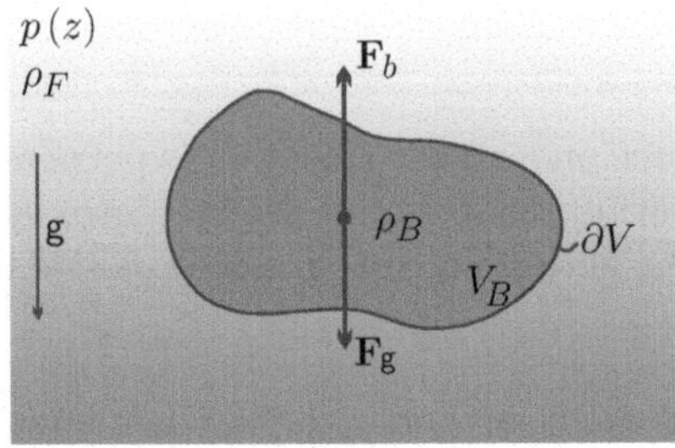

Fig. 3.4 Body B immersed in a fluid (liquid or gas). Pressure gradient induced by gravity leads to the buoyancy force $\mathbf{F}_b$ upwards, which competes with the gravitational force on the body, $\mathbf{F}_\mathrm{g}$.

$$F_{\mathrm{g},i} = \int_{V_B} \rho_B f_i dV = \rho_B V_B f_i \tag{3.56}$$

and the surface force due to the pressure $p(z)$ of the surrounding fluid, the buoyancy force,

$$F_{b,i} = \oint_{\partial V_B} t_{ij} n_j dA = -\oint_{\partial V_B} p n_i dA \,. \tag{3.57}$$

With the Gauss theorem, the surface integral can be converted into a volume integral, so that

$$F_{b,i} = -\int_{V_B} \frac{\partial p}{\partial x_i} dV = -\int_{V_B} \rho_F f_i dV = -\rho_F V_B f_i \,, \tag{3.58}$$

where we have used the reduced balance of momentum for the surrounding fluid (3.52) to replace the pressure gradient. Here, $\rho_F V_B$ is the mass of fluid that is displaced by the body,[7] or, in other words, the mass of fluid that would fill the volume, if the body was not present—this is Archimedes' principle (Archimedes of Syracuse, 287-212 BCE).

The net force on the body is the sum

$$F_i = F_{\mathrm{g},i} + F_{b,i} = (\rho_F - \rho_B) V_B \{0, 0, \mathrm{g}\}_i \,. \tag{3.59}$$

Hence, if the density of the fluid is larger than the density of the body, $\rho_F > \rho_B$, the net force is upwards, and a helium balloon in air, or a bubble in water, rise. If $\rho_F < \rho_B$, the body will fall, e.g., a stone in air or water.

[7] If the fluid is a gas, density ρ_F varies with height. However, the body is assumed to be small, so that variation of density ρ_F over the region of volume V_B can be ignored.

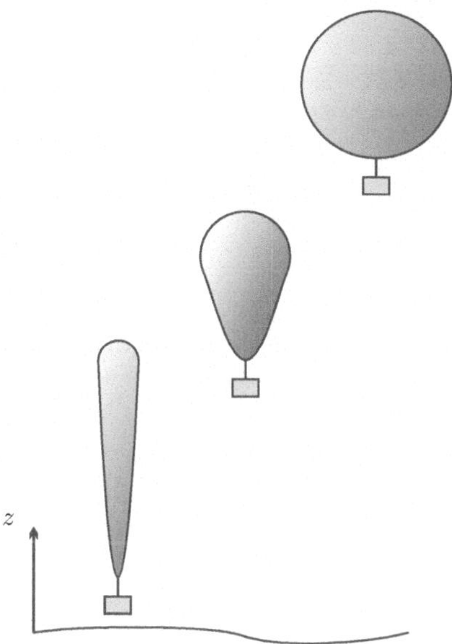

Fig. 3.5 Helium balloon rising in the atmosphere.

3.8.5 Membrane Stress

Figure 3.5 shows a helium filled balloon rising in the atmosphere. In early stages, the balloon shell is relaxed, and the helium pressure in the balloon is equal to the atmospheric pressure. As the balloon rises, the atmospheric pressure decreases (⇒ barometric formula) and the helium expands.

Above a certain height the shell is completely filled, and the volume of the helium, and hence its pressure, will not change further as the balloon keeps rising. Now, helium pressure in the balloon and atmospheric pressure outside are different, that is the shell is stressed.

To determine surface stresses σ_S, we consider a spherical balloon of radius r, where we balance forces for one half of the balloon as shown in Fig. 3.6. The pressure difference $p_{\text{He}} - p_{\text{air}}$ acts locally normal to the balloon shell. Due to symmetry, the horizontal contributions cancel, hence we only need to consider the contributions of the pressure forces in vertical direction, which are balanced by the surface stresses σ_S in the shell.

Using spherical coordinates, the pressure force in vertical direction results from integration of the vertical component $\cos\theta$ of the normal vector over the half sphere,

$$F_p = \int_{\varphi=0}^{2\pi} \int_{\theta=0}^{\pi/2} (p_{\text{He}} - p_{\text{air}}) \cos\theta \; r^2 \sin\theta d\theta d\varphi = (p_{\text{He}} - p_{\text{air}}) \, \pi r^2 \tag{3.60}$$

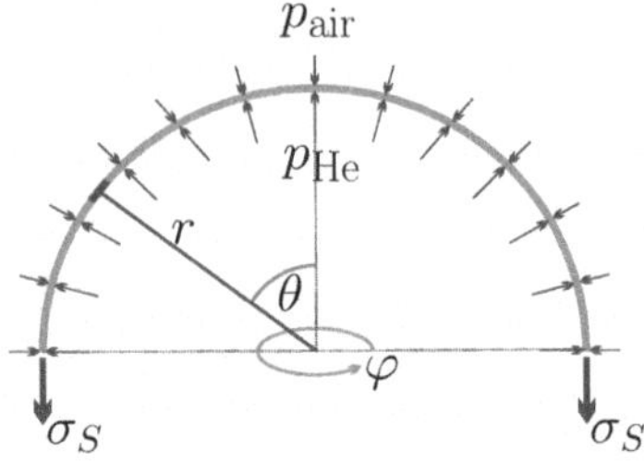

Fig. 3.6 Free body diagram for the upper half of the balloon shell. The tension $\sigma^{(s)}$ balances the pressure difference between helium filling and external air.

Effectively, the pressure difference is projected on the surface πr^2 and surface stresses act on the circumference $2\pi r$, so that

$$(p_{\rm He} - p_{\rm air})\,\pi r^2 = \sigma_S 2\pi r \quad \Rightarrow \quad (p_{\rm He} - p_{\rm air}) = \frac{2\sigma_S}{r}\,. \tag{3.61}$$

Considering the shell as a singular surface and applying (3.45) with $v_i = w_i = 0$ and stress tensor $t_{ij} = -p\delta_{ij}$, we find the reduced equation

$$[[p]]\,n_i = (p_{\rm air} - p_{\rm He})\,n_i = \zeta_i^S \tag{3.62}$$

Hence, we can link the surface supply to the stresses as[8]

$$\zeta_i^S = -\frac{2\sigma_S}{r} n_i\,. \tag{3.63}$$

Before the balloon is completely filled, there are no stresses in the shell, $\sigma_S = 0$, and $p_{\rm He} = p_{\rm air}$, $\zeta_i^S = 0$. After the balloon is filled, stresses σ_S in the shell balance the pressure difference, and induce the supply ζ_i^S.

3.9 Energy Balances

3.9.1 Forms of Energy

We have discussed energy in earlier chapters. In the present context, potential energy arises as a supply, and we have to consider only the kinetic energy E_{kin} of macroscopic motion, and the internal energy U.

The internal energy accounts for energy on the microscopic level which we cannot resolve with our eyes or tools. In particular it accounts for: kinetic

[8] We only consider spherical shells; for non-sperical shells one must use a more detailed description with two local radii of curvature. The normal vector points outwards.

energy of the atomic or molecular motion relative to the average velocity v_i of the material; microscopic potential energies between atoms or molecules; chemical energy stored or released in molecular bonds; quantum states of atoms and molecules; nuclear energy.

We write for the total internal and kinetic energies

$$U = \int_V \rho u dV \quad , \quad E_{kin} = \int_V \frac{\rho}{2} v^2 dV \tag{3.64}$$

From our experience in thermodynamics we know that the conservation law for energy—the First Law of Thermodynamics—for a closed system reads

$$\frac{d}{dt}\left(U + E_{kin}\right) = \dot{Q} - \dot{W} \ , \tag{3.65}$$

where $\dot{Q}$ and $\dot{W}$ are heating rate and power, respectively.

Heating and cooling is possible either by convective heat transfer through the surface, or by absorption and emission of radiation in the bulk. We write

$$\dot{Q} = -\oint q_k n_k dA + \int_V \rho \mathfrak{r} dV \ , \tag{3.66}$$

where we introduced q_i as the non-convective energy flux, or *heat flux vector*, and $\mathfrak{r}$ as the specific energy supply rate through absorption and emission of radiation. The signs are chosen such that heat transfer $\dot{Q}$ into the system is positive; then, the heat flux points into the system, opposite to the normal vector n_i.

Mechanical power is force times velocity of the point of application. The forces must be the same as in the momentum balance, that is the stresses $\sigma_i = t_{ij} n_j$ that act on the matter at the boundary of the system, and the body forces f_i that act in the bulk. Accordingly, we have surface and bulk contributions to power,

$$\dot{W} = -\oint_{\partial V} t_{ik} n_k v_i dA - \int_V \rho f_i v_i dV \ , \tag{3.67}$$

where the signs follow the convention that work is positive when the system does work on its surroundings. The body force term is a contribution to energy supply.

Finally, we have to account for convective energy transfer, that is energy entering or leaving the system with mass flow. Combining all, we find the integral energy balance as

$$\frac{d}{dt}\int_V \rho\left(u+\frac{1}{2}v^2\right)dV + \oint_{\partial V}\left[\rho\left(u+\frac{1}{2}v^2\right)(v_k-w_k)+q_k-t_{ik}v_i\right]n_k dA = \int_V \rho\left(\mathfrak{r}+f_iv_i\right)dV \quad (3.68)$$

By comparison with the general balance laws (3.6, 3.10, 3.15), we identify the local balance in regular points as

$$\frac{\partial}{\partial t}\left[\rho\left(u+\frac{1}{2}v^2\right)\right]+\frac{\partial}{\partial x_k}\left[\rho\left(u+\frac{1}{2}v^2\right)v_k+q_k-t_{ik}v_i\right]=\rho\left(\mathfrak{r}+f_iv_i\right) \quad (3.69)$$

or, with the convective derivative,

$$\rho\frac{D}{Dt}\left(u+\frac{1}{2}v^2\right)+\frac{\partial}{\partial x_k}\left[q_k-t_{ik}v_i\right]=\rho\left(\mathfrak{r}+f_iv_i\right) \ . \quad (3.70)$$

At singular surfaces we find

$$\left[\left[\rho\left(u+\frac{1}{2}v^2\right)(v_k-w_k)+q_k-t_{ik}v_i\right]\right]n_k=\mathfrak{r}_S \ , \quad (3.71)$$

where $\mathfrak{r}_S$ is energy supply to the singular surface, e.g., by absorption or emission of radiation at a wall or membrane.

3.9.2 Balance of Kinetic Energy

The balance of kinetic energy $E_{kin}=\int_V \frac{\rho}{2}v^2 dV$ is obtained from scalar multiplication of the momentum balance with the local velocity v_i which gives $(\psi=\frac{1}{2}v^2)$

$$\rho\frac{D}{Dt}\left(\frac{1}{2}v^2\right)-\frac{\partial v_i t_{ik}}{\partial x_k}=\rho f_i v_i - t_{ij}\frac{\partial v_i}{\partial x_j} \ . \quad (3.72)$$

We identify the non-convective flux of kinetic energy as $\phi_k=-v_it_{ik}$, the supply as $\rho\zeta=\rho f_iv_i$, and the production as $\rho\pi=-t_{ij}\frac{\partial v_i}{\partial x_j}$. Note that, while total energy is conserved, kinetic energy is not conserved. The production describes conversion of kinetic energy into internal energy through volume change and friction. Global (integral) form, and the balance on singular surfaces will not be considered.

3.9.3 Balance of Internal Energy

We have just seen that the balance of kinetic energy follows from the momentum balance as (3.72). Hence, having kinetic energy in the energy balance does not give additional information. The balance of internal energy results from subtracting the balance of kinetic energy from the balance of total energy. In local form we have

$$\rho \frac{Du}{Dt} + \frac{\partial q_k}{\partial x_k} = t_{ij} \frac{\partial v_i}{\partial x_j} + \rho \mathfrak{r} \ . \tag{3.73}$$

where $t_{ij} \frac{\partial v_i}{\partial x_j}$ is interpreted as the production of internal energy.

The corresponding integral balance reads

$$\frac{d}{dt} \int_V \rho u dV + \oint_{\partial V} \rho u \left(v_k - w_k\right) n_k dA = \\ - \oint_{\partial V} q_k n_k dA + \int_V \rho \mathfrak{r} dV + \int_V t_{ij} \frac{\partial v_i}{\partial x_j} dV \ . \tag{3.74}$$

In a reversible process the stress tensor reduces to $t_{ij} = -p\delta_{ij}$ and the pressure p is homogeneous at all times. For reversible processes in a material volume (where $w_i = v_i$) this equation reduces to the usual energy balance for a closed system,[9]

$$\frac{dU}{dt} = \dot{Q} - p \frac{dV}{dt} \ . \tag{3.75}$$

3.9.4 Potential Energy

Above, the body force f_i appears in the energy supply $\rho f_i v_i$. If the body force results from a time independent potential φ, we have $f_i = -\frac{\partial \varphi}{\partial x_i}$. For a fixed volume V_0 (that is for $w_i = 0$), the integral energy balance then assumes the form

$$\frac{d}{dt} \int_{V_0} \rho \left(u + \frac{1}{2} v^2 + \varphi\right) dV \\ + \oint_{\partial V_0} \left[\rho \left(u + \frac{1}{2} v^2 + \varphi\right) v_k - t_{ik} v_i + q_k\right] n_k dA = \int_V \rho \mathfrak{r} dV \tag{3.76}$$

where potential energy $E_{pot} = \int_{V_0} \rho \varphi dV$ appears as an additional form of energy. It is left to the reader to find the corresponding local balances for regular points and singular surfaces.

---

[9] Prob. 3.2 shows that for a material volume $\int_{V(t)} \frac{\partial v_i}{\partial x_j} dV = \frac{dV}{dt}$.

The usual textbook form of the energy balance for open systems (control volumes), Eq. (1.8), results from the above under some simplifying assumptions, such as taking average values of properties at inlets and outlets, ignoring dissipative stresses (so that $t_{ij} = -p\delta_{ij}$) at inlets and outlets; this is left to the reader as Prob. 3.10.

3.10 Conservation Laws in Alternative Notations

There are many alternative ways to write balance laws, using different variants of index or direct notation. While we rely mostly on the former, it is a useful exercise to translate equations from one form to another, not least since presentation in the literature differs widely, and one should be able to read equations in any of these.

3.10.1 Index Notation

The local balances for mass, mass fraction, momentum, internal energy read in index notation (as above)

$$\frac{\partial \rho}{\partial t} + v_k \frac{\partial \rho}{\partial x_k} + \rho \frac{\partial v_k}{\partial x_k} = 0 \,, \tag{3.77}$$

$$\rho \left(\frac{\partial \mathfrak{c}_\alpha}{\partial t} + v_k \frac{\partial \mathfrak{c}_\alpha}{\partial x_k} \right) + \frac{\partial J_k^\alpha}{\partial x_k} = \sum_{a=1}^{n} \gamma_\alpha^a M_\alpha \lambda^a \,, \tag{3.78}$$

$$\rho \left(\frac{\partial v_i}{\partial t} + v_k \frac{\partial v_i}{\partial x_k} \right) - \frac{\partial t_{ik}}{\partial x_k} = \rho f_i \,, \tag{3.79}$$

$$\rho \left(\frac{\partial u}{\partial t} + v_k \frac{\partial u}{\partial x_k} \right) + \frac{\partial q_k}{\partial x_k} = t_{ij} \frac{\partial v_i}{\partial x_j} + \rho \mathfrak{r} \,. \tag{3.80}$$

3.10.2 Index Notation with Nabla

The local balances for mass, mass fraction, momentum, internal energy read in index notation using the nabla symbol

$$\frac{\partial \rho}{\partial t} + v_k \nabla_k \rho + \rho \nabla_k v_k = 0 \,, \tag{3.81}$$

$$\rho\left(\frac{\partial c_\alpha}{\partial t} + v_k\nabla_k c_\alpha\right) + \nabla_k J_k^\alpha = \sum_{a=1}^{n}\gamma_\alpha^a M_\alpha \lambda^a \ , \tag{3.82}$$

$$\rho\left(\frac{\partial v_i}{\partial t} + v_k\nabla_k v_i\right) - \nabla_k t_{ik} = \rho f_i \ , \tag{3.83}$$

$$\rho\left(\frac{\partial u}{\partial t} + v_k\nabla_k u\right) + \nabla_k q_k = t_{ij}\nabla_j v_i + \rho\mathfrak{r} \ . \tag{3.84}$$

3.10.3 Direct Notation with Nabla

The local balances for mass, mass fraction, momentum, internal energy read in direct notation using the nabla symbol

$$\frac{\partial \rho}{\partial t} + \mathbf{v}\cdot\boldsymbol{\nabla}\rho + \rho\boldsymbol{\nabla}\cdot\mathbf{v} = 0 \ , \tag{3.85}$$

$$\rho\left(\frac{\partial c_\alpha}{\partial t} + \mathbf{v}\cdot\boldsymbol{\nabla}c_\alpha\right) + \boldsymbol{\nabla}\cdot\mathbf{J}^\alpha = \sum_{a=1}^{n}\gamma_\alpha^a M_\alpha \lambda^a \ , \tag{3.86}$$

$$\rho\left(\frac{\partial \mathbf{v}}{\partial t} + \mathbf{v}\cdot\boldsymbol{\nabla}\mathbf{v}\right) - \boldsymbol{\nabla}\cdot\mathbf{t}^T = \rho\mathbf{f} \ , \tag{3.87}$$

$$\rho\left(\frac{\partial u}{\partial t} + \mathbf{v}\cdot\boldsymbol{\nabla}u\right) + \boldsymbol{\nabla}\cdot\mathbf{q} = \mathbf{t}:(\boldsymbol{\nabla}\mathbf{v}) + \rho\mathfrak{r} \ . \tag{3.88}$$

3.10.4 Direct Notation with div *and* grad

The local balances for mass, mass fraction, momentum, internal energy read in direct notation using div and grad

$$\frac{\partial \rho}{\partial t} + \mathbf{v}\cdot \operatorname{grad}\rho + \rho \operatorname{div}\mathbf{v} = 0 \ , \tag{3.89}$$

$$\rho\left(\frac{\partial c_\alpha}{\partial t} + \mathbf{v}\cdot \operatorname{grad} c_\alpha\right) + \operatorname{div}\mathbf{J}^\alpha = \sum_{a=1}^{n}\gamma_\alpha^a M_\alpha \lambda^a \ , \tag{3.90}$$

$$\rho\left(\frac{\partial \mathbf{v}}{\partial t} + \mathbf{v}\cdot (\operatorname{grad}\mathbf{v})^T\right) - \operatorname{div}\mathbf{t}^T = \rho\mathbf{f} \ , \tag{3.91}$$

$$\rho\left(\frac{\partial u}{\partial t} + \mathbf{v}\cdot \operatorname{grad} u\right) + \operatorname{div}\mathbf{q} = \mathbf{t}:(\operatorname{grad}\mathbf{v}) + \rho\mathfrak{r} \ . \tag{3.92}$$

Problems

3.1. General Balance Law

1. Write the general balance law in regular points in the forms (3.10) and (3.29) in direct tensor notation using both nabla and div/grad symbols.
2. Write the general balance law in singular points (3.15) in direct tensor notation.

3.2. Volume Change

Show that the rate of change of a material volume is given by

$$\frac{dV}{dt} = \int_{V(t)} \frac{\partial v_i}{\partial x_i} dV = \int_{V(t)} \operatorname{div} \mathbf{v} dV \,.$$

3.3. Mass Balance for a Control Volume

The mass balance for a control volume is typically written as

$$\frac{dm}{dt} = \sum_{\text{in}} \dot{m}_i - \sum_{\text{out}} \dot{m}_e \,,$$

where $\dot{m} = \rho v A$ is the mass flux across an inlet or exit of the system. Stating all assumptions and simplifications used, reduce the integral mass balance (3.25) to this form.

3.4. Buoyancy

A spherical balloon with radius $r = 5\,\text{m}$ is filled with hydrogen (ideal gas, $M_{H_2} = 2\frac{\text{kg}}{\text{kmol}}$) at $p_{H_2} = 2\,\text{bar}$, $T_0 = 278\,\text{K}$. The total mass of balloon shell and payload is $m_B = 400\,\text{kg}$.

1. Determine the mass of hydrogen in the balloon.
2. Determine the height z_{max} to which the balloon will rise in the atmosphere, where the pressure obeys the barometric law $p_{air}(z) = p_0 \exp\left(-\frac{gz}{R_{air}T_0}\right)$, with $p_0 = 1\,\text{bar}$, $g = 9.81\frac{\text{m}}{\text{s}^2}$, $M_{air} = 29\frac{\text{kg}}{\text{kmol}}$.
3. Determine the surface stress σ_S in the balloon shell at z_{max}.

3.5. High-Altitude Balloon

For his jump from 39 km above sea level, Baumgartner used a helium filled balloon. Mass of balloon and load was 1680 kg, the maximum volume of the balloon shell was $850,000\,\text{m}^3$. The balloon was filled at the ground with $180,000\,\text{ft}^3$ of helium at $p_0 = 1\,\text{bar}$ and $T_0 = -3\,°\text{C}$. Assume the temperature of the atmosphere as constant, and validity of the barometric formula.

1. Determine the volume of the balloon as function of height before the balloon is fully expanded.
2. At what height is the shell fully expanded?
3. Determine the maximum height that the balloon can reach.

4. Determine the forces acting on the balloon as function of height (for buoyancy, distinguish before and after the shell is fully expanded).

3.6. Balance of Momentum
The balance of momentum reads

$$\rho\frac{Dv_i}{Dt}-\frac{\partial t_{ik}}{\partial x_k}=\rho f_i$$

Fully expand this equation in vector and tensor components.

3.7. Balance of Kinetic Energy
Kinetic energy is given by $E_{kin}=\int_V \frac{\varrho}{2}v^2dV$. The local balance for kinetic energy results from scalar multiplication of velocity v_i on the conservation law for momentum,

$$\frac{\partial\rho v_i}{\partial t}+\frac{\partial}{\partial x_{kk}}\left[\rho v_iv_k-t_{ik}\right]-\rho f_i\;.$$

1. Perform all the necessary steps to bring the balance of kinetic energy into standard form

$$\frac{\partial\left(\frac{\varrho}{2}v^2\right)}{\partial t}+\frac{\partial}{\partial x_k}\left[\frac{\rho}{2}v^2v_k+\phi_k\right]=\rho\left(\pi+\zeta\right)$$

Identify ϕ_i, π and ζ. Discuss physical meaning of production and supply.
2. Introduce the convective derivative and write in the form $\rho\frac{D\frac{1}{2}v^2}{Dt}+\cdots$
3. Write the integral form of the balance of kinetic energy.
4. Write the balance of kinetic energy on singular surfaces.

3.8. Balance of Internal Energy

1. Subtract the balance of kinetic energy from the full energy balance, to find the balance of internal energy,

$$\frac{\partial\rho u}{\partial t}+\frac{\partial}{\partial x_k}\left[\rho uv_k+q_k\right]=t_{ij}\frac{\partial v_i}{\partial x_j}+\rho\mathfrak{r}$$

2. To find the balance of energy for reversible processes in closed systems, $\frac{dU}{dt}=\dot{Q}-p\frac{dV}{dt}$, set $t_{ij}=-p\delta_{ij}$, and integrate the above over a material volume. Assume homogeneous pressure. Note: you will have to use that $\frac{dV}{dt}=\int_{V(t)}\frac{\partial v_i}{\partial x_i}dV$.

3.9. Potential Energy
In the local balance of total energy (3.69) the term ρf_kv_k describes the power of body forces, which enters the equation as a supply term. Consider the case that the specific body force f_k follows from a time-independent potential Ψ (e.g., $\Psi=gz$) as $f_k=-\frac{\partial\Psi}{\partial x_k}$and show that the balance can be written as

$$\frac{\partial \rho \left(u + \frac{1}{2}v^2 + \Psi\right)}{\partial t} + \frac{\partial \left[\rho \left(u + \frac{1}{2}v^2 + \Psi\right) v_k - t_{jk}v_j + q_k\right]}{\partial x_k} = \rho \mathfrak{r}$$

3.10. The First Law for a Control Volume

The first law for an open system is commonly written as (g is gravitational acceleration))

$$\frac{dE}{dt} = \sum_{in} \dot{m}_i \left(h + \frac{1}{2}v^2 + \mathrm{g}z\right)_i - \sum_{out} \dot{m}_e \left(h + \frac{1}{2}v^2 + \mathrm{g}z\right)_e + \dot{Q} - \dot{W}$$

This equation follows from the balance given in the previous problem through integration over the control volume. Identify the quantities E, $\dot{m}$, $\dot{Q}$, $\dot{W}$, and state all simplifications required to obtain this form.

3.11. Direct Tensor Notation

1. Write the momentum and energy balances in direct tensor notation. From these, find the balances of kinetic and internal energy as in Probs. 3.7, 3.8, but using direct notation.
2. Use direct notation only to determine the balance of moment of momentum from the momentum balance.
3. Write the balances at singular surfaces for mass, momentum, energy in direct notation.
4. Write partial mass balances and mass fraction balances in direct notation.

Chapter 4
Linear Irreversible Thermodynamics

The conservation laws for (partial) mass, momentum and (total, internal, kinetic) energy contain non-convective fluxes as well as chemical reaction rates. These quantities depend on the material and the nonequilibrium flow state of the system through constitutive equations that must be specified based on rational arguments. In this chapter we employ the assumption of Local Thermodynamic Equilibrium (LTE) to construct the second law of thermodynamics—the balance law for entropy—accompanying the conservation laws. The constitutive equations are obtained from the requirement that the local entropy generation rate must be non-negative, and vanishes in equilibrium.

4.1 Conservation Laws and Variables

In the previous chapter we have found conservation laws for mass, momentum and energy in various forms. Here, we list the equations in regular points which are required to describe processes in reacting mixtures with ν components and n independent reactions:

Total Mass Balance

$$\frac{D\rho}{Dt} + \rho\frac{\partial v_k}{\partial x_k} = 0 \tag{4.1}$$

Mass Fraction Balances

$$\rho\frac{D\mathsf{c}_\alpha}{Dt} + \frac{\partial J_k^\alpha}{\partial x_k} = \sum_{a=1}^{n} \gamma_\alpha^a M_\alpha \lambda^a \qquad \text{for} \quad \alpha = 1, \ldots, \nu - 1 \ , \tag{4.2}$$

with $\sum_{\alpha=1}^{\nu} \mathsf{c}_\alpha = 1$, $\sum_{\alpha=1}^{\nu} J_i^\alpha = 0$, $\sum_{\alpha=1}^{\nu} \gamma_\alpha^a M_\alpha = 0$.

Momentum Balance

$$\rho\frac{Dv_i}{Dt} - \frac{\partial t_{ik}}{\partial x_k} = \rho f_i \tag{4.3}$$

H. Struchtrup, *A Thermodynamic Introduction to Transport Phenomena*,
https://doi.org/10.1007/978-3-031-61868-0_4

Balance of Internal Energy

$$\rho\frac{Du}{Dt}+\frac{\partial q_k}{\partial x_k}=t_{ij}\frac{\partial v_i}{\partial x_j}+\rho\mathfrak{r} \tag{4.4}$$

The intuitive variables are the measurable properties mass density ρ, mass fractions c_α, flow velocity v_i, temperature T. With this, the number of variables equals the number of equations listed. Note that temperature T does not appear explicitly in the equations; it replaces internal energy, in order to have a measurable property as variable.

The above conservation laws do not form a closed system for the variables, since they contain in addition to the variables the following quantities: diffusion fluxes J_i^α, reaction rates λ^a, stress tensor t_{ij}, heat flux vector q_i, and internal energy u. Moreover, knowledge of the supply terms, i.e., the body force f_i and the radiative absorption/emission supply $\mathfrak{r}$ is required.

To close the system of conservation laws, we need to develop *constitutive relations* that relate the unknown quantities $\{J_i^\alpha,\lambda^a,t_{ij},q_i,u\}$ to the local state $\{\rho,\mathsf{c}_\alpha,v_i,T\}$ of the system, through the variables themselves, and their gradients in space (and possibly time).

The caloric equation of state, $u\,(T,\rho,\mathsf{c}_\alpha)$ is one example for a constitutive relation that is known from the consideration of equilibrium states and reversible processes. We already extended its use to nonequilibrium states for processes that can be described under the assumption of local thermodynamic equilibrium (LIT).

The other properties, $\{J_i^\alpha,\lambda^a,t_{ij},q_i\}$ are fluxes and production terms that describe nonequilibrium processes, and we have to devise a strategy to find meaningful constitutive relations for these.

Different approaches to this end can be found in the literature, and here we use the most prominent, and most fundamental, method, which is known as *Linear Irreversible Thermodynamics*, or short, as *LIT*.

4.2 Entropy Balance

In *Linear Irreversible Thermodynamics* the constitutive relations are constructed such that the second law of thermodynamics is obeyed at all times, that is for all solutions to the closed conservation laws.

Since entropy $S=\int\rho s dV$ is an extensive property of the system, it must obey a balance law in the general forms outlined before. Therefore, we write the entropy balance in integral, local and jump form as

$$\frac{d}{dt}\int_{V(t)}\rho s dV+\oint_{\partial V(t)}\left[\rho s\left(v_k-w_k\right)+\varphi_k\right]n_k dA=\int_{V(t)}\left(\Sigma+\rho\xi\right)dV \tag{4.5}$$

$$\rho \frac{Ds}{Dt} + \frac{\partial \varphi_k}{\partial x_k} = \Sigma + \rho \xi \tag{4.6}$$

$$[[\rho s (v_k - w_k) + \varphi_k]] \, n_k = \Sigma_S + \xi_S \, . \tag{4.7}$$

Here we have introduced the non-convective entropy flux φ_i, density of entropy generation rate Σ, surface production rate Σ_S, and entropy supplies ξ, ξ_S in bulk and at singular surfaces, respectively.

Both entropy production rates must be non-negative,

$$\Sigma \geq 0 \quad , \quad \Sigma_S \geq 0 \, , \tag{4.8}$$

with $\Sigma = \Sigma_S = 0$ in equilibrium. Entropy supply, ξ and ξ_S, are not restricted in sign.

From equilibrium thermodynamics it is known that entropy is not an independent property of the system, but depends on other properties through the Gibbs equation (1.44),

$$T ds = du - \frac{p}{\rho^2} d\rho - \sum_{\alpha=1}^{\nu-1} (\mu_\alpha - \mu_\nu) \, d\mathrm{c}_\alpha \, . \tag{4.9}$$

The main assumption of LIT is that of *Local Thermodynamic Equilibrium (LTE)*, as discussed in Sec. 1.6.3. In LTE, the thermal and caloric equations of state and the Gibbs equation are valid locally, at *any* continuum point $\mathbf{x}$, at all times t.

Considering the Gibbs equation for the change of properties of a material volume links the rate of change of entropy to those appearing in the conservation laws,

$$T \frac{Ds}{Dt} = \frac{Du}{Dt} - \frac{p}{\rho^2} \frac{D\rho}{Dt} - \sum_{\alpha=1}^{\nu-1} (\mu_\alpha - \mu_\nu) \frac{D\mathrm{c}_\alpha}{Dt} \tag{4.10}$$

With the help of the conservation laws, we bring the above into standard balance law form. Specifically, we multiply with ρ, divide by T, and replace the rates on the right by means of the conservation laws, so that

$$\begin{aligned} \rho \frac{Ds}{Dt} = \frac{1}{T} \left[-\frac{\partial q_k}{\partial x_k} + t_{ij} \frac{\partial v_i}{\partial x_j} + \rho \mathfrak{r} \right] - \frac{p}{\rho T} \left[-\rho \frac{\partial v_k}{\partial x_k} \right] \\ - \sum_{\alpha=1}^{\nu-1} \frac{\mu_\alpha - \mu_\nu}{T} \left[-\frac{\partial J_k^\alpha}{\partial x_k} + \sum_{a=1}^{n} \gamma_\alpha^a M_\alpha \lambda^a \right] \end{aligned} \tag{4.11}$$

Some work is required to bring this into the desired form (4.6). Using the product rule backwards, we have

$$-\frac{1}{T}\frac{\partial q_j}{\partial x_j} = -\frac{\partial}{\partial x_j}\left(\frac{q_j}{T}\right) + q_j\frac{\partial}{\partial x_j}\left(\frac{1}{T}\right) \tag{4.12}$$

and

$$\sum_{\alpha=1}^{\nu-1}\frac{\mu_\alpha - \mu_\nu}{T}\frac{\partial J_k^\alpha}{\partial x_i} = \frac{\partial}{\partial x_k}\left(\sum_{\alpha=1}^{\nu-1}\frac{\mu_\alpha - \mu_\nu}{T}J_k^\alpha\right) - \sum_{\alpha=1}^{\nu-1}J_k^\alpha\frac{\partial}{\partial x_k}\left(\frac{\mu_\alpha - \mu_\nu}{T}\right) \tag{4.13}$$

Next, we decompose the (symmetric!) stress tensor and velocity gradient into their irreducible parts

$$t_{ij}\frac{\partial v_i}{\partial x_j} = \frac{1}{3}t_{kk}\frac{\partial v_l}{\partial x_l} + t_{\langle ij\rangle}\frac{\partial v_{\langle i}}{\partial x_{j\rangle}} \tag{4.14}$$

Here, indices in angular brackets denote the trace-free and symmetric part of a tensor, see Appendix A. With all this, we find[1]

$$\begin{aligned}\rho\frac{Ds}{Dt} + \frac{\partial}{\partial x_k}\left(\frac{q_k}{T} - \sum_{\alpha=1}^{\nu-1}\frac{\mu_\alpha - \mu_\nu}{T}J_k^\alpha\right) &= \\ = \frac{1}{T}\left[\frac{1}{3}t_{kk} + p\right]\frac{\partial v_i}{\partial x_i} &+ \sum_{a=1}^{n}\frac{\lambda^a}{T}\left[-\sum_{\alpha=1}^{\nu}\gamma_\alpha^a M_\alpha\mu_\alpha\right] \\ + q_k\frac{\partial}{\partial x_k}\left(\frac{1}{T}\right) + \sum_{\alpha=1}^{\nu-1}J_k^\alpha\left[-\frac{\partial}{\partial x_k}\left(\frac{\mu_\alpha - \mu_\nu}{T}\right)\right] &+ \frac{1}{T}t_{\langle ij\rangle}\frac{\partial v_{\langle i}}{\partial x_{j\rangle}} + \frac{1}{T}\rho\mathfrak{r}\end{aligned} \tag{4.15}$$

Comparison with the general form (4.6) allows us to identify:
non-convective entropy flux

$$\varphi_i = \frac{q_k}{T} - \sum_{\alpha=1}^{\nu-1}\frac{\mu_\alpha - \mu_\nu}{T}J_k^\alpha \tag{4.16}$$

entropy generation rate

$$\begin{aligned}\Sigma = \left[\frac{1}{3}t_{kk} + p\right]\frac{1}{T}\frac{\partial v_i}{\partial x_i} &+ \sum_{a=1}^{n}\lambda^a\left[-\frac{1}{T}\sum_{\alpha=1}^{\nu}\gamma_\alpha^a M_\alpha\mu_\alpha\right] \\ &+ q_k\frac{\partial}{\partial x_k}\left(\frac{1}{T}\right) + \sum_{\alpha=1}^{\nu-1}J_k^\alpha\left[-\frac{\partial}{\partial x_k}\left(\frac{\mu_\alpha - \mu_\nu}{T}\right)\right] \\ &+ \frac{1}{T}t_{\langle ij\rangle}\frac{\partial v_{\langle i}}{\partial x_{j\rangle}}\end{aligned} \tag{4.17}$$

and entropy supply

[1] Recall that $\sum_{\alpha=1}^{\nu}\gamma_\alpha^a M_\alpha = 0$, hence $\sum_{\alpha=1}^{\nu-1}\gamma_\alpha^a M_\alpha\left(\mu_\alpha - \mu_\nu\right) = \sum_{\alpha=1}^{\nu}\gamma_\alpha^a M_\alpha\mu_\alpha$.

$$\xi = \frac{\mathfrak{r}}{T} \,. \tag{4.18}$$

For simple substances, where composition does not change, the entropy balance reduces to

$$\rho \frac{Ds}{Dt} + \frac{\partial \frac{q_k}{T}}{\partial x_k} = \Sigma \tag{4.19}$$

and, as shown in Prob. 4.1, integration over system volume leads to the second law in the form (1.10). The balance on singular surfaces reads

$$\left[\left[\rho s \left(v_i - w_i\right) + \frac{q_i}{T}\right]\right] n_i = \Sigma_S + \xi_S \,, \tag{4.20}$$

where Σ_S and ξ_S are surface entropy generation and supply, respectively.

4.3 Thermodynamic Forces and Fluxes

We recognize the entropy production rate (4.17) as a sum of products of thermodynamic forces $\mathcal{F}_A$, which describe the deviation from equilibrium states, and thermodynamic fluxes $\mathcal{J}_A$, which move the system towards equilibrium, that is

$$\Sigma = \sum_A \mathcal{F}_A \mathcal{J}_A \,. \tag{4.21}$$

The following table lists the forces and their associated fluxes:

Force $\mathcal{F}_A$	**Flux $\mathcal{J}_A$**	**Description**
$\frac{1}{T}\frac{\partial v_i}{\partial x_i}$	$\left[\frac{1}{3}t_{kk} + p\right]$	stresses drive flow towards homogeneous velocity
$\frac{1}{T}\frac{\partial v_{\langle i}}{\partial x_{j\rangle}}$	$t_{\langle ij\rangle}$	stresses drive flow towards homogeneous velocity
$\frac{\partial}{\partial x_k}\left(\frac{1}{T}\right)$	q_k	heat flux drives system towards homogeneous temperature
$-\frac{\partial}{\partial x_k}\left(\frac{\mu_\alpha - \mu_\nu}{T}\right)$	J_k^α	diffusion flux drives system towards equilibrium mixture
$-\frac{1}{T}\sum_{\alpha=1}^{\nu} \gamma_\alpha^a M_\alpha \mu_\alpha$	λ^a	reactions at rate λ^a drive system towards reactive equilibrium (l.o.m.a.)

Some comments are in order:

In Chap. 2 it was found that in equilibrium velocity, temperature, and chemical potentials are homogeneous, that is their gradients vanish. Hence, gradients of velocity, temperature and chemical potential describe deviations from mechanical, thermal and chemical equilibrium, that is nonequilibrium states. These gradients are interpreted as thermodynamic forces, in the entropy generation they are paired with stress tensor, heat flux and diffusion fluxes, which are interpreted as fluxes that aim to reduce the gradients.

In chemical equilibrium the Law of Mass Action holds [TEC, Chap. 23],

$$\left[\sum_{\alpha=1}^{\nu} \gamma_\alpha^a M_\alpha \mu_\alpha\right]_{|E} = 0 \quad , \quad a = 1, \ldots, n \ , \tag{4.22}$$

which yields the equilibrium composition of a reacting mixture in dependence on temperature and pressure. We denote the expressions

$$\bar{\mathcal{A}}^a = \sum_{\alpha=1}^{\nu} \gamma_\alpha^a M_\alpha \mu_\alpha = \sum_{\alpha=1}^{\nu} \gamma_\alpha^a \bar{\mu}_\alpha \tag{4.23}$$

as affinities, so that the law of mass action states that the affinities $\bar{\mathcal{A}}^a$ must vanish in equilibrium. Non-zero affinities indicate nonequilibrium states, and in extension of the above are interpreted as thermodynamic forces. Reaction rates λ^a are interpreted as fluxes that drive the system towards chemical equilibrium. The overbar indicates molar properties, affinity and molar chemical potential $\bar{\mu}_\alpha = M_\alpha \mu_\alpha$ have the unit $\left[\frac{\mathrm{kJ}}{\mathrm{mol}}\right]$.

4.4 Phenomenological Equations

Recall that we need constitutive equations for the fluxes $\{J_i^\alpha, \lambda^a, q_i, t_{ij}\}$ in terms of the variables $\{\rho, \mathsf{c}_\alpha, v_i, T\}$. In equilibrium, when the forces vanish, the fluxes must vanish as well so that the equilibrium state is maintained. Moreover, entropy production must be non-negative.

These requirements are fulfilled when one writes the thermodynamic fluxes as linear functions of the forces,

$$\mathcal{J}_A = \sum_B L_{AB} \mathcal{F}_B \tag{4.24}$$

where $L_{AB}\left(\rho, \mathsf{c}_\alpha, T\right)$ is a matrix of coefficients that depend on the local state. The elements of L_{AB} are known as the *phenomenological coefficients*. The non-convective fluxes are all measured in the center of mass system of the material, and are independent of the flow velocity, which depends on the motion of the observer relative to the flow. Accordingly, the L_{AB} must be independent of velocity.

The phenomenological coefficients must be determined from measurements, or found from underlying microscopic theories. As discussed below, the matrix of phenomenological coefficients is subject to several conditions which reduce the number of measurements required to obtain their values.

With the linear relation, the entropy generation becomes

$$\Sigma = \sum_{A} \mathcal{F}_A \mathcal{J}_A = \sum_{A,B} \mathcal{F}_A L_{AB} \mathcal{F}_B \geq 0 \,. \tag{4.25}$$

The sign requirement is guaranteed when the matrix of phenomenological coefficients L_{AB} is non-negative definite. The linear relations ensure that all forces and fluxes, and the entropy generation rate, vanish in equilibrium.

Linearity of the force-flux relations implies a restriction to processes not too far from equilibrium states. For most applications in fluid mechanics and heat and mass transfer, linear relations provide accurate models.

4.4.1 Curie Principle

Notably, the linear ansatz for the fluxes pairs a flux not only with its own force—the one with which it is appears in the entropy generation rate—but also with other forces. A reduction of the phenomenological relations, related to tensor structure of fluxes and forces, is due to Pierre Curie (1859-1906), the husband of double Nobel laureate Marie Curie (1867-1934).

The *Curie Principle* states that for isotropic media, such as a liquid or a fluid,

- scalar fluxes depend on scalar forces,
- vectorial fluxes depend on vectorial forces, and
- tensor forces depend on tensor forces.

The Curie principle ensures proper transformation of thermodynamic forces and fluxes between different observer frames.

4.4.2 Onsager-Casimir Relations

The phenomenological coefficients L_{AB} must be obtained from measurements, as functions of the variables $(\rho, \mathsf{c}_\alpha, T)$. This task is somewhat simplified by the *Onsager-Casimir Reciprocity Relations*, which state that the matrix of phenomenological coefficients is symmetric, if either all forces are invariant in time reversal, or change sign in time reversal (Lars Onsager, 1903-1976, Nobel Prize 1968; Hendrik B.G. Casimir, 1909-2000). If forces are paired in (4.25) where one changes sign, and the other not, then the corresponding matrix elements are antisymmetric.

Here, time reversal means running backward in time, i.e., as in rewinding a movie. Examples for invariant forces are gradients of temperature or chemical potential, which keep their direction in rewinding. However, velocity v_i changes its sign in reversal, and hence the forces $\frac{\partial v_i}{\partial x_i}$ and $\frac{\partial v_{\langle i}}{\partial x_{j\rangle}}$ change their signs as well.

The Onsager-Casimir relations are derived from arguments on statistics of microscopic states, see, e.g., the classic monograph by de Groot & Mazur: Non-equilibrium Thermodynamics, 1969.

4.4.3 Scalar Force-Flux Relations

Looking at the scalar force-flux relations first, we find (with the common factor T absorbed into the L_{AB})

$$\frac{1}{3}t_{kk} + p = \lambda \frac{\partial v_i}{\partial x_i} + \sum_{b=1}^{n} L_b \left[- \sum_{\alpha=1}^{\nu} \gamma_\alpha^a M_\alpha \mu_\alpha \right] \tag{4.26}$$

$$\lambda^a = \tilde{L}_a \frac{\partial v_i}{\partial x_i} + \sum_{b=1}^{n} L_{ab} \left[- \sum_{\alpha=1}^{\nu} \gamma_\alpha^b M_\alpha \mu_\alpha \right] \quad , \quad a = 1, \ldots, n \tag{4.27}$$

with the $([1+n] \times [1+n])$ matrix of phenomenological coefficients

$$\begin{bmatrix} \lambda & L_b \\ \\ \tilde{L}_a & L_{ab} \end{bmatrix} \text{—positive definite} \tag{4.28}$$

The coefficient λ is known as bulk viscosity, or volume viscosity, and describes the resistance to volume change.

For the relations between scalar fluxes and forces, the Onsager relations imply that the submatrix L_{ab} is symmetric,

$$L_{ab} = L_{ba} \tag{4.29}$$

while the mixed coefficients must be anti-symmetric,

$$\tilde{L}_b = -L_b \tag{4.30}$$

The cross coupling between the mechanical and chemical effects is often ignored, so that $\tilde{L}_b = -L_b \simeq 0$. For most chemical reactions, the deviation from equilibrium states is large, so that a linear relation between reaction rates and affinities, i.e., $\lambda^a = -\sum_b L_{ab}\mathcal{A}^b$, does not suffice. Then, non-linear relations, such as the Arrhenius law, must be used, see Sec. 9.3.3.

4.4.4 Vectorial Force-Flux Relations

For vector fluxes and forces we find

$$q_k = \kappa T^2 \frac{\partial}{\partial x_k}\left(\frac{1}{T}\right) + \sum_{\beta=1}^{\nu-1} B_\beta \left[-\frac{\partial}{\partial x_k}\left(\frac{\mu_\beta - \mu_\nu}{T}\right)\right] \tag{4.31}$$

$$J_k^\alpha = \tilde{B}_\alpha \frac{\partial}{\partial x_k}\left(\frac{1}{T}\right) + \sum_{\beta=1}^{\nu-1} B_{\alpha\beta} \left[-\frac{\partial}{\partial x_k}\left(\frac{\mu_\beta - \mu_\nu}{T}\right)\right] \tag{4.32}$$

with phenomenological coefficients κ, B_β, $\tilde{B}_\alpha$, $B_{\alpha\beta}$. Since all forces are invariant against time reversal, we have for the coefficient matrix

$$\begin{bmatrix} \kappa T^2 & B_\beta \\ \\ \tilde{B}_\alpha & B_{\alpha\beta} \end{bmatrix} \text{ — symmetric and positive definite }, \tag{4.33}$$

in particular

$$\tilde{B}_\alpha = B_\alpha \quad , \quad B_{\alpha\beta} = B_{\beta\alpha} \; . \tag{4.34}$$

The coefficient κ is a heat conductivity. The diagonal elements of a positive definite matrix are positive, which implies $\kappa > 0$. Since $\kappa T^2 \frac{\partial}{\partial x_k}\left(\frac{1}{T}\right) = -\kappa \frac{\partial T}{\partial x_k}$, the first term in the relation for heat flux describes transfer of heat opposite to the gradient of temperature—heat, by itself will go from warm to cold.

Diffusion fluxes are induced by gradients of the chemical potentials, the coefficients $B_{\alpha\beta}$ are related to diffusion coefficients.

But there is more: a temperature gradient induces diffusion fluxes, with the coefficients $\tilde{B}_\alpha$. This is known as the Soret, or thermo-diffusion, effect (Charles Soret 1854-1904). The off-diagonal elements $\tilde{B}_\alpha$ have no sign restrictions, so that thermo-diffusion might be in direction of the temperature gradient, or in the opposite direction.

Correspondingly, concentration gradients imply gradients of the chemical potential, which induce heat flux (coefficients $B_\beta = \tilde{B}_\alpha$). This is known as Dufour, or diffusion-thermo effect (Louis Dufour 1832-1892).

4.4.5 Tensorial Force-Flux Relation

Finally, we have only one tensor force and one corresponding flux, which yields the (trace-free part of the) Newton stress tensor

$$t_{\langle ij\rangle} = 2\eta \frac{\partial v_{\langle i}}{\partial x_{j\rangle}} . \tag{4.35}$$

Here, the shear viscosity η must be positive; the factor 2 appears due to convention.

4.4.6 Determination of Phenomenological Coefficients

Even with the reduction through the Curie and Onsager-Casimir principles, the cross-coupling of forces and fluxes through the phenomenological equations (4.24) introduces a large number of coefficients, at least when mixtures, with or without reaction, are considered. These coefficients depend on the local state $\{\rho, \mathsf{c}_\alpha, T\}$ of the material, and must either be measured through careful experiments, or determined from molecular theories and simulations. This provides a formidable task, due to the possibly complicated interplay of effects in experiments, such as buoyancy in inhomogeneous mixtures, and the limited capability of measuring exact property data within the flow field. Moreover, the variable space $\{\rho, \mathsf{c}_\alpha, T\}$ is possibly quite large, which significantly increases the number of required measurements. Light scattering experiments offer an interesting alternative for the measurement of nonequilibrium transport coefficients from the fluctuations of a system in equilibrium.

4.5 Equilibrium in a Reacting Mixture under Gravity

As a first application, we determine the equilibrium state in mixtures from the balance laws and the phenomenological relations.

Indeed, in equilibrium the local entropy generation rate Σ must vanish, which in LIT implies that all thermodynamic forces $\mathcal{F}_A$, and hence all thermodynamic fluxes $\mathcal{J}_A$, vanish. Vanishing forces yields

$\dfrac{\partial T}{\partial x_i} = 0$	$\Rightarrow T = T_0$	homogeneous temperature
$\dfrac{\partial v_{\langle i}}{\partial x_{j\rangle}} = 0 \ , \ \dfrac{\partial v_k}{\partial x_k} = 0$	$\Rightarrow v_i = 0$	homogeneous velocity
$-\dfrac{\partial}{\partial x_i}\left(\dfrac{\mu_\alpha - \mu_\nu}{T}\right) = 0$	$\Rightarrow \dfrac{\partial \mu_\alpha}{\partial x_i} = \dfrac{\partial \mu_\nu}{\partial x_i} = K_i$	K_i independent of α
$-\sum_{\alpha=1}^{\nu} \gamma_\alpha^a M_\alpha \mu_\alpha = 0$	$\Rightarrow$ l.o.m.a.	law of mass action

With all convective and conductive fluxes vanishing, the transport equations (4.1 - 4.4) reduce to

$$\frac{\partial \rho}{\partial t} = 0 \quad , \quad \frac{\partial \mathsf{c}_\alpha}{\partial t} = 0 \quad , \quad \frac{\partial p}{\partial x_i} = \rho f_i \quad , \quad \frac{\partial u}{\partial t} = 0 \ , \tag{4.36}$$

that is all properties remain in equilibrium.

To further evaluate the conditions for chemical equilibrium, we use the Gibbs-Duhem equation (1.48), which after multiplication with mixture density ρ reads

$$\sum_{\alpha=1}^{\nu} \rho_\alpha d\mu_\alpha = -\rho s dT + dp \ . \tag{4.37}$$

Hence, with $T = T_0$, $dT = 0$,

$$\sum_{\alpha=1}^{\nu} \rho_\alpha \frac{\partial \mu_\alpha}{\partial x_i} = \frac{\partial p}{\partial x_i} \ . \tag{4.38}$$

Since the gradients of the chemical potentials of all components have the same value, K_i, we find by replacing the pressure gradient from the balance of momentum,

$$\sum_{\alpha=1}^{\nu} \rho_\alpha K_i = \rho K_i = \frac{\partial p}{\partial x_i} = \rho f_i \quad \text{so that} \quad K_i = f_i \ . \tag{4.39}$$

In the gravitational field $f_i = \{0, 0, -\mathsf{g}\}_i$, hence $\frac{\partial \mu_\alpha}{\partial x_i} = \{0, 0, -\mathsf{g}\}_i$ for all α, which upon integration gives

$$\mu_\alpha + \mathsf{g} z = \mu_\alpha^0 \ , \tag{4.40}$$

where μ_α^0 are the chemical potentials at $z = 0$, which appear as integration constants. These equations yield the barometric and hydrostatic laws for mixtures, see Probs. 2.3, 2.4.

Using the equilibrium result (4.40) for μ_α in the law of mass action, we have as the requirement for reactive equilibrium in the gravitational field

$$0=\sum_{\alpha=1}^{\nu}\gamma_\alpha^a M_\alpha\mu_\alpha=\sum_{\alpha=1}^{\nu}\gamma_\alpha^a M_\alpha\left(\mu_\alpha^0-\mathrm{g}z\right)=\sum_{\alpha=1}^{\nu}\gamma_\alpha^a M_\alpha\mu_\alpha^0-\mathrm{g}z\sum_{\alpha=1}^{\nu}\gamma_\alpha^a M_\alpha\,. \tag{4.41}$$

Due to mass conservation $\sum_{\alpha=1}^{\nu}\gamma_\alpha^a M_\alpha=0$, hence the reactive equilibrium requires

$$\sum_{\alpha=1}^{\nu}\gamma_\alpha^a M_\alpha\mu_\alpha^0=0\,. \tag{4.42}$$

That is, in equilibrium, when the law of mass action is fulfilled at height $z=0$ (i.e., for the μ_α^0) it is fulfilled at all heights. Composition of the reacting mixture depends on height (e.g., barometric formula for ideal gases), but due to the pressure dependence of the chemical potentials, reactive equilibrium is maintained at all heights.

4.6 Maxwell-Stefan Equations

An alternative approach to diffusion fluxes is given through the Maxwell-Stefan equations, which we briefly motivate here, for the special case of isothermal diffusion in a system of homogeneous pressure. We start with the diffusion fluxes (4.32),

$$J_k^\alpha=-\sum_{\beta=1}^{\nu-1}\frac{B_{\alpha\beta}}{T}\frac{\partial\left(\mu_\beta-\mu_\nu\right)}{\partial x_i}\quad,\quad\alpha=1,\ldots,\nu-1 \tag{4.43}$$

The symmetric tensor $B_{\alpha\beta}$ $(\alpha,\beta=1,\ldots,\nu-1)$ has $\frac{(\nu-1)\nu}{2}$ independent components. We introduce the extension $\breve{B}_{\alpha\beta}$ of this tensor, by adding a row and a column for component ν, so that

$$\breve{B}_{\alpha\beta}=B_{\alpha\beta}\quad,\quad\breve{B}_{\alpha\nu}=-\sum_{\beta=1}^{\nu-1}B_{\alpha\beta}=-\sum_{\beta=1}^{\nu-1}B_{\beta\alpha}=\breve{B}_{\nu\alpha}\quad,\quad\breve{B}_{\nu\nu}=0\,. \tag{4.44}$$

With this we can express the diffusion fluxes as

$$J_k^\alpha=-\sum_{\beta=1}^{\nu}\frac{\breve{B}_{\alpha\beta}}{T}\frac{\partial\mu_\beta}{\partial x_i}\quad,\quad\alpha=1,\ldots,\nu \tag{4.45}$$

where $\sum_{\alpha=1}^{\nu}J_i^\alpha=0$ results from the definition of the added elements $\breve{B}_{\alpha\nu}$.

By inversion of the matrix $\breve{B}_{\alpha\beta}$ we find

$$-\frac{1}{T}\frac{\partial \mu_\beta}{\partial x_i} = \sum_{\alpha=1}^{\nu-1} \breve{B}^{-1}_{\beta\alpha} J^\alpha_k \quad , \quad \beta = 1, \ldots, \nu \, . \tag{4.46}$$

Inserting this into the Gibbs-Duhem equation (1.48) for $dp = dT = 0$ gives

$$0 = -\frac{1}{T}\sum_{\beta=1}^{\nu} \mathsf{c}_\beta \frac{\partial \mu_\beta}{\partial x_i} = \sum_{\alpha=1}^{\nu}\left(\sum_{\beta=1}^{\nu} \mathsf{c}_\beta \breve{B}^{-1}_{\beta\alpha}\right) J^\alpha_k \, , \tag{4.47}$$

hence

$$\sum_{\beta=1}^{\nu} \mathsf{c}_\beta \breve{B}^{-1}_{\beta\alpha} = 0 \quad \Longrightarrow \quad \sum_{\substack{\alpha=1 \\ (\alpha\neq\beta)}}^{\nu} \mathsf{c}_\alpha \breve{B}^{-1}_{\alpha\beta} + \mathsf{c}_\beta \breve{B}^{-1}_{\beta\beta} = 0 \tag{4.48}$$

so that

$$-\frac{1}{T}\frac{\partial \mu_\beta}{\partial x_i} = \sum_{\alpha,\alpha\neq\beta} \mathsf{c}_\alpha \breve{B}^{-1}_{\alpha\beta}\frac{J^\alpha_k}{\mathsf{c}_\alpha} + \mathsf{c}_\beta \breve{B}^{-1}_{\beta\beta}\frac{J^\beta_k}{\mathsf{c}_\beta} = \sum_{\alpha,\alpha\neq\beta} \mathsf{c}_\alpha \breve{B}^{-1}_{\alpha\beta}\left(\frac{J^\alpha_k}{\mathsf{c}_\alpha} - \frac{J^\beta_k}{\mathsf{c}_\beta}\right) \tag{4.49}$$

Finally, with[2] $\frac{J^\alpha_k}{\mathsf{c}_\alpha} = \rho\left(v^\alpha_i - v_i\right)$, $\rho_\alpha = M_\alpha \bar{\rho}_\alpha = M_\alpha \bar{\rho} X_\alpha$ and $M_\beta \mu_\beta = \bar{\mu}_\beta$ we find the Maxwell-Stefan equations (for isothermal and isobaric diffusion) as

$$-\frac{1}{\bar{R}T}\frac{\partial \bar{\mu}_\beta}{\partial x_i} = \sum_{\substack{\alpha=1 \\ (\alpha\neq\beta)}}^{\nu} \frac{X_\alpha}{\mathfrak{D}_{\alpha\beta}}\left(v^\beta_i - v^\alpha_i\right) \, , \quad \beta = 1, \ldots, \nu - 1 \tag{4.50}$$

It suffices to consider only $(\nu - 1)$ Maxwell-Stefan equations, since the gradients of chemical potentials are coupled by (4.47). The symmetric Maxwell-Stefan diffusion coefficients are given by

$$\mathfrak{D}_{\alpha\beta} = -\frac{\bar{R}}{\bar{\rho} M_\alpha M_\beta \breve{B}^{-1}_{\alpha\beta}} \, , \tag{4.51}$$

where only coefficients $\mathfrak{D}_{\alpha\beta}$ for $\alpha \neq \beta$ appear, so that $\frac{\nu(\nu-1)}{2}$ independent coefficients are required (and must be determined). These correspond to the same number of $B_{\alpha\beta}$ in the original formulation (4.43). Experience shows that the $\mathfrak{D}_{\alpha\beta}$ depend only weakly on concentrations and often can be considered as constants.

For an ideal mixture, $\bar{\mu}_\beta = \bar{g}_\beta(T, p) + \bar{R}T \ln X_\beta$ and the Maxwell-Stefan relations assume the compact form

$$\frac{\partial X_\beta}{\partial x_i} = \sum_{\substack{\alpha=1 \\ (\alpha\neq\beta)}}^{\nu} \frac{X_\alpha X_\beta}{\mathfrak{D}_{\alpha\beta}}\left(v^\alpha_i - v^\beta_i\right) \, , \quad \beta = 1, \ldots, \nu - 1 \, . \tag{4.52}$$

[2] The overbar denotes molar properties, e.g. $\bar{\rho}$ is molar density with units $\mathrm{kmol/m^3}$.

Problems

4.1. The Second Law for a Control Volume

In undergraduate thermodynamics, the second law for an open system is commonly written as

$$\frac{dS}{dt}+\sum_{out}\dot{m}_e s_e-\sum_{in}\dot{m}_i s_i-\sum_{A}\frac{\dot{Q}_K}{T_K}=\dot{S}_{gen}\geq 0\ ,$$

where $\dot{Q}_K$ is the heat crossing a system boundary at temperature T_K. Show how this form is obtained from the LIT entropy balance for a simple substance, $\rho\frac{Ds}{Dt}+\frac{\partial}{\partial x_k}\left(\frac{q_k}{T}\right)=\sigma$.

4.2. Frictional Heating

1 litre of water (incompressible fluid, mass density $\rho=1\frac{\mathrm{kg}}{\mathrm{litre}}$, specific heat $c_w=4.18\frac{\mathrm{kJ}}{\mathrm{kg\,K}}$) is heavily stirred in an adiabatic container, so that the average velocity of the swirling water is $v=10\frac{\mathrm{m}}{\mathrm{s}}$.

When the stirrer is removed at $t=0$, the temperature is 300 K.

1. Use the first law of thermodynamics to determine the change of temperature as the water comes to rest.
2. Use the second law of thermodynamics to determine the amount of entropy generated, S_{gen}, and the associated work loss $T_0 S_{gen}$.

4.3. Variants for Newton's Law of Cooling

Show that the entropy generation associated with heat transfer between two temperature reservoirs A and B is $\Sigma_T=\left(\frac{1}{T_A}-\frac{1}{T_B}\right)\dot{Q}$.

Which of the following suggestions for the heat flux $\dot{Q}$ are acceptable choices? State your reasons, and give the range of allowed values for the constants β_α (e.g., positive, negative, either sign).

1. $\dot{Q}=\beta_1\left(\frac{1}{T_B}-\frac{1}{T_A}\right)$
2. $\dot{Q}=\beta_2\left(T_B-T_A\right)$
3. $\dot{Q}=\beta_3\left(T_B-T_A\right)^2$
4. $\dot{Q}=\beta_4\left(\ln T_B-\ln T_A\right)$
5. $\dot{Q}=\frac{\beta_5}{\frac{1}{T_B}-\frac{1}{T_A}}$
6. $\dot{Q}=\beta_6 e^{T_B-T_A}$
7. $\dot{Q}=\beta_7\left(e^{1/T_B-1/T_A}-1\right)$

Chapter 5
Navier-Stokes-Fourier Models

For simple substances, i.e., substances where no change of composition occurs, the transport equations of the previous chapter reduce to the Navier-Stokes-Fourier (NSF) equations of classical hydrodynamics. This chapter presents the NSF equations for compressible flows and their reduced form for incompressible flows. The Boussinesq approximation for quasi-incompressible liquids with thermal expansion is discussed in some detail. The equations are solved for Couette, Poiseuille and Boussinesq flows. The chapter closes with a discussion of co- and counterflow heat exchangers.

5.1 Simple Substances

For a simple substance, where no chemical changes occur, the force-flux relations of the last chapter reduce to the Fourier law for heat flux (Jean-Baptiste Joseph Fourier 1768-1830)

$$q_k = -\kappa \frac{\partial T}{\partial x_k} , \tag{5.1}$$

where $\kappa > 0$ is the heat conductivity (unit $\frac{\mathrm{W}}{\mathrm{m\,K}} = \frac{\mathrm{N}}{\mathrm{K\,s}}$), and the Newton stress tensor

$$t_{ij} = -p\delta_{ij} + \lambda \frac{\partial v_k}{\partial x_k}\delta_{ij} + 2\eta \frac{\partial v_{\langle i}}{\partial x_{j\rangle}} , \tag{5.2}$$

where $\lambda > 0$ is the bulk viscosity, or volume viscosity, which describes the resistance to volume change, and $\eta > 0$ is the shear viscosity (both with unit $\frac{\mathrm{N\,s}}{\mathrm{m}^2} = \mathrm{Pa\,s}$).

With the definition of the trace-free tensor, the stress can be rewritten in the more explicit form

$$t_{ij} = -p\delta_{ij} + \left(\lambda - \frac{2}{3}\eta\right) \frac{\partial v_k}{\partial x_k}\delta_{ij} + \eta \left(\frac{\partial v_i}{\partial x_j} + \frac{\partial v_j}{\partial x_i}\right) . \tag{5.3}$$

H. Struchtrup, *A Thermodynamic Introduction to Transport Phenomena*,
https://doi.org/10.1007/978-3-031-61868-0_5

5.2 Compressible Navier-Stokes-Fourier Equations

Insertion of Fourier law and Newton stress tensor into the conservation laws for mass, momentum and energy yields the Navier-Stokes-Fourier (NSF) equations (Claude-Louis Navier, 1785-1836; George Stokes, 1819-1903). The resulting equations are the compressible NSF equations,

$$\frac{D\rho}{Dt}+\rho\frac{\partial v_k}{\partial x_k}=0 \tag{5.4}$$

$$\rho\frac{Dv_i}{Dt}+\frac{\partial p}{\partial x_i}-\frac{\partial}{\partial x_i}\left[\left(\lambda-\frac{2}{3}\eta\right)\frac{\partial v_k}{\partial x_k}\right]-\frac{\partial}{\partial x_j}\left[\eta\left(\frac{\partial v_i}{\partial x_j}+\frac{\partial v_j}{\partial x_i}\right)\right]=\rho f_i \tag{5.5}$$

$$\rho\frac{Du}{Dt}-\frac{\partial}{\partial x_k}\left[\kappa\frac{\partial T}{\partial x_k}\right]=-p\frac{\partial v_k}{\partial x_k}+\lambda\left(\frac{\partial v_k}{\partial x_k}\right)^2+2\eta\frac{\partial v_{\langle i}}{\partial x_{j\rangle}}\frac{\partial v_{\langle i}}{\partial x_{j\rangle}}+\rho\mathfrak{r} \tag{5.6}$$

Note that the equations also require the thermal and caloric equations of state, $p(\rho,T)$, $u(\rho,T)$, and that the transport coefficients $\lambda(\rho,T)$, $\eta(\rho,T)$, $\kappa(\rho,T)$ depend on the local state.

These five equations are appropriate for the description of liquid and gas flows in macroscopic regimes, i.e., when the typical length scale of the system is much larger than the mean free path of the gas molecules. The five variables of the system are mass density ρ, fluid velocity v_i, and temperature T.

The entropy balance for the NSF equations reads

$$\rho\frac{Ds}{Dt}-\frac{\partial}{\partial x_k}\left(\frac{\kappa}{T}\frac{\partial T}{\partial x_k}\right)=\Sigma+\frac{1}{T}\rho\mathfrak{r} \tag{5.7}$$

with the local entropy generation rate

$$\Sigma=\frac{\lambda}{T}\left(\frac{\partial v_k}{\partial x_k}\right)^2+\frac{2\eta}{T}\frac{\partial v_{\langle i}}{\partial x_{j\rangle}}\frac{\partial v_{\langle i}}{\partial x_{j\rangle}}+\frac{\kappa}{T^2}\frac{\partial T}{\partial x_k}\frac{\partial T}{\partial x_k}\geq 0\ . \tag{5.8}$$

While the second law is satisfied for all solutions of the NSF equations, it remains important since thermodynamic work losses are related to the entropy generation within the system as $T_0\dot{S}_{gen}$, with the overall entropy generation rate resulting from integration over the system volume,

$$\dot{S}_{gen}=\int_V \Sigma dV\ . \tag{5.9}$$

The local entropy generation rate Σ provides deeper insight into irreversible losses within the system, e.g., it can be used to identify regions of larger losses, which invite redesign of the system.

5.3 A Change of Variables: $\rho \rightarrow p$

As an educational exercise, and in preparation for the equations for incompressible flows, we perform a change of variables such that pressure p replaces density ρ. Hence, we need to consider the thermal equation of state as $\rho(T,p)$ with the differential

$$d\rho = \left(\frac{\partial \rho}{\partial T}\right)_p dT + \left(\frac{\partial \rho}{\partial p}\right)_T dp = -\rho\alpha dT + \rho\kappa_T dp\ , \tag{5.10}$$

where

$$\alpha = -\frac{1}{\rho}\left(\frac{\partial \rho}{\partial T}\right)_p \quad , \quad \kappa_T = \frac{1}{\rho}\left(\frac{\partial \rho}{\partial p}\right)_T \ . \tag{5.11}$$

denote coefficient of thermal expansion and isothermal compressibility. Specific heats at constant pressure and volume are related by

$$c_p - c_v = \frac{T}{\rho}\frac{\alpha^2}{\kappa_T} \ . \tag{5.12}$$

The differential of internal energy $u(T,p)$ becomes

$$du = \left(\frac{\partial u}{\partial T}\right)_p dT + \left(\frac{\partial u}{\partial p}\right)_T dp \tag{5.13}$$

where we use thermodynamic property relations, including $\left(\frac{\partial h}{\partial p}\right)_T = \frac{T}{\rho^2}\left(\frac{\partial \rho}{\partial T}\right)_p + \frac{1}{\rho}$, to identify

$$\left(\frac{\partial u}{\partial T}\right)_p = \left(\frac{\partial\left(h - \frac{p}{\rho}\right)}{\partial T}\right)_p = c_p + \frac{p}{\rho^2}\left(\frac{\partial \rho}{\partial T}\right)_p = c_p - \frac{p}{\rho}\alpha \tag{5.14}$$

$$\left(\frac{\partial u}{\partial p}\right)_T = \left(\frac{\partial\left(h - \frac{p}{\rho}\right)}{\partial p}\right)_T = \left(\frac{\partial h}{\partial p}\right)_T - \frac{1}{\rho} + \frac{p}{\rho^2}\left(\frac{\partial \rho}{\partial p}\right)_T = -\frac{T}{\rho}\alpha + \frac{p}{\rho}\kappa_T \tag{5.15}$$

With the new variables, the compressible NSF equations for $\{p, v_i, T\}$ become

$$-\alpha\frac{DT}{Dt} + \kappa_T\frac{Dp}{Dt} + \frac{\partial v_k}{\partial x_k} = 0 \tag{5.16}$$

$$\rho\frac{Dv_i}{Dt} + \frac{\partial p}{\partial x_i} - \frac{\partial}{\partial x_i}\left[\left(\lambda - \frac{2}{3}\eta\right)\frac{\partial v_k}{\partial x_k}\right] - \frac{\partial}{\partial x_j}\left[\eta\left(\frac{\partial v_i}{\partial x_j} + \frac{\partial v_j}{\partial x_i}\right)\right] - \rho f_i \tag{5.17}$$

$$\left[\rho c_p\frac{DT}{Dt} - T\alpha\frac{Dp}{Dt}\right] - \frac{\partial}{\partial x_k}\left[\kappa\frac{\partial T}{\partial x_k}\right] = \lambda\left(\frac{\partial v_k}{\partial x_k}\right)^2 + 2\eta\frac{\partial v_{\langle i}}{\partial x_{j\rangle}}\frac{\partial v_{\langle i}}{\partial x_{j\rangle}} + \rho\mathfrak{r} \tag{5.18}$$

The NSF equations for variables $\{p, v_i, T\}$ for an ideal gas are obtained by setting $T\alpha = p\kappa_T = 1$ and $c_p = c_v + R$.

5.4 Incompressible Liquids

For a truly incompressible liquid, $\alpha = \kappa_T = 0$, and $c_p = c_v = c$, as well as $\rho = \rho_0 = const$, so that the above reduce to the incompressible NSF equations for the variables $\{p, v_i, T\}$:

$$\frac{\partial v_k}{\partial x_k} = 0 \tag{5.19}$$

$$\rho_0 \frac{Dv_i}{Dt} + \frac{\partial p}{\partial x_i} - \eta \frac{\partial^2 v_i}{\partial x_j \partial x_j} = \rho_0 f_i \tag{5.20}$$

$$\rho_0 c \frac{DT}{Dt} - \kappa \frac{\partial^2 T}{\partial x_k \partial x_k} = 2\eta \frac{\partial v_{(i}}{\partial x_{j)}} \frac{\partial v_{(i}}{\partial x_{j)}} + \rho_0 \mathfrak{r} \ . \tag{5.21}$$

Here, for simplicity, we have assumed constant viscosities and heat transfer coefficient, but in general these will be functions of temperature. The mass balance states that the velocity is divergence free, which implies $\frac{\partial v_{\langle i}}{\partial x_{j\rangle}} = \frac{\partial v_{(i}}{\partial x_{j)}}$.

Often, thermal effects can be ignored, that is temperature is assumed to be constant, or varying so little that it has no influence on the flow. Then the energy balance (5.21) can be ignored and mass and momentum balances (5.19, 5.20) form the incompressible Navier-Stokes equations, which are extensively studied in the field of classical fluid dynamics. Note that in this case the viscosities are true constants.

5.5 Boussinesq Approximation

The incompressible NSF equations can be found directly from setting $\rho = \rho_0$ in the NSF equations, and then identifying pressure as the required variable to replace density. The detailed approach chosen here becomes valuable when true incompressibility breaks down due to thermal effects, specifically inhomogeneous densities induced by thermal expansion of the liquid. An important application is natural convection, where buoyancy effects induce convection flows. The Boussinesq approximation provides a reduced model, where only the leading effect of thermal expansion is included.

In the case of interest the dimensionless specific heat

$$\hat{c}_p = c_p \frac{\rho T}{p} \gg 1 \tag{5.22}$$

is rather large, while the dimensionless thermal expansion

$$\hat{\alpha} = T\alpha = -\frac{T}{\rho}\left(\frac{\partial\rho}{\partial T}\right)_p \ll 1 \tag{5.23}$$

and the dimensionless isothermal compressibility

$$\hat{\kappa}_T = p\kappa_T = \frac{p}{\rho}\left(\frac{\partial\rho}{\partial p}\right)_T \ll 1 \tag{5.24}$$

are rather small, with $\hat{\kappa}_T \ll \hat{\alpha}$. For water at $p_0 = 1\,\mathrm{MPa}$, $T_0 = 298.15\,\mathrm{K}$ the values of these properties are $\hat{c}_p = 1243$, $\hat{\alpha} = 0.077$, $\hat{\kappa}_T = 4.5 \times 10^{-4}$.

We now use the relative size of these coefficients to reduce the equations (5.16-5.18). Writing the mass balance as

$$\frac{\partial v_k}{\partial x_k} = -\frac{1}{\rho}\frac{D\rho}{Dt} = \hat{\alpha}\frac{D\ln T}{Dt} - \hat{\kappa}_T\frac{D\ln p}{Dt} \tag{5.25}$$

we recognize that the divergence of velocity is small to the order of $\hat{\alpha}$. With that, in the balances for momentum (5.17) and energy (5.18) the viscous divergence contributions are considerably smaller than the other viscous terms, and can be ignored against these. This amounts to effectively setting the divergence to zero, $\frac{\partial v_k}{\partial x_k} = 0$.

Due to the large difference between $\hat{c}_p \gg 1$ and $\hat{\alpha} \ll 1$, the leading term in the energy balance reduces to

$$\left[\rho c_p\frac{DT}{Dt} - T\alpha\frac{Dp}{Dt}\right] = p\left[\hat{c}_p\frac{D\ln T}{Dt} - \hat{\alpha}\frac{D\ln p}{Dt}\right] \simeq p\left[\hat{c}_p\frac{D\ln T}{Dt}\right] = \rho c_p\frac{DT}{Dt} \tag{5.26}$$

Finally, by integrating (5.10), we write the mass density relative to a reference state as

$$\rho = \rho_0\left[1 - \hat{\alpha}\frac{T - T_0}{T_0} + \hat{\kappa}_T\frac{p - p_0}{p_0}\right] \simeq \rho_0\left[1 - \alpha\left(T - T_0\right)\right] \ , \tag{5.27}$$

where the pressure term is ignored since[1] $\hat{\kappa}_T \ll \hat{\alpha}$. In the terms with $\rho\frac{D(\bullet)}{Dt}$ in momentum and energy balance, the density correction from thermal expansion is so small, that we can write $\rho_0\frac{D}{Dt}$ instead. In the body force, however, we retain the thermal contribution, to find the NSF equations in Boussinesq

[1] On the side, we recall that by (5.12) $c_p - c_v = \frac{T}{\rho}\frac{\alpha^2}{\kappa_T} > 0$, and that thermodynamic stability demands $c_v > 0$. Combining these two conditions yields $0 < \frac{\frac{\hat{\alpha}^2}{\hat{c}_p}}{\hat{\kappa}_T} = 1 - \frac{c_v}{c_p} < 1$, hence $\hat{\kappa}_T > \frac{\hat{\alpha}^2}{\hat{c}_p}$ which is fulfilled by the data.

approximation as[2]

$$\frac{\partial v_k}{\partial x_k} = 0 \tag{5.28}$$

$$\rho_0 \frac{Dv_i}{Dt} + \left[\frac{\partial p}{\partial x_i} - \rho_0 f_i\right] - \eta \frac{\partial^2 v_i}{\partial x_j \partial x_j} = -\alpha \left(T - T_0\right) \rho_0 f_i \tag{5.29}$$

$$\rho_0 c_p \frac{DT}{Dt} - \frac{\partial}{\partial x_k}\left[\kappa \frac{\partial T}{\partial x_k}\right] = 2\eta \frac{\partial v_{\langle i}}{\partial x_{j\rangle}} \frac{\partial v_{\langle i}}{\partial x_{j\rangle}} + \rho_0 \mathfrak{r} \tag{5.30}$$

In case of an equilibrium rest state, with $T = T_0$, $v_i = 0$, the mass and energy balances are trivially fulfilled, and the momentum balance reduces to $\left[\frac{\partial p}{\partial x_i} - \rho_0 f_i\right] = 0$, which yields the hydrostatic pressure.

For nonequilibrium processes, the term $\left[\frac{\partial p}{\partial x_i} - \rho_0 f_i\right]$ gives the deviation of pressure from the hydrostatic rest state, which for problems of natural convection is expected to be small. With the body force $\rho_0 f_i$ compensated by the pressure gradient, the buoyancy contribution $\left[-\alpha \left(T - T_0\right) \rho_0 f_i\right]$ stands alone against the other terms in the momentum balance, and cannot be ignored.

The Boussinesq equations are valid for a quasi-incompressible liquid with thermal expansion. For most forced flows, the buoyancy correction can be ignored, $\left[-\rho_0 \alpha \left(T - T_0\right) f_i\right] \simeq 0$, so that the flow is described by the incompressible NSF equations (5.19-5.21). For free convection flows one typically can assume validity of the hydrostatic pressure law for the flow, $\left[\frac{\partial p}{\partial x_i} - \rho_0 f_i\right] = 0$.

Moreover, for slow flows the friction term in the energy balance can be ignored. While the variable change leads to the appearance of the specific heat c_p, within the approximation there is only one specific heat, $c_p = c_{\mathsf{v}} = c$.

5.6 Plane Flows in Steady State

5.6.1 Geometry and Equations

The NSF equations are the basic equations in the fields of fluid mechanics and heat transfer, where they are considered in great detail analytically as well as numerically. In general, that is for problems in complex geometries, and possibly with time dependence, the NSF equations must be solved numerically, that is using the methods of Computational Fluid Mechanics (CFD). Analytical solutions are possible only for simple geometries, but offer good insight into the behavior of NSF (or other transport equations).

[2] For simplicity, in the momentum balance we ignore the non-linear term $\frac{\partial \eta}{\partial x_j}\left[\left(\frac{\partial v_i}{\partial x_j} + \frac{\partial v_j}{\partial x_i}\right)\right] = \frac{d\eta}{dT}\frac{\partial T}{\partial x_j}\left[\left(\frac{\partial v_i}{\partial x_j} + \frac{\partial v_j}{\partial x_i}\right)\right]$, since in free convection gradients of velocity are typically small.

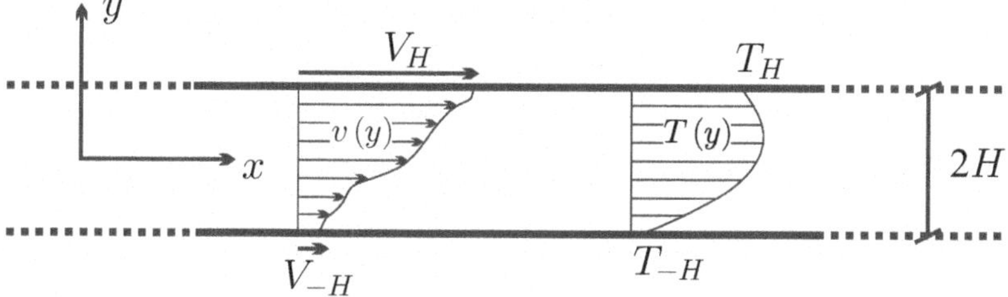

Fig. 5.1 Geometry of plane flows between two parallel plates, with velocities V_H, V_{-H}, temperatures T_H, T_{-H}.

Plane flows in steady state offer the simplest examples to get acquainted with the NSF equations. A number of end-of-chapter problems examine steady state flow and heat transfer of an incompressible liquid confined between infinite parallel plates at distance $2H$, at boundary locations $y_B = \pm H$, as depicted in Fig. 5.1. Flow velocity points strictly in direction x parallel to the plates, while heat transfer is perpendicular to the plates, so that velocity and temperature depend only on the normal coordinate y,

$$v_i = (v(y), 0, 0) \quad , \quad T = T(y) \ . \tag{5.31}$$

We assume that there are no jump and slip effects at the plates, i.e., at $\pm H$ the liquid sticks to the plates and assumes their temperature. For further simplification, specific heat, heat conductivity and viscosity are assumed to be constants.

The first step in the analysis is to reduce the full set of equations to adapt to geometry and assumptions on property relations.

Under the specified conditions we have for derivatives of T, v:

$$\frac{\partial}{\partial t} = 0 \quad \text{and} \quad v_k \frac{\partial}{\partial x_k} = v_x \frac{\partial}{\partial x} = 0 \quad \Rightarrow \quad \frac{D}{Dt} = \frac{\partial}{\partial t} + v_k \frac{\partial}{\partial x_k} = 0 \ . \tag{5.32}$$

The velocity gradient reduces to

$$\frac{\partial v_i}{\partial x_j} = \begin{pmatrix} \frac{\partial v_x}{\partial x} & \frac{\partial v_x}{\partial y} & \frac{\partial v_x}{\partial z} \\ \frac{\partial v_y}{\partial x} & \frac{\partial v_y}{\partial y} & \frac{\partial v_y}{\partial z} \\ \frac{\partial v_z}{\partial x} & \frac{\partial v_z}{\partial y} & \frac{\partial v_z}{\partial z} \end{pmatrix} = \begin{pmatrix} 0 & \frac{dv}{dy} & 0 \\ 0 & 0 & 0 \\ 0 & 0 & 0 \end{pmatrix} , \tag{5.33}$$

while the divergence vanishes by

$$\frac{\partial v_k}{\partial x_k} = \frac{\partial v_x}{\partial x} + \frac{\partial v_y}{\partial y} + \frac{\partial v_z}{\partial z} = \frac{\partial v(y)}{\partial x} = 0 \ . \tag{5.34}$$

With this, the stress tensor (5.3) reads

$$t_{ij} = -p \begin{pmatrix} 1 & 0 & 0 \\ 0 & 1 & 0 \\ 0 & 0 & 1 \end{pmatrix} + \eta \begin{pmatrix} 0 & \frac{dv}{dy} & 0 \\ \frac{dv}{dy} & 0 & 0 \\ 0 & 0 & 0 \end{pmatrix} . \tag{5.35}$$

With temperature depending only on the normal coordinate, the heat flux vector becomes

$$q_i = -\kappa \frac{\partial T}{\partial x_i} = -\kappa \left(0, \frac{dT}{dy}, 0 \right) . \tag{5.36}$$

From these, the local entropy generation rate (5.8) for the plane flow processes considered is

$$\Sigma = \frac{\eta}{T} \left(\frac{dv}{dy} \right)^2 + \frac{\kappa}{T^2} \left(\frac{dT}{dy} \right)^2 . \tag{5.37}$$

Finally, we use all of the above to identify the corresponding Boussinesq equations, where we assume that the body force f_i points only in flow direction, and no radiative transfer takes place,

$$f_i = (f_x, 0, 0) \quad , \quad \mathfrak{r} = 0 . \tag{5.38}$$

The mass balance (5.28) is identically fulfilled. The balance of momentum in the y-direction gives $\frac{\partial p}{\partial y} = 0$, which implies that pressure is only a function of flow direction x. The momentum balance (5.29) in x-direction, and the energy balance (5.30) finally become

$$\left[\frac{dp}{dx} - \rho_0 f_x \right] - \eta \frac{d^2 v}{dy^2} = -\alpha \left(T - T_0 \right) \rho_0 f_x , \tag{5.39}$$

$$-\kappa \frac{d^2 T}{dy^2} = \eta \left(\frac{dv}{dy} \right)^2 . \tag{5.40}$$

Velocity $v\left(y\right)$ and temperature $T\left(y\right)$ can now be determined for a variety of processes, with the detailed calculations considered in the end-of-chapter problems. While the solutions and plots are given below, it is left to the interested reader to solve the equations step by step, and plot the resulting curves for velocity, temperature, and entropy generation for a variety of parameters. The solutions give indication how these processes can be used to identify transport coefficients. Transient processes in this simple geometry are considered in the next chapter.

5.6.2 Couette Flow

Couette flow (Prob. 5.2) is driven by motion of the boundary plates against each other. The plates move in opposite directions with velocities $\pm V$, and both have the same temperature T_0, while $\frac{dp}{dx} = f_x = \alpha = 0$, so that the balances reduce to

$$\eta \frac{d^2 v}{dy^2} = 0 \quad , \quad -\kappa \frac{d^2 T}{dy^2} = \eta \left(\frac{dv}{dy} \right)^2 \, . \tag{5.41}$$

Solving first for velocity, then for temperature, and last for entropy generation rate, one finds

$$v\left(y\right) = V \frac{y}{H} \, , \tag{5.42}$$

$$T\left(y\right) = T_0 + \frac{1}{2} \frac{\eta}{\kappa} V^2 \left(1 - \frac{y^2}{H^2}\right) \, , \tag{5.43}$$

$$\Sigma\left(y\right) = \frac{\eta}{T_0} \frac{V^2}{H^2} \left(1 + \frac{\eta}{\kappa} V^2 \frac{y^2}{H^2}\right) \, . \tag{5.44}$$

The average entropy generation density for the channel is

$$\bar{\Sigma} = \frac{1}{2H} \int_{-H}^{H} \Sigma dy = \frac{\eta}{T_0} \frac{V^2}{H^2} \left(1 + \frac{1}{3} \frac{\eta}{\kappa} V^2\right) \, . \tag{5.45}$$

The entropy generation grows quadratically with the velocity gradient $\frac{V}{H}$, with a further quadratic increase in velocity alone due to frictional heating. The local entropy generation is largest at the boundaries. Curves for velocity, temperature and entropy generation are shown as solid lines in Fig. 5.2.

As an extension of this setting, Prob. 5.5 considers Couette flow with temperature difference between the boundaries.

5.6.3 Poiseuille Flow

In Poiseuille flow (Prob. 5.3) the plates are at rest and have the same temperature T_0, and the flow is driven by a fixed pressure gradient $\frac{dp}{dx} = P$ while $f_x = \alpha = 0$. In this case, the balances reduce to

$$\eta \frac{d^2 v}{dy^2} = P \quad , \quad -\kappa \frac{d^2 T}{dy^2} = \eta \left(\frac{dv}{dy} \right)^2 \, . \tag{5.46}$$

with the solution

$$v\left(y\right) = -\frac{1}{2} \frac{P}{\eta} H^2 \left(1 - \frac{y^2}{H^2}\right) \, , \tag{5.47}$$

$$T\left(y\right) = T_0 + \frac{1}{12} \frac{P^2 H^4}{\kappa \eta} \left(1 - \frac{y^4}{H^4}\right) \, , \tag{5.48}$$

$$\Sigma\left(y\right) = \frac{P^2 H^2}{T \eta} \left(\frac{y^2}{H^2} + \frac{1}{9} \frac{P^2 H^4}{T \kappa \eta} \frac{y^6}{H^6}\right) \, . \tag{5.49}$$

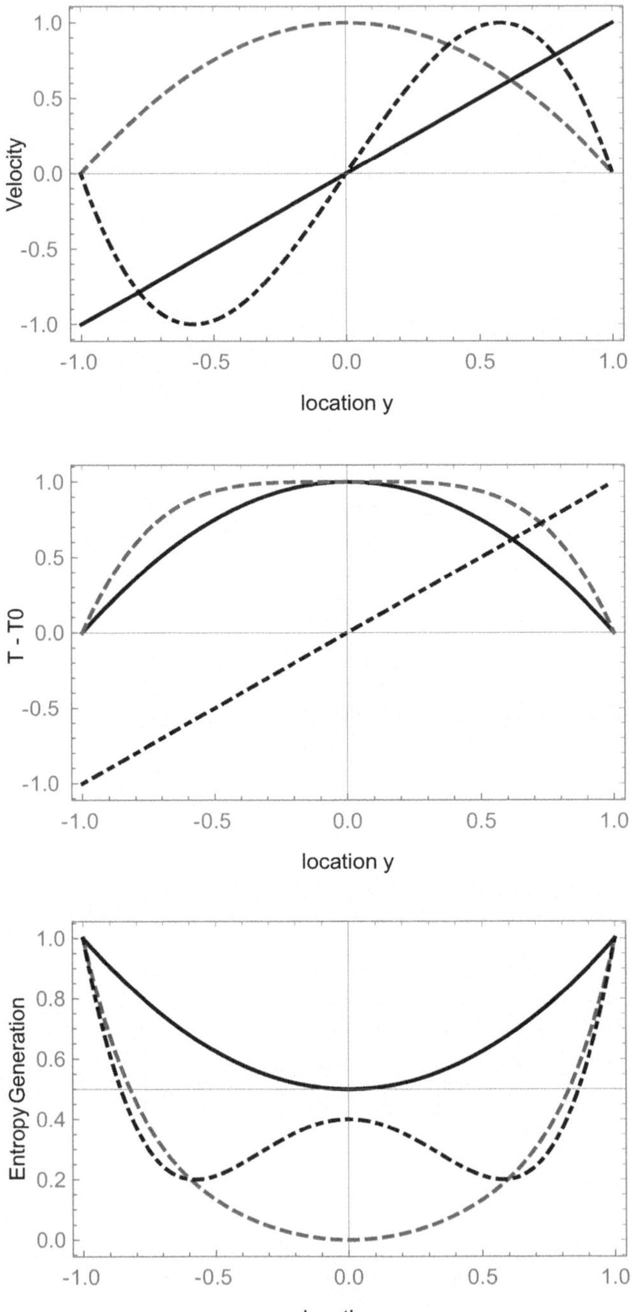

Fig. 5.2 Velocity $v(y)$, temperature deviation $T(y) - T_0$, and entropy generation rate $\Sigma(y)$ for Couette flow (solid line), Poiseuille flow (dashed), and Boussinesq convection flow (dot-dashed); in suitable dimensionless units.

The average velocity through the channel,

$$\bar{v} = \frac{1}{2H}\int_{-H}^{H} v dy = -\frac{1}{3}\frac{P}{\eta}H^2 , \tag{5.50}$$

is proportional and opposite in direction to the pressure gradient, so that the flow is from high to low pressure. The velocity is growing quadratically with H, hence a desired flow velocity $\bar{v}$ requires significantly smaller pressure gradients in wider channels. Correspondingly, the average density of entropy generation, and hence the work loss, is smaller for a wide channel (at same $\bar{v}$), with

$$\bar{\Sigma} = \frac{1}{3}\frac{P^2 H^2}{T\eta}\left(1 + \frac{1}{21}\frac{P^2 H^4}{T\kappa\eta}\right) = \frac{3}{4}\frac{\eta\bar{v}^2}{TH^2}\left(1 + \frac{3}{28}\frac{\eta\bar{v}^2}{T\kappa}\right) . \tag{5.51}$$

Curves for velocity, temperature and entropy generation are shown as dashed lines in Fig. 5.2.

5.6.4 Boussinesq Flow

For simple Boussinesq flow (Prob. 5.4) the plates are at rest and maintained at different temperatures $T_0 \pm \Delta T$, and the gravitational field $f_x = -\mathrm{g}$ induces a convection flow. The velocity is so small that frictional heating can be ignored, and it is assumed that $\left[\frac{dp}{dx} - \rho_0 f_x\right] \simeq 0$, which gives momentum and energy balance as

$$-\eta\frac{d^2 v}{dy^2} = -\alpha\left(T - T_0\right)\rho_0 f_x \quad , \quad -\kappa\frac{d^2 T}{dy^2} = 0 . \tag{5.52}$$

Here, one must first solve for temperature, and then use the result to find velocity and then entropy generation rate, as

$$T\left(y\right) = T_0 + \Delta T\frac{y}{H} , \tag{5.53}$$

$$v\left(y\right) = \frac{\alpha\rho_0 \mathrm{g}\Delta T H^2}{6\eta}\left(\frac{y}{H} - \frac{y^3}{H^3}\right) , \tag{5.54}$$

$$\Sigma\left(y\right) = \frac{\kappa}{T^2}\left(\frac{\Delta T}{H}\right)^2\left[1 + \frac{T\alpha^2\rho_0^2 \mathrm{g}^2 H^4}{36\kappa\eta}\left(1 - 3\frac{y^2}{H^2}\right)^2\right] \tag{5.55}$$

On the cold side of the flow, for $y < 0$, the velocity points downward, in the direction of the gravitational field, while on the warm side, for $v > 0$, the velocity is upwards—warm fluid rises, and cold fluid falls (assuming $\alpha > 0$).

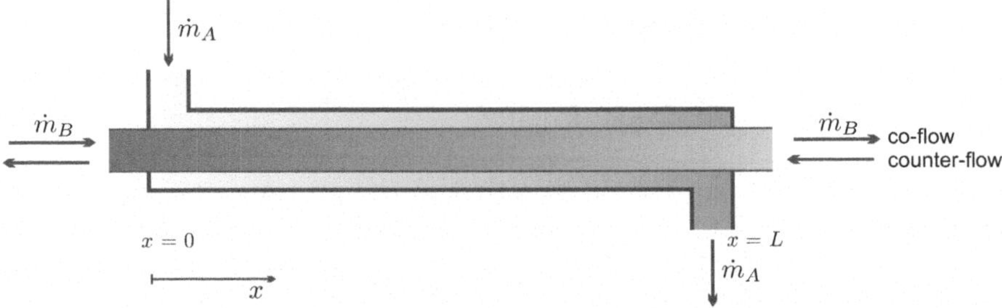

Fig. 5.3 Heat exchange between two flows A and B. Co- and counter flow settings are indicated by the arrows that show the direction of flows.

Curves for velocity, temperature and entropy generation are shown as dot-dashed lines in Fig. 5.2.

5.7 Entropy Generation in 1-D Heat Exchangers

In technical applications of thermodynamics one should always aim for systems of high efficiencies. Inefficient systems are those with significant generation of entropy, that is one must aim for systems with low entropy generation rate. The discussion of 1-D heat exchangers presented below relies on a number of simplifying assumptions, and is included mainly to show the usefulness of using entropy based arguments for system design.[3]

5.7.1 Basic Equations

We discuss the principles of simple heat exchange between two flows, which are either running in the same direction (co-flow), or in opposite directions (counter-flow), see Fig. 5.3 for a basic sketch of the set-up. Heat exchange is assumed to be a one-dimensional process, where the temperatures $T_A(x)$ and $T_B(x)$ of the two flows depend only on the space coordinate in flow direction, x. In particular, temperature profiles perpendicular to the flow are ignored, that is the given temperatures are cross-sectional averages. The heat exchangers are assumed to be adiabatic to the outside, that is heat losses to the surroundings are ignored. Friction losses are ignored as well.

[3] The material in this section is taken from [TEC, Chap. 15]. The approach differs from the other parts of this book in that the modelling is rather direct, instead of simplifying from full balance laws.

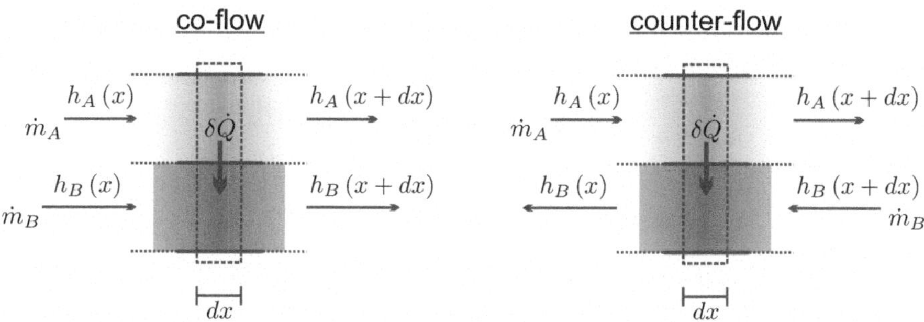

Fig. 5.4 A volume element dx of the heat exchanger, and the corresponding fluxes for co- and counter-flow heat exchangers.

Heat is transferred between the two flows due to difference in temperatures $T_A(x)$ and $T_B(x)$. Typically, the inflow temperatures of the flows are given, and one aims at determining the outflow temperatures.

Since the temperatures depend on the location x, we need to balance energy locally. Figure 5.4 shows small elements of infinitesimal width dx and the corresponding flows and properties for co- and counter-flow settings, where energy flux is only convective, that is conductive heat flux in flow direction, q_x, is ignored.

The task at hand is to determine the temperature profiles $T(x)$ in both flows. Applying the first law to each flow in the element dx gives

$$\dot{m}_A\left[h_A(x+dx)-h_A(x)\right]=-\delta\dot{Q}\,, \tag{5.56}$$

$$\pm\dot{m}_B\left[h_B(x+dx)-h_B(x)\right]=\delta\dot{Q}\,, \tag{5.57}$$

where in the second equation the upper sign refers to co-flow, and the lower sign refers to counter-flow.

Heat is transferred by conduction through the solid walls separating the two flows. The heat exchange between the flows within the element dx is denoted by $\delta\dot{Q}$, which is the heat transferred through a wall element of area $H\cdot dx$ where H is the width of the duct. With Fourier's law for the heat flux in the wall (direction z), we have

$$\delta\dot{Q}=\int_A^B q_z dz\cdot H\cdot dx=-\kappa\int_A^B\frac{dT}{dz}dz\cdot H\cdot dx=-\kappa H\int_A^B dT\cdot dx \tag{5.58}$$

which simplifies to Newton's law of cooling

$$\delta\dot{Q}=\alpha\left(T_A-T_B\right)dx\,. \tag{5.59}$$

The heat transfer coefficient $\alpha=\kappa H$ is assumed to be constant.

Taylor expansion relates the enthalpy differences to specific heats and temperature gradients as

$$h_A(x+dx) - h_A(x) = \frac{dh_A}{dx}dx = c_p^A \frac{dT_A}{dx}dx \ , \tag{5.60}$$

$$h_B(x+dx) - h_B(x) = \frac{dh_B}{dx}dx = c_p^B \frac{dT_B}{dx}dx \ ; \tag{5.61}$$

for simplicity specific heats are assumed to be constant.

Combining all of the above yields two coupled differential equations for the temperatures,

$$\frac{dT_A}{dx} = \hat{\alpha}_A (T_B - T_A) \quad , \quad \frac{dT_B}{dx} = \mp \hat{\alpha}_B (T_B - T_A) \ , \tag{5.62}$$

with the abbreviations

$$\hat{\alpha}_A = \frac{\alpha}{\dot{m}_A c_p^A} \quad , \quad \hat{\alpha}_B = \frac{\alpha}{\dot{m}_B c_p^B} \ . \tag{5.63}$$

The coupled equations (5.62) can be integrated easily,[4] and the solutions read

$$T_A(x) = K_2 \exp\left[-(\hat{\alpha}_A \pm \hat{\alpha}_B) x\right] + \frac{\hat{\alpha}_A}{\hat{\alpha}_A \pm \hat{\alpha}_B} K_1 \ , \tag{5.64}$$

$$T_B(x) = \mp \frac{\hat{\alpha}_B}{\hat{\alpha}_A} K_2 \exp\left[-(\hat{\alpha}_A \pm \hat{\alpha}_B) x\right] + \frac{\hat{\alpha}_A}{\hat{\alpha}_A \pm \hat{\alpha}_B} K_1 \ , \tag{5.65}$$

where K_1 and K_2 are integrating constants, and, as in all equations in this section, the upper sign is for co-flow, and the lower sign is for counter-flow exchangers.

5.7.2 Co-Flow Heat Exchangers

For co-flow heat exchangers (upper sign) the known inflow conditions are the temperatures $T_A(0)$ and $T_B(0)$. Evaluating of (5.64, 5.65) at the two inlets gives

$$\text{inflow of A:} \quad T_A(0) = K_2 + \frac{\hat{\alpha}_A}{\hat{\alpha}_A + \hat{\alpha}_B} K_1 \tag{5.66}$$

$$\text{inflow of B:} \quad T_B(0) = -\frac{\hat{\alpha}_B}{\hat{\alpha}_A} K_2 + \frac{\hat{\alpha}_A}{\hat{\alpha}_A + \hat{\alpha}_B} K_1 \ . \tag{5.67}$$

Solving for the constants K_1 and K_2 and inserting these into (5.64) gives the temperature curves for the co-flow heat exchanger as

[4] Take the difference to find the equation $\frac{d(T_A - T_B)}{dx} = -(\hat{\alpha}_A \pm \hat{\alpha}_B)(T_A - T_B)$ which can be integrated. Use the result to eliminate T_B in the equation for T_A, and solve for T_A.

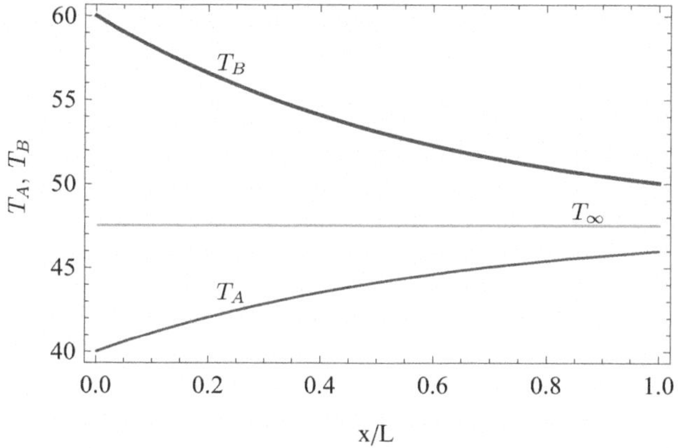

Fig. 5.5 Temperatures T_A and T_B in a co-flow heat exchanger with inflow temperatures $T_A(0) = 40\,°\mathrm{C}$ and $T_B(0) = 60\,°\mathrm{C}$, for $\hat{\alpha}_A/L = 0.7$ and $\hat{\alpha}_B/L = 1$. The asymptotic value $T_A(\infty) = T_B(\infty) = 47.5\,°\mathrm{C}$ is shown as well.

$$T_A(x) = \frac{T_A(0) + \frac{\hat{\alpha}_A}{\hat{\alpha}_B} T_B(0)}{1 + \frac{\hat{\alpha}_A}{\hat{\alpha}_B}} + \frac{T_A(0) - T_B(0)}{1 + \frac{\hat{\alpha}_B}{\hat{\alpha}_A}} \exp\left[-\left(\hat{\alpha}_A + \hat{\alpha}_B\right) x\right] , \quad (5.68)$$

$$T_B(x) = \frac{T_A(0) + \frac{\hat{\alpha}_A}{\hat{\alpha}_B} T_B(0)}{1 + \frac{\hat{\alpha}_A}{\hat{\alpha}_B}} - \frac{T_A(0) - T_B(0)}{1 + \frac{\hat{\alpha}_A}{\hat{\alpha}_B}} \exp\left[-\left(\hat{\alpha}_A + \hat{\alpha}_B\right) x\right] . \quad (5.69)$$

According to these equations, the two flows exponentially approach a common temperature value, which is obtained in the limit $x \longrightarrow \infty$ as

$$T_A(\infty) = T_B(\infty) = \frac{T_A(0)}{1 + \frac{\hat{\alpha}_A}{\hat{\alpha}_B}} + \frac{T_B(0)}{1 + \frac{\hat{\alpha}_B}{\hat{\alpha}_A}} . \quad (5.70)$$

In a heat exchanger of finite length L, the exit temperatures $T_A(L)$ and $T_B(L)$ differ from this value as seen in Fig. 5.5.

Since the heat exchanger is adiabatic to the outside, the entropy generation is just the difference between entropy flows leaving and entering the system, for simple incompressible fluids it is computed as

$$\dot{S}_{gen} = \dot{m}_A c_p^A \ln \frac{T_A(L)}{T_A(0)} + \dot{m}_B c_p^B \ln \frac{T_B(L)}{T_B(0)} > 0 . \quad (5.71)$$

In a co-flow heat exchanger, all heat transfer takes places over finite temperature differences and the entropy generation is always finite.

5.7.3 Counter-Flow Heat Exchangers

For counter-flow heat exchangers (lower sign) the known inflow conditions are the temperatures $T_A(0)$ and $T_B(L)$, for which (5.64) gives

$$\text{inflow of A: } T_A(0) = K_2 + \frac{\hat{\alpha}_A}{\hat{\alpha}_A - \hat{\alpha}_B} K_1 \,, \tag{5.72}$$

$$\text{inflow of B: } T_B(L) = \frac{\hat{\alpha}_B}{\hat{\alpha}_A} K_2 \exp\left[-\left(\hat{\alpha}_A - \hat{\alpha}_B\right) L\right] + \frac{\hat{\alpha}_A}{\hat{\alpha}_A - \hat{\alpha}_B} K_1 \,. \tag{5.73}$$

Solving for the constants K_1 and K_2 and inserting these into (5.64) gives the temperature curves for the counter-flow heat exchanger as

$$T_A(x) = T_A(0) + \left[T_B(L) - T_A(0)\right] \frac{\exp\left[\left(\hat{\alpha}_B - \hat{\alpha}_A\right) x\right] - 1}{\frac{\hat{\alpha}_B}{\hat{\alpha}_A} \exp\left[\left(\hat{\alpha}_B - \hat{\alpha}_A\right) L\right] - 1} \,, \tag{5.74}$$

$$T_B(x) = T_B(L) + \left[T_B(L) - T_A(0)\right] \frac{\exp\left[\left(\hat{\alpha}_B - \hat{\alpha}_A\right) x\right] - \exp\left[\left(\hat{\alpha}_B - \hat{\alpha}_A\right) L\right]}{\exp\left[\left(\hat{\alpha}_B - \hat{\alpha}_A\right) L\right] - \frac{\hat{\alpha}_A}{\hat{\alpha}_B}} \,. \tag{5.75}$$

This solution becomes singular for the special case $\hat{\alpha}_A = \hat{\alpha}_B = \hat{\alpha}$. L'Hôpital's rule gives the temperature curves for this case as

$$T_A(x) = T_A(0) + \left[T_B(L) - T_A(0)\right] \frac{\hat{\alpha} x}{1 + \hat{\alpha} L} \,, \tag{5.76}$$

$$T_B(x) = T_B(L) + \left[T_B(L) - T_A(0)\right] \frac{\hat{\alpha}(x - L)}{1 + \hat{\alpha} L} \,. \tag{5.77}$$

Thus, in general, one observes exponential curves for the temperatures, but straight lines when $\hat{\alpha}_A = \hat{\alpha}_B$, as shown in Fig. 5.6.

For counter-flow heat exchangers the exit temperatures are not limited by a common mean value as for co-flow exchange, but can be close to the inlet temperature of the other stream. For an exchanger of length L, the exit temperatures of the two streams are

$$T_A(L) = T_A(0) + \left[T_B(L) - T_A(0)\right] \frac{\exp\left[\left(\hat{\alpha}_B - \hat{\alpha}_A\right) L\right] - 1}{\frac{\hat{\alpha}_B}{\hat{\alpha}_A} \exp\left[\left(\hat{\alpha}_B - \hat{\alpha}_A\right) L\right] - 1} \,, \tag{5.78}$$

$$T_B(0) = T_B(L) + \left[T_B(L) - T_A(0)\right] \frac{1 - \exp\left[\left(\hat{\alpha}_B - \hat{\alpha}_A\right) L\right]}{\exp\left[\left(\hat{\alpha}_B - \hat{\alpha}_A\right) L\right] - \frac{\hat{\alpha}_A}{\hat{\alpha}_B}} \,. \tag{5.79}$$

Considering the case where $\hat{\alpha}_B < \hat{\alpha}_A$, the limiting exit temperatures for infinite length, $L \to \infty$ are

$$T_A(L_\infty) = T_B(L_\infty) \,, \tag{5.80}$$

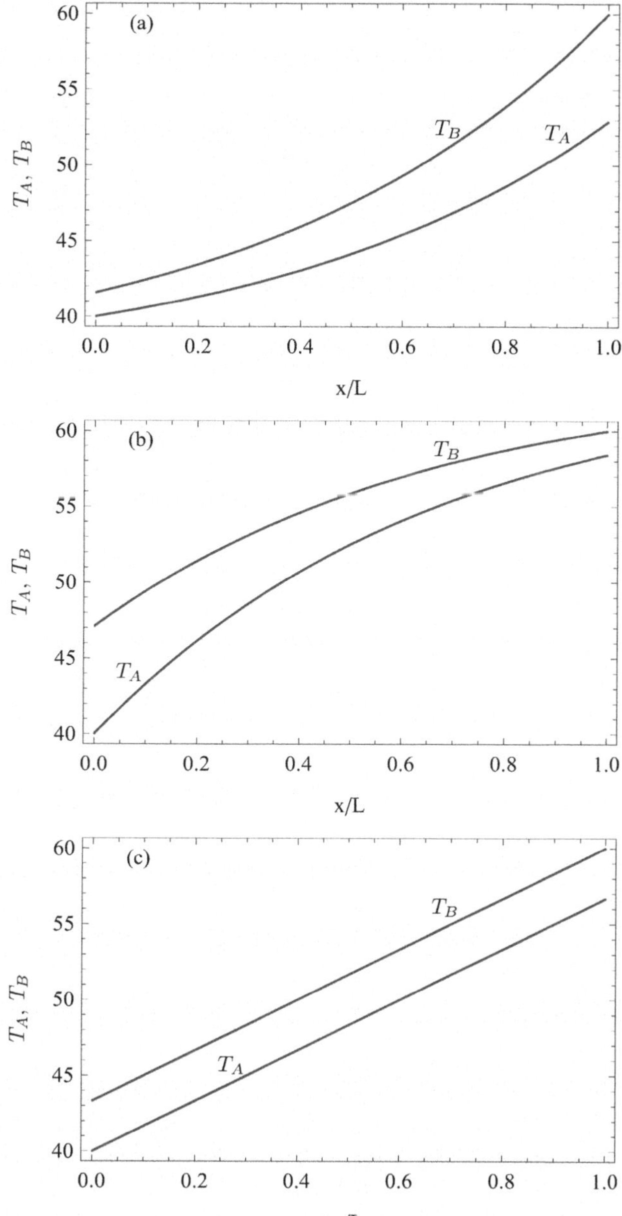

Fig. 5.6 Temperature curves for a counter-flow heat exchanger with $T_A\,(0) = 40\,°\mathrm{C}$ and $T_B\,(0) = 60\,°\mathrm{C}$, for the three cases (a) $\hat{\alpha}_A/L = 3.5$ and $\hat{\alpha}_B/L = 5$; (b) $\hat{\alpha}_A/L = 5$ and $\hat{\alpha}_B/L = 3.5$, (c) $\hat{\alpha}_A/L = \hat{\alpha}_B/L = 5$.

$$T_B(0) = T_A(0)\frac{\hat{\alpha}_B}{\hat{\alpha}_A} + \left(1 - \frac{\hat{\alpha}_B}{\hat{\alpha}_A}\right) T_B(L_\infty) \ . \tag{5.81}$$

Thus, stream A exits in thermal equilibrium with incoming stream B at $x = L_\infty$, but exiting stream B cannot achieve equilibrium with the incoming stream A at $x = 0$. When $\hat{\alpha}_B > \hat{\alpha}_A$, the behavior is opposite. The only case where both exiting streams are in equilibrium with the incoming streams, in the case $L \to \infty$, is for $\hat{\alpha}_B = \hat{\alpha}_A = \hat{\alpha}$.

The above discussion shows that a counter-flow heat exchanger works particularly well when $\hat{\alpha}_B = \hat{\alpha}_A = \hat{\alpha}$, from (5.63) this is the case when the mass flows are matched such that

$$\dot{m}_A c_p^A = \dot{m}_B c_p^B \ . \tag{5.82}$$

The discussion of the entropy generation rate of the heat exchanger sheds more light on this. Still ignoring heat loss to the exterior, the entropy generation is

$$\dot{S}_{gen} = \dot{m}_A c_p^A \ln \frac{T_A(L)}{T_A(0)} + \dot{m}_B c_p^B \ln \frac{T_B(0)}{T_B(L)} \ . \tag{5.83}$$

Figure 5.7 shows the reduced entropy generation rate

$$\frac{\dot{S}_{gen}}{\sqrt{\dot{m}_A c_p^A \dot{m}_B c_p^B}} = \ln\left[\left(\frac{T_A(L)}{T_A(0)}\right)^{\sqrt{\frac{\hat{\alpha}_B}{\hat{\alpha}_A}}}\left(\frac{T_B(0)}{T_B(L)}\right)^{\sqrt{\frac{\hat{\alpha}_A}{\hat{\alpha}_B}}}\right] , \tag{5.84}$$

as a function of the ratio $\frac{\hat{\alpha}_A}{\hat{\alpha}_B} = \frac{\dot{m}_B c_p^B}{\dot{m}_A c_p^A}$ for various total lengths L. If the heat exchanger is sufficiently long, the entropy generation develops a minimum for $\frac{\hat{\alpha}_A}{\hat{\alpha}_B} = 1$, which therefore is the optimum condition for running counter-flow heat exchangers.

5.7.4 Discussion

In an infinitely long counter-flow heat exchanger running at optimum condition $\frac{\hat{\alpha}_A}{\hat{\alpha}_B} = 1$, both streams have the same temperatures at all locations (with an infinitesimal difference for heat transfer, of course), and no entropy is generated. In co-flow heat exchangers, the entropy generation is always non-zero, since the temperature difference at the common inlet is given by the temperatures of the incoming streams.

Realistic heat exchangers have finite length, and thus some generation of entropy, which is always less in counter-flow heat exchangers as compared to co-flow systems. In the latter, both flows approach an intermediate temperature, while in the counter-flow case, the flows approach an exchange of their temperatures: the exit temperature of one flow is close to the inlet tem-

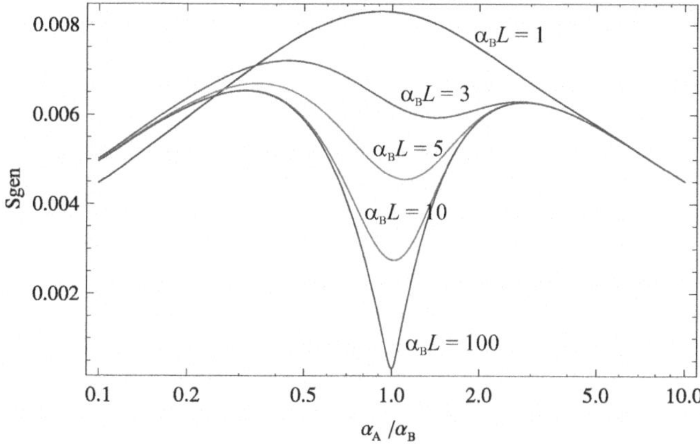

Fig. 5.7 Counter-flow heat exchanger: Reduced entropy generation rate over the ratio $\frac{\hat{\alpha}_A}{\hat{\alpha}_B} - \frac{\dot{m}_B c_p^B}{\dot{m}_A c_p^A}$ for various values of $\hat{\alpha}_B L$; note the logarithmic scale.

perature of the other. In most applications this will be the more desirable outcome.

Entropy generation and work loss arguments should be used in the design of heat exchangers, which should be build such that the overall entropy generation in the heat exchanger is as small as possible. Full 3-D simulation of the flow and temperature details based on the NSF equations is required for an accurate estimate of entropy generation that will account for losses due to heat transfer and viscosity effects.

Problems

5.1. NSF Equations

Write the 3-dimensional NSF equations in direct tensor notation and in index notation.

5.2. Plane Couette Flow

In steady state Couette flow, the plates move with constant velocities $\pm V$ in x-direction and are kept at temperature T_0, the pressure is constant, body force and thermal expansion can be ignored.

1. Solve the momentum balance for velocity $v(y)$ and use boundary conditions to determine the constants of integration.
2. Solve the energy balance for temperature, use boundary conditions to determine the constants of integration.
3. Determine stress tensor t_{ij}, heat flux q_i and entropy generation rate Σ.

4. Sketch the temperature, velocity and entropy generation as functions of y.

5.3. Plane Poiseuille Flow

In steady state Poiseuille flow, the plates are at rest and are kept at temperature T_0, the flow is driven by a constant pressure gradient $P = \frac{dp}{dx}$, body force and thermal expansion can be ignored.

1. Solve the momentum balance for velocity $v(y)$, use boundary conditions to determine the constants of integration.
2. Solve the energy balance for temperature $T(y)$, use boundary conditions to determine the constants of integration.
3. Determine stress tensor t_{ij}, heat flux q_i and entropy generation rate Σ.
4. Sketch the temperature, velocity and entropy generation as functions of y.

5.4. Boussinesq Flow: Plane Heat Transfer with Convection

Consider the case of natural convection with the following details: The plates are oriented such that the flow is in direction of the body force $f_i = (-\mathrm{g}, 0, 0)$. The plates are at rest and kept at temperatures $T_0 \pm \Delta T$. Pressure gradient and body force at T_0 cancel, $\left[\frac{\partial p}{\partial x_i} - \rho_0 f_i\right] = 0$, and the flow is so slow that frictional heating can be ignored.

1. Solve the simplified energy balance for temperature $T(y)$, use boundary conditions to determine the constants of integration.
2. Solve the momentum balance for velocity $v(y)$, use boundary conditions to determine the constants of integration. Note that the flow is driven by the buoyancy term $-\alpha(T - T_0)\rho_0 f_i$
3. Determine stress tensor t_{ij}, heat flux q_i and entropy generation rate Σ.
4. Sketch the temperature, velocity and entropy generation as functions of x.

5.5. Shear Flow with Heat Transfer

Simplify the conservation laws for mass, momentum and energy of a Navier-Stokes-Fourier fluid for plane shear flow (Couette flow) with heat transfer, at constant pressure.

Assume flow in x-direction between two parallel plates of distance $2H$ that move with velocities $v(-H) = -V$, $v(H) = +V$ in the x-direction, and have the temperatures $T(-H) = T_0 - \Delta T$, $T(H) = T_0 + \Delta T$.

Assume constant viscosity and heat conductivity and show that the equation for velocity decouples.

Solve with no-slip boundary conditions at steady state for velocity and temperature as functions of the space variable y between the plates. Solve symbolically, do not insert data.

Discuss how this process can be used to measure the transport coefficients.

Plot temperature and velocity profiles with the following data (water): viscosity $\eta = 10^{-3}\frac{\mathrm{kg}}{\mathrm{m\,s}}$, $\kappa = 0.6\frac{\mathrm{J}}{\mathrm{m\,s\,K}}$, $V = 100\frac{\mathrm{m}}{\mathrm{s}}$, $T_0 = 300\,\mathrm{K}$, $T_L = 310\,\mathrm{K}$, $L = 10\,\mathrm{cm}$

5.6. Temperature Dependence of Heat Conductivity

An ideal gas is confined between two walls of temperatures T_{W_1}, T_{W_2}, the distance of the walls is L and the gas is at rest. Kinetic theory shows that the heat conductivity of an ideal gas is given by $\kappa = \kappa_0 T^{\varpi}$, where κ_0 and ϖ are constants. For most gases ϖ is between 0.5 and 1.

Consider a steady state heat transfer process for a gas at rest. Reduce the balance of internal energy for this process, and then solve for temperature $T(x)$. Assume that there are no temperature jumps at the walls. Plot your results for $T_{W_2} = 1.1T_{W_1}$, $T_{W_2} = 2T_{W_1}$ and $T_{W_2} = 5T_{W_1}$, for $\varpi = 0, \varpi = 0.5$ and $\varpi = 1$, and interpret the results.

5.7. Fourier's Law in Anisotropic Bodies

1. Consider heat transfer in an anisotropic solid of constant density. Argue that the in this case Fourier's law must read

$$q_i = -\kappa_{ij}(T)\frac{\partial T}{\partial x_i}$$

where κ_{ij} is a positive definite matrix.

2. The body is made up of very thin layers (thickness δ much smaller than length of volume element ΔV) of two different substances A and B with heat conductivities κ_A, κ_B. Find expressions for the overall heat conductivity perpendicular to the layers, and along the layers. Assume continuous temperature between layers (no T-jumps).

Chapter 6
Linear Problems and Waves

Solutions of transport equations in simple settings are invaluable for the deeper understanding of the various transport phenomena. In this chapter we linearize the transport equations for small deviations from an equilibrium ground state, and study one-dimensional transport behavior. We focus on wave solutions, and study dispersion and damping for wave and diffusion equations. The Cattaneo equation is derived and discussed as an example of non-local transport, where the system is not in local equilibrium states.

6.1 Euler Equations and the Wave Equation

As a first simple application for time dependent problems, we consider the Euler equations for a simple fluid (no chemical changes). In the Euler equations, all irreversible fluxes are ignored, and mass, momentum and entropy balance reduce to

$$\begin{aligned} \frac{D\rho}{Dt} + \rho\frac{\partial v_k}{\partial x_k} &= 0 \\ \rho\frac{Dv_i}{Dt} + \frac{\partial p}{\partial x_i} &= 0 \\ \frac{Ds}{Dt} &= 0 \end{aligned} \tag{6.1}$$

Here, we ignore body forces as well. Note that we use the entropy balance instead of the reversible energy balance $\rho\frac{Du}{Dt} = -p\frac{\partial v_k}{\partial x_k} = \frac{p}{\rho}\frac{D\rho}{Dt}$. By means of Gibbs equation $Tds = du - \frac{p}{\rho^2}$ and mass balance $\frac{D\rho}{Dt} = -\rho\frac{\partial v_k}{\partial x_k}$ it is easy to see that both equations are equivalent.

If entropy is homogeneous initially, then the last equation states that entropy will remain at a constant homogenous value through the process.

H. Struchtrup, *A Thermodynamic Introduction to Transport Phenomena*,
https://doi.org/10.1007/978-3-031-61868-0_6

To proceed, it is convenient to consider density and entropy as variables, hence the thermal equation of state is $p(\rho, s)$. Since entropy remains constant, we have for the pressure gradient

$$\frac{\partial p}{\partial x_i} = \left(\frac{\partial p}{\partial \rho}\right)_s \frac{\partial \rho}{\partial x_i} + \left(\frac{\partial p}{\partial s}\right)_\rho \frac{\partial s}{\partial x_i} = \left(\frac{\partial p}{\partial \rho}\right)_s \frac{\partial \rho}{\partial x_i} . \tag{6.2}$$

The undisturbed fluid assumes a time-independent equilibrium ground state ρ_E, s_E, $v_k^E = 0$. Sound waves are rather small deviations from the ground state, and we linearize the transport equations accordingly. Specifically, we write the variables as

$$\rho = \rho_E + \tilde{\rho} \quad , \quad s = s_E + \tilde{s} \quad , \quad v_k = \tilde{v}_k \ , \tag{6.3}$$

where the properties with tilde are small disturbances of the ground state, i.e., $\tilde{\rho} \ll \rho_E$, etc. Inserting these into the transport equations, and deleting all contributions which are non-linear in the deviations, we find the linearized equations for the deviations as

$$\frac{\partial \tilde{\rho}}{\partial t} + \rho_E \frac{\partial \tilde{v}_k}{\partial x_k} = 0 \tag{6.4}$$

$$\rho_E \frac{\partial \tilde{v}_i}{\partial t} + \left(\frac{\partial p}{\partial \rho}\right)_{s|E} \frac{\partial \tilde{\rho}}{\partial x_i} = 0 \tag{6.5}$$

The entropy balance simply states that entropy remains constant $\frac{\partial \tilde{s}}{\partial t} = 0$.

To proceed, we can either eliminate density, or velocity. To do the former, we take space and time derivatives of the equations,

$$\frac{\partial^2 \tilde{\rho}}{\partial x_i \partial t} + \rho_E \frac{\partial^2 \tilde{v}_k}{\partial x_i \partial x_k} = 0 \ , \tag{6.6}$$

$$\rho_E \frac{\partial^2 \tilde{v}_i}{\partial t^2} + \left(\frac{\partial p}{\partial \rho}\right)_{s|E} \frac{\partial^2 \tilde{\rho}}{\partial t \partial x_i} = 0 \ , \tag{6.7}$$

and take the difference to find the wave equation for velocity

$$\frac{\partial^2 \tilde{v}_i}{\partial t^2} - a^2 \frac{\partial^2 \tilde{v}_k}{\partial x_i \partial x_k} = 0 \ . \tag{6.8}$$

Here we have introduced the speed of sound as

$$a = \sqrt{\left(\frac{\partial p}{\partial \rho}\right)_s} \ . \tag{6.9}$$

For an ideal gas, the speed of sound is $a = \sqrt{\gamma R T}$ with the ratio of specific heats $\gamma = \frac{c_p}{c_v}$ (Prob. 6.1).

Alternatively, we can take time and space derivatives, and eliminate velocity to find the wave equation for density,

$$\frac{\partial^2 \tilde{\rho}}{\partial t^2} - a^2 \frac{\partial^2 \tilde{\rho}}{\partial x_i \partial x_i} = 0 \,. \tag{6.10}$$

One distinguishes between longitudinal waves, where velocity oscillates in the direction of propagation, so that $\tilde{v}_k = \tilde{v} n_k$, and transversal waves, where velocity oscillation is perpendicular to direction of propagation, $\tilde{v}_k n_k = 0$.

For a transversal wave propagating in $x_1 = x$ direction, the velocity is $v_i = \{0, v(y), 0\}_i$ so that $\frac{\partial \tilde{v}_k}{\partial x_k} = 0$. Accordingly, the Euler equations describe only longitudinal waves.

These equations will be further evaluated in the discussion below, and in problems.

6.2 Diffusion Equations

6.2.1 Momentum Diffusion (Incompressible Liquid)

We consider simple shear flow for the incompressible NSF equations (5.19–5.21), ignoring the mild dependence of shear viscosity η on temperature. Then, the flow is decoupled from the temperature field, and the flow field is obtained from the *incompressible Navier-Stokes* equations

$$\frac{\partial v_k}{\partial x_k} = 0 \quad , \quad \rho_0 \frac{D v_i}{D t} + \frac{\partial p}{\partial x_i} - \eta \frac{\partial^2 v_i}{\partial x_j \partial x_j} = \rho_0 f_i \,. \tag{6.11}$$

with pressure p and velocity v_i as variables.

A simple example for the equations is given by laminar shear flow between two infinite parallel plates as sketched in Fig. 5.1. When the plates move in x-direction, and their distance is in y-direction, we expect velocity to be of the form

$$v_i = \{v(y,t), 0, 0\}_i \,. \tag{6.12}$$

This ansatz already is divergence free,

$$\frac{\partial v_k}{\partial x_k} = \frac{\partial v(y,t)}{\partial x} = 0, \tag{6.13}$$

that is the mass balance is guaranteed. We find for the momentum balance in x- and y-directions[1]

[1] For this geometry: $\frac{D v_i}{D t} = \frac{\partial v_i}{\partial t} + v_k \frac{\partial v_i}{\partial x_k} = \frac{\partial v_i}{\partial t}$.

$$\rho_0 \frac{\partial v}{\partial t} - \eta \frac{\partial^2 v}{\partial y^2} = \left(\rho_0 f_x - \frac{\partial p}{\partial x} \right) \quad , \quad \left(\frac{\partial p}{\partial y} - \rho_0 f_y \right) = 0 \ . \tag{6.14}$$

The second equation gives the hydrostatic pressure as a response to the body force. The first equation gives the development of velocity $v(y,t)$ in space-time as a response to a net force $\rho_{0A} f_x - \frac{\partial p}{\partial x}$, and boundary conditions that specify the interaction between the fluid and the plates.

Typical cases are *Couette flow*, where the net force vanishes ($\rho_0 f_x = \frac{\partial p}{\partial x}$), and motion of the fluid is only in response to movement of the plates, and *Poiseuille flow*, in which the plates are at rest, and motion is due to body force and pressure gradient.

For Couette flow, the velocity evolution is given by the diffusion equation for shear velocity

$$\frac{\partial v}{\partial t} - \frac{\eta}{\rho_0} \frac{\partial^2 v}{\partial y^2} = 0 \ , \tag{6.15}$$

with the momentum diffusivity $\frac{\eta}{\rho_0}$, which is also known as kinematic viscosity. In steady state the velocity is linear, with no-slip boundary conditions for plates in distance $2H$ we have $v = \frac{V_H}{2}\left(1 + \frac{y}{H}\right) + \frac{V_{-H}}{2}\left(1 - \frac{y}{H}\right)$ with the corresponding constant stress $t_{xy} = -\eta \frac{V_H - V_{-H}}{2H}$. Measurement of velocities and shear stress allows determination of viscosity η.[2]

6.2.2 Heat Equation

Next, we consider the balance of internal energy,

$$\rho \frac{Du}{Dt} + \frac{\partial q_j}{\partial x_j} = t_{ij} \frac{\partial v_i}{\partial x_j} + \rho \mathfrak{r} \tag{6.16}$$

For a simple incompressible fluid, the internal energy depends only on temperature, $u(T)$, the heat flux obeys Fourier's law,

$$q_k = -\kappa \frac{\partial T}{\partial x_k} \ , \tag{6.17}$$

and the stress tensor is given by (5.3) with $\frac{\partial v_k}{\partial x_k} = 0$, so that the temperature follows from the equation (compare with 5.21),

$$\rho_0 \mathrm{c_v} \frac{DT}{Dt} - \frac{\partial}{\partial x_k} \left(\kappa \frac{\partial T}{\partial x_k} \right) = 2\eta \frac{\partial v_{(i}}{\partial x_{j)}} \frac{\partial v_{(i}}{\partial x_{j)}} + \rho \mathfrak{r} \tag{6.18}$$

[2] We consider plane Couette flow for simplicity, actual viscosity measurements are performed with rotating Couette viscometers.

This equation describes the evolution of temperature in space-time due to convective transfer (through the material derivative), heat conduction (through the heat flux vector), frictional heating (through the square of velocity gradient) and radiative transfer (through $\mathfrak{r}$).

If viscosity is independent of temperature, the energy balance can be solved after the Navier-Stokes equations. However, if temperature dependence of shear viscosity must be considered, than one has to solve the equations simultaneously.

For a fluid at rest, the energy balance reduces to

$$\rho_0 \mathsf{c}_\mathsf{v} \frac{\partial T}{\partial t} - \frac{\partial}{\partial x_k}\left(\kappa \frac{\partial T}{\partial x_k}\right) = 0 \,, \tag{6.19}$$

which for constant heat conductivity κ reduces to the classical heat equation,

$$\frac{\partial T}{\partial t} - \frac{\kappa}{\rho_0 \mathsf{c}_\mathsf{v}} \frac{\partial^2 T}{\partial x_k \partial x_k} = 0 \,. \tag{6.20}$$

In 1-D, this reduces to the diffusion equation for temperature,

$$\frac{\partial T}{\partial t} - \frac{\kappa}{\rho_0 \mathsf{c}_\mathsf{v}} \frac{\partial^2 T}{\partial y^2} = 0 \tag{6.21}$$

with the thermal diffusivity $\frac{\kappa}{\rho_0 c_v}$.

In steady state heat transfer between two plates one finds $T(y) = \frac{T_H}{2}\left(1+\frac{y}{H}\right)+\frac{T_{-H}}{2}\left(1-\frac{y}{H}\right)$ with the corresponding heat flux $q_y = -\kappa\frac{T_H - T_{-H}}{2H}$. Measurement of temperature and heat flux allows determination of heat conductivity κ.

6.2.3 Mass Diffusion

As another example, we study a binary mixture with mass fractions $\mathsf{c}_1 = (1-\mathsf{c}_2) = \mathsf{c}$, and diffusion fluxes $J_i^1 = -J_i^2 = J_i$. Considering a mixture at rest ($v_i = 0$), in an isothermal ($T = const$) and isobaric ($p = const$) process, mass, momentum and energy balance are fulfilled, and the single mass fraction balance reduces to

$$\rho \frac{\partial \mathsf{c}}{\partial t} + \frac{\partial J_k}{\partial x_k} = 0 \,. \tag{6.22}$$

From the phenomenological relation, the diffusion flux is

$$J_k = -B_{11} \frac{\partial}{\partial x_k}\left(\frac{\mu_1 - \mu_2}{T}\right) = -B_{11} \frac{\partial}{\partial \mathsf{c}}\left(\frac{\mu_1 - \mu_2}{T}\right) \frac{\partial \mathsf{c}}{\partial x_k} = -D \frac{\partial \mathsf{c}}{\partial x_k} \tag{6.23}$$

where D is the diffusion coefficient. Note that generally the chemical potentials depend also on temperature and pressure, but there are no gradients of these in the diffusion flux due to the assumptions.

Assuming constant diffusion coefficient, we once more find the diffusion equation

$$\frac{\partial \mathsf{c}}{\partial t} - \frac{D}{\rho} \frac{\partial^2 \mathsf{c}}{\partial x_k \partial x_k} = 0 \,, \tag{6.24}$$

with the diffusivity $\frac{D}{\rho}$.

6.2.4 Discussion

For some simplified flow configurations, we have now found the diffusion equation for shear flow, heat transfer, and mass transfer. This is a reflection of the fact that all three transport mechanisms are tied to transport of molecular properties by random molecular motion, and exchange of momentum and energy by molecular interactions.

6.3 Plane Wave Solutions

6.3.1 Plane Harmonic Waves

When we consider processes in which the variables differ only slightly from the equilibrium ground states, we can linearize the transport equations as was done above. This leads to sets of linear partial differential equations (pde's), with wave and diffusion equations as the simplest examples.

The discussion of plane harmonic waves offers excellent insight into the general transport characteristics of linear pde's, and their solutions. Indeed, solutions of linear pde's can often be written as superpositions of waves of different frequencies and/or wavelengths. The formal approach to this is the theory of Fourier series and Fourier integrals, of which we only provide some elements below.

Figure 6.1 shows a damped harmonic wave for a signal $\mathsf{s}(x_i, t)$ travelling in direction n_i. The wave is one-dimensional, that is in a plane perpendicular to n_i there is no variation of s, hence we speak of plane waves. One might, for instance, think of a signal induced by an harmonically oscillating plane, that travels away from the source, while being damped, such as the pressure wave induced by a loudspeaker. Mathematically, the wave can be described as

$$\mathsf{s}(x_i, t) = \check{\mathsf{s}} e^{k^{(i)} n_r x_r} \cos\left(\omega t - k^{(r)} n_r x_r + \varphi\right) \tag{6.25}$$

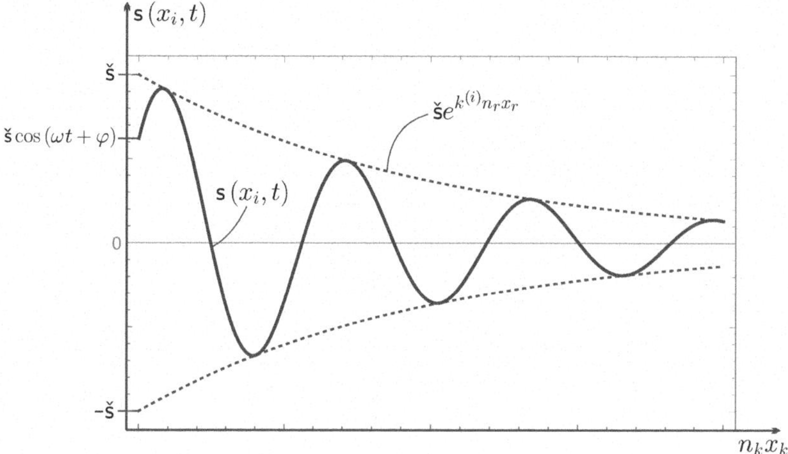

Fig. 6.1 Plane harmonic wave $\mathsf{s}(x, t)$.

where $\check{\mathsf{s}}$ is the amplitude of the wave, $\omega = 2\pi f$ is its angular frequency, where f is ordinary frequency, $k^{(i)}$ is the damping, $k^{(r)} = \frac{2\pi}{\lambda}$ is the wave number,[3] where λ is the wavelength, and φ denotes the phase shift.

The argument of the cosine function is the phase of the wave. A point of constant phase, for instance the maximum of the signal, moves with the phase velocity v_{ph}. Constant phase implies

$$\omega t - k^{(r)} n_r x_r^{ph} + \varphi = const. \tag{6.26}$$

where $x_r^{ph}(t)$ is the associated location of the moving point. Taking the time derivative, we find,

$$\omega - k^{(r)} n_r \frac{dx_r^{ph}}{dt} = 0 \ . \tag{6.27}$$

The phase velocity in direction of propagation is therefore

$$v_{ph} = n_r \frac{dx_r^{ph}}{dt} = \frac{\omega}{k^{(r)}} \ . \tag{6.28}$$

To proceed, we switch to complex number notation, where the cosine is the real part of a complex exponential,[4]

$$\cos\alpha = \mathrm{Re}\left(e^{i\alpha}\right) \ . \tag{6.29}$$

With $k^{(i)} n_r x_r = i\left(-ik^{(i)} n_r x_r\right)$, we can thus write instead of (6.25)

[3] The resaons for the notation will become clear soon.

[4] Here, $i = \sqrt{-1}$ is the imaginary unit.

$$\mathsf{s}(x_i,t) = \mathrm{Re}\left[\hat{\mathsf{s}}e^{i(\omega t - kn_r x_r)}\right] \tag{6.30}$$

where

$$k = k^{(r)} + ik^{(i)} \tag{6.31}$$

is the complex wave number, which combines information on wavelength and damping, and

$$\hat{\mathsf{s}} = \check{\mathsf{s}}e^{i\varphi} \tag{6.32}$$

is the complex amplitude, which combines information on amplitude $\check{\mathsf{s}}$ and phase shift φ.

With the understanding that only the real part of the wave is relevant, we can consider a complex plane wave,

$$\mathsf{s}(x_i,t) = \hat{\mathsf{s}}e^{i(\omega t - kn_r x_r)} . \tag{6.33}$$

For time and space derivatives of the complex wave we find

$$\frac{\partial \mathsf{s}}{\partial t} = i\omega\hat{\mathsf{s}}e^{i(\omega t - kn_r x_r)} = i\omega\mathsf{s} \quad , \quad \frac{\partial \mathsf{s}}{\partial x_k} = -ikn_k\hat{\mathsf{s}}e^{i(\omega t - kn_r x_r)} = -ikn_k\mathsf{s} \tag{6.34}$$

that is in complex notation derivatives become simple algebraic expressions.

6.3.2 Wave Equation

We consider the one-dimensional wave equation for a variable $w(x,t)$

$$\frac{\partial^2 w}{\partial t^2} - a^2\frac{\partial^2 w}{\partial x^2} = 0 , \tag{6.35}$$

where a is the speed of sound. For the solution, we assume a one-dimensional complex wave ($n_r x_r = x$),

$$w(x_i,t) = \hat{w}e^{i(\omega t - kx)} \tag{6.36}$$

and insert this into the wave equation to find

$$\left(-\omega^2 + a^2k^2\right)\hat{w} = 0 . \tag{6.37}$$

We are interested in non-trivial solutions, where $\tilde{w} \neq 0$, hence the bracket must vanish, so that wave number and frequency are not independent of each other,

$$-\omega^2 + a^2k^2 = 0 \quad \Rightarrow \quad k = \pm\frac{\omega}{a} . \tag{6.38}$$

Such a relation between frequency ω and wave number is known as *dispersion relation*. For the wave equation, the dispersion relation states that k is real, and phase velocity and damping are given by

$$v_{ph} = \frac{\omega}{k^{(r)}} = \pm a \quad , \quad k^{(i)} = 0 \ . \tag{6.39}$$

Hence, damping vanishes, and waves of any frequency move with the speed of sound a, in positive or negative direction. Realistic signals are superpositions of many waves. A locally induced signal will split into two signals travelling in opposite directions, undisturbed and undamped (Prob. 6.10).

6.3.3 Diffusion Equation

Next, we study wave solutions of the one-dimensional diffusion equation with diffusivity a,

$$\frac{\partial w}{\partial t} - \mathsf{a}\frac{\partial^2 w}{\partial x^2} = 0 \ , \tag{6.40}$$

for which the wave ansatz (6.36) yields

$$\left(i\omega + \mathsf{a}k^2\right)\hat{w} = 0 \tag{6.41}$$

which gives the dispersion relation[5]

$$i\omega + \mathsf{a}k^2 = 0 \quad \Rightarrow \quad k = \pm\sqrt{-i}\sqrt{\frac{\omega}{\mathsf{a}}} = \pm(1-i)\sqrt{\frac{\omega}{2\mathsf{a}}} \ . \tag{6.42}$$

Hence, k is complex,

$$k^{(r)} = \pm\sqrt{\frac{\omega}{2\mathsf{a}}} \quad , \quad k^{(i)} = \mp\sqrt{\frac{\omega}{2a}} \ , \tag{6.43}$$

with phase velocity

$$v_{ph} = \frac{\omega}{k^{(r)}} = \pm\sqrt{2\mathsf{a}\omega} \ . \tag{6.44}$$

Waves travel in positive and negative direction with a frequency dependent phase speed $v_{ph} = \sqrt{2\mathsf{a}\omega}$, that is the waves show dispersion. Indeed, a locally induced signal, that is a superposition of many waves, will split into two signals travelling in opposite directions, which deform, since all contributing waves move at different speed $\sqrt{2\mathsf{a}\omega}$ (waves of higher frequency are faster), and experience different damping $\sqrt{\frac{\omega}{2\mathsf{a}}}$ (waves of higher frequency are damped more strongly).

Note that for waves travelling in positive direction ($v_{ph} > 0$) the damping coefficient is negative ($k^{(i)} < 0$), while for waves travelling into negative direction, the damping coefficient is positive. Thus, the damping term $e^{k^{(i)}x}$

[5] The square root of the the imaginary element is $\sqrt{\pm i} = \frac{1 \pm i}{\sqrt{2}}$.

decreases in the direction of travel, that is wave amplitude decays, as one would expect.

6.3.4 Boundary Value Problems

The full problem statement for the solution of pde's requires integration, in time and space, which introduces constants of integration. These must be determined from initial and boundary data.

Wave solutions, as discussed here, can normally be used either for pure initial value problems, where boundaries are infinitely far away, or for pure boundary value problems, where initial data is long forgotten, but not for full initial *and* boundary value problems.[6]

Our approach so far was appropriate for *boundary value problems*, where disturbances $\mathsf{s}(0,t)$ of real frequency ω at one location ($x = n_r x_r = 0$) induce damped waves that travel in positive and negative direction x, with amplitude decreasing spatially.[7] Specifically, frequency is real, while wave number is complex, so that the complex representation of the wave reads

$$\mathsf{s}(x,t) = \hat{\mathsf{s}} e^{i(\omega t - k(\omega) x)} \; . \tag{6.45}$$

The wave solution we are studying is the real part of this,

$$\mathsf{s}(x,t) = \check{\mathsf{s}} e^{k^{(i)}(\omega) x} \cos\left(\omega\left(t - \frac{x}{v_{ph}(\omega)}\right) + \varphi\right) \tag{6.46}$$

with the phase velocity

$$v_{ph}(\omega) = \frac{\omega}{k^{(r)}(\omega)} \; . \tag{6.47}$$

6.3.5 Initial Value Problems–Stability

For *initial value problems*, a harmonic spatial disturbance $\mathsf{s}(x,0)$ of wavelength λ is prescribed in space initially (at $t = 0$). In this case the wave number is real, $k = \pm\frac{2\pi}{\lambda}$, where we allow positive and negative values of k which refer to waves travelling in opposite directions. The dispersion relation typically yields complex frequencies that depend on wave number k,

---

[6] Our (complex) wave solutions can be interpreted as (complex) Fourier transforms. To incorporate inital and boundary conditions one has to perform Laplace transforms instead.

[7] Some transport models yield waves with amplitude increasing in the direction of travel, see H. Struchtrup, B. Nadler, Wave Motion **97**, 102612 (2020), doi: 10.1016/j.wavemoti.2020.102612

$$\omega = \omega(k) = \omega^{(r)} + i\omega^{(i)} . \tag{6.48}$$

Hence, the complex wave now is

$$\mathsf{s}(x,t) = \hat{\mathsf{s}} e^{i(\omega(k)t - kx)} \tag{6.49}$$

with real part

$$\mathsf{s}(x,t) = \check{\mathsf{s}} e^{-\omega^{(i)}t} \cos\left(k\left(v_{ph}t - x\right) + \varphi\right) \tag{6.50}$$

An initially prescribed wave travels with the phase velocity

$$v_{ph}(k) = \frac{\omega^{(r)}(k)}{k} . \tag{6.51}$$

The amplitude of the wave changes over time due to the temporal damping described by $\omega^{(i)}(k)$. Recall that the second law of thermodynamics ensures that a system left to itself will approach a stable equilibrium state. For meaningful solutions of transport equations, we thus expect decay of arbitrary initial disturbances over time, i.e., to obey the second law we must have positive damping,

$$\omega^{(i)}(k) \geq 0 . \tag{6.52}$$

If this condition is violated, any disturbance with arbitrary small amplitude will over time blow-up to infinite amplitude—the system is unstable. Positive temporal damping guarantees stability.

For the wave equation, we find $\omega = ak$, hence no damping at all, $\omega^{(i)} = 0$, and phase speeds $v_{ph} = \frac{\omega^{(r)}}{k} = \pm a$. Note that typically one will be interested in positive frequencies only, and consider positive and negative wave numbers to distinguish between waves travelling into positive and negative directions.

The dispersion relation is more interesting for the diffusion equation, from which we find a purely imaginary frequency,

$$\omega(k) = \omega^{(i)} = +iak^2 . \tag{6.53}$$

Accordingly, $\omega^{(r)} = 0$, that is the phase velocity vanishes, $v_{ph} = 0$. Hence, for initial value problems, the diffusion equation describes waves that do not move in space, but decay over time.

6.4 Heat Pulse

The general solution of the diffusion equation on an infinite domain as initial value problem is a superposition of waves with all possible wave numbers k,

$$u(x,t) = \int_{-\infty}^{\infty} \hat{u}(k)\, e^{i(\omega t - kx)} dk \tag{6.54}$$

where, from the dispersion relation

$$\omega(k) = i\mathsf{a}k^2 \; ; \tag{6.55}$$

a denotes diffusivity. The amplitudes $\tilde{u}(k)$ of the waves must be found from the initial distribution

$$u(x,0) = u_0(x) = \int_{-\infty}^{\infty} \hat{u}(k)\, e^{-ikx} dk \; . \tag{6.56}$$

To proceed, we use some knowledge from the theory of Fourier transforms. The delta distribution $\delta(x-x')$ is defined such that its integral gives the value of a function f at one location,[8]

$$\int_{-\infty}^{\infty} f(x)\,\delta(x-x')\,dx = f(x') \; , \tag{6.57}$$

and as well, simply writing k instead of x,

$$\int_{-\infty}^{\infty} f(k)\,\delta(k-k')\,dk = f(k') \; . \tag{6.58}$$

The delta distribution can be expressed as

$$\delta(x-x') = \frac{1}{2\pi}\int_{-\infty}^{\infty} e^{i(x'-x)k} dk \quad , \quad \delta(k-k') = \frac{1}{2\pi}\int_{-\infty}^{\infty} e^{i(k'-k)x} dx \; . \tag{6.59}$$

Hence, multiplication of the initial condition (6.56) with $\frac{1}{2\pi}e^{-ik'x}$ and integration over space gives

$$\begin{aligned}\frac{1}{2\pi}\int_{-\infty}^{\infty} e^{ik'x} u_0(x)\,dx &= \frac{1}{2\pi}\int_{-\infty}^{\infty}\int_{-\infty}^{\infty} \hat{u}(k)\, e^{i(k'-k)x} dk dx \\ &= \int_{-\infty}^{\infty} \hat{u}(k)\,\delta(k-k')\,dk = \tilde{u}(k') \; . \end{aligned} \tag{6.60}$$

Thus, after renaming $k' \to k$ and $x \to x'$, the amplitudes are determined from the initial conditions as

$$\hat{u}(k) = \frac{1}{2\pi}\int_{-\infty}^{\infty} e^{ikx'} u_0(x')\,dx' \; . \tag{6.61}$$

The general solution of the initial value problem results from inserting amplitudes and dispersion relation into (6.54) as

$$u(x,t) = \frac{1}{2\pi}\int_{-\infty}^{\infty}\int_{-\infty}^{\infty} u_0(x')\, e^{-\mathsf{a}k^2 t} e^{i(x'-x)k} dk dx' \; . \tag{6.62}$$

[8] Roughly speaking, the delta distribution is infinite for $x = x'$, and zero elsewhere.

The further evaluation of this double integral depends on the initial condition $u_0(x)$. We consider the sudden localized heating such that the initial condition is A peak of strength U_0 at $x = 0$ which is expressed through a delta distribution

$$u_0(x) = U_0 \delta(x) \ . \tag{6.63}$$

When this is inserted into the general solution (6.62), space integration gives

$$u(x,t) = \frac{U_0}{2\pi} \int_{-\infty}^{\infty} e^{-ak^2 t} e^{-ikx} dk \ , \tag{6.64}$$

where only the real part is of interest.

The integration over wave number k is accessible through the following steps: The real part of the integral is

$$I(x) = \int_{-\infty}^{\infty} e^{-ak^2 t} \cos(kx)\, dk \tag{6.65}$$

where the value at $x = 0$ is[9]

$$I(0) = \int_{-\infty}^{\infty} e^{-ak^2 t} dk = \sqrt{\frac{\pi}{at}} \ . \tag{6.66}$$

The space derivative of the integral is

$$\frac{dI(x)}{dx} = -\int_{-\infty}^{\infty} e^{-ak^2 t} k \sin(kx)\, dk = \frac{1}{2at} \int_{-\infty}^{\infty} \sin(kx) \frac{d}{dk}\left(e^{-ak^2 t}\right) dk \ , \tag{6.67}$$

or, with integration by parts,

$$\frac{dI(x)}{dx} = \frac{1}{2at}\left[\int_{-\infty}^{\infty} \frac{d}{dk}\left(\sin(kx)\, e^{-ak^2 t}\right) dk - \int_{-\infty}^{\infty} e^{-ak^2 t} x \cos(kx)\, dk\right] . \tag{6.68}$$

The first integral vanishes, since $e^{-ak^2 t} \to 0$ for $k \to \pm\infty$, and in the second integral we identify $I(x)$ itself,

$$\frac{dI(x)}{dx} = -\frac{x}{2at} \int_{-\infty}^{\infty} e^{-ak^2 t} \cos(kx)\, dk = -\frac{x}{2at} I(x) \ . \tag{6.69}$$

[9] The integral is solved by considering its square, and use of polar coordinates:

$$I(0) = 2\int_0^{\infty} e^{-ak^2 t} dk = 2\sqrt{\int_0^{\infty}\int_0^{\infty} e^{-at(k^2+j^2)} dk dj} = 2\sqrt{\int_0^{\pi/2}\int_0^{\infty} e^{-atr^2} r dr d\varphi}$$

$$= 2\sqrt{\frac{\pi}{2}\int_0^{\infty} e^{-atr^2} \frac{1}{2at} d(atr^2)} = \sqrt{\frac{\pi}{at}\left[-e^{-atr^2}\right]_0^{\infty}} = \sqrt{\frac{\pi}{at}}$$

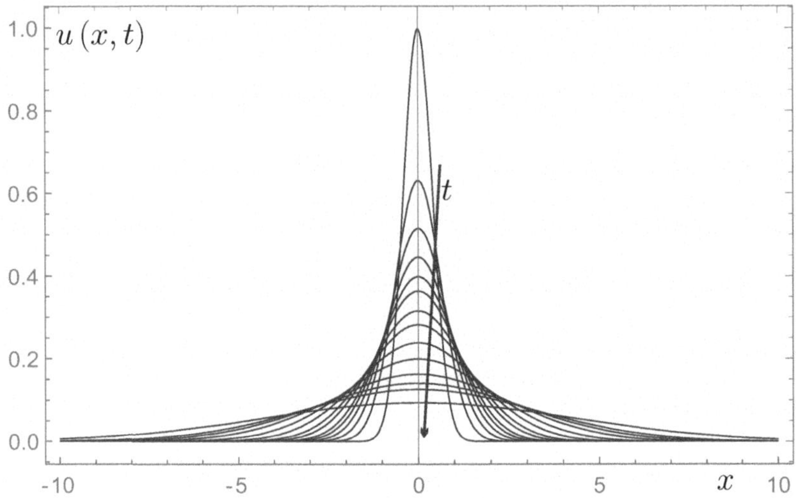

Fig. 6.2 Space-time evolution $u(x,t)$ of initial heat pulse at $x = 0$ as solution of the diffusion equation, with $U_0 = 1$, $\mathsf{a} = 1$, $t = \{0.08, 0.2, 0.3, 0.4, 0.5, 0.6, 0.8, 1, 1.4, 2, 3, 4, 5, 9\}$.

This simple differential equation is solved by separation of variables,

$$\frac{dI(x)}{I(x)} = -\frac{x dx}{2\mathsf{a}t} \quad \Rightarrow \quad I(x) = I(0)\, e^{-\frac{x^2}{4\mathsf{a}t}} = \sqrt{\frac{\pi}{\mathsf{a}t}} e^{-\frac{x^2}{4\mathsf{a}t}} \tag{6.70}$$

Putting everything together, we find the space time evolution of the initial localized pulse as Green's function for the diffusion equation,

$$u(x,t) = \frac{U_0}{2\pi} I(x) = \frac{U_0}{\sqrt{4\pi \mathsf{a}t}} e^{-\frac{x^2}{4\mathsf{a}t}} \tag{6.71}$$

Figure 6.2 shows the distribution of the initial peak over space and time. Note that, as a reflection of energy conservation, the area below the curve is constant,

$$\int_{-\infty}^{\infty} u(x,t)\, dx = U_0 \ . \tag{6.72}$$

Recall that for the solution of initial value problems the diffusion equation yields vanishing phase velocity. The overall result is the superposition of (infinitely many) waves that decay over time, but do not move. The expansion in space is a result of superposition, and the widely different damping of the waves.

An initial pulse centered at location x' yields the shifted response

$$u(x,t;x') = \frac{U_0(x')}{\sqrt{4\pi \mathsf{a}t}} e^{-\frac{(x-x')^2}{4\mathsf{a}t}} \ . \tag{6.73}$$

Since the diffusion equation is linear, a general solution is obtained from superposition of solutions. Considering the initial condition as a sum of delta peaks, the solution becomes an integral over Green's function,

$$u(x,t) = \int_{-\infty}^{\infty} \frac{U_0(x')}{\sqrt{4\pi \mathsf{a} t}} e^{-\frac{(x-x')^2}{4\mathsf{a}t}} dx' \; , \tag{6.74}$$

where $U_0(x)$ is the initial condition.

6.5 Damping of Sound Waves

The Euler equations describe undamped waves without dispersion. We now study the linearized Navier-Stokes and Fourier laws for ideal gases, for which mass, momentum and energy balance reduce to

$$\frac{\partial \tilde{\rho}}{\partial t} + \rho_E \frac{\partial \tilde{v}_k}{\partial x_k} = 0 \tag{6.75}$$

$$\rho_E \frac{\partial \tilde{v}_i}{\partial t} + RT_E \frac{\partial \tilde{\rho}}{\partial x_i} + \rho_E R \frac{\partial \tilde{T}}{\partial x_i} - \left(\lambda + \frac{\eta}{3}\right) \frac{\partial^2 \tilde{v}_k}{\partial x_i \partial x_k} - \eta \frac{\partial^2 \tilde{v}_i}{\partial x_k \partial x_k} = 0 \tag{6.76}$$

$$\rho_E c_{\mathsf{v}} \frac{\partial \tilde{T}}{\partial t} - \kappa \frac{\partial^2 \tilde{T}}{\partial x_k \partial x_k} + p_E \frac{\partial \tilde{v}_k}{\partial x_k} = 0 \tag{6.77}$$

Here, we consider small deviations $\left\{\tilde{\rho}, \tilde{v}_i, \tilde{T}\right\}$ from the equilibrium rest state ρ_E, T_E, $p_E = \rho_E R T_E$. Specific heat c_{v}, viscosities η, λ, and heat conductivity κ are taken at the rest state.

For the solution of the system, we assume that the solutions for all variables are plane waves of the form discussed above,

$$\left\{\tilde{\rho},\, \tilde{v}_i,\, \tilde{T}\right\} = \left\{\tilde{\rho} e^{i(\omega t - k n_r x_r)},\, \tilde{v}_i e^{i(\omega t - k n_r x_r)},\, \tilde{T} e^{i(\omega t - k n_r x_r)}\right\} \tag{6.78}$$

Insertion of the waves leads to an algebraic system for the complex amplitudes,

$$i\omega\tilde{\rho} - ik\rho_E \tilde{v}_k n_k = 0 \tag{6.79}$$

$$i\omega\rho_E \tilde{v}_i - ikn_i RT_E \tilde{\rho} - ikn_i \rho_E R\tilde{T} + k^2\eta\tilde{v}_i + k^2\left(\lambda + \frac{\eta}{3}\right) n_i \tilde{v}_k n_k = 0 \tag{6.80}$$

$$i\omega\rho_E c_{\mathsf{v}} \tilde{T} + k^2\kappa\tilde{T} - ikp_E \tilde{v}_k n_k = 0 \tag{6.81}$$

To proceed, we distinguish between longitudinal waves, for which the velocity amplitude points in the direction of propagation, so that $\tilde{v}_k n_k = \tilde{v} n_k$, and transversal waves, for which the velocity amplitude is perpendicular to the direction of propagation, so that $\tilde{v}_k n_k = 0$.

For transversal waves, the above system reduces to

$$i\omega\tilde{\rho} = 0 \tag{6.82}$$

$$i\omega\rho_E\tilde{v}_i - ikn_iRT_E\tilde{\rho} - ikn_i\rho_E R\tilde{T} + k^2\eta\tilde{v}_i = 0 \tag{6.83}$$

$$i\omega\rho_E c_\mathsf{v}\tilde{T} + k^2\kappa\tilde{T} = 0 \tag{6.84}$$

We see immediately from the first equation that $\tilde{\rho} = 0$, that is density remains undisturbed. Scalar product of the middle equation with the direction vector n_i shows that also $\tilde{T} = 0$, which fulfills the last equation. Hence, for transversal waves, the system reduces to the diffusion equation,

$$\left[i\omega\rho_E + k^2\eta\right]\tilde{v}_i = 0\ . \tag{6.85}$$

Transversal waves, or shear waves, diffuse, and do not affect density or temperature.

Sound propagation is only possible for longitudinal waves. Again, we multiply the middle equation with n_i, to find a linear system for the amplitudes ($\tilde{v} = \tilde{v}_k n_k$)

$$\begin{bmatrix} i\omega & -ik\rho_E & 0 \\ -ikRT_E & i\omega\rho_E + k^2\left(\lambda + \frac{4}{3}\eta\right) & -ik\rho_E R \\ 0 & -ikp_E & i\omega\rho_E c_\mathsf{v} + k^2\kappa \end{bmatrix} \begin{bmatrix} \tilde{\rho} \\ \tilde{v} \\ \tilde{T} \end{bmatrix} = 0\ . \tag{6.86}$$

Non-trivial solutions require vanishing determinant, which gives the dispersion relation as a fourth order complex bi-quadratic polynomial in k,

$$\begin{aligned}\left[\kappa p_E + i\omega\left(\lambda + \frac{4}{3}\eta\right)\kappa\right]k^4 \\ + \left[i\omega\left(c_\mathsf{v} + R\right)p_E\rho_E - \omega^2\left(\rho_E\kappa + \rho_E c_\mathsf{v}\left(\lambda + \frac{4}{3}\eta\right)\right)\right]k^2 \\ - i\omega^3\rho_E^2 c_\mathsf{v} = 0\ .\end{aligned} \tag{6.87}$$

Considering boundary value problems such as the transmission of sound emerging from a loudspeaker, we need to solve for $k(\omega)$. The bi-quadratic equation provides two pairs of solutions $k_{1,2}(\omega)$ and $k_{3,4}(\omega)$, where each pair describes damped waves travelling in positive and negative direction.

Damping and phase speed depend on the three transport coefficients bulk viscosity λ, shear viscosity η, and heat conductivity κ. Shear viscosity and heat conductivity are measured in shear flow and heat transfer experiments, hence wave attenuation data can be used to determine bulk viscosity λ.

Figures 6.3, 6.4 show damping and phase speed for the four modes $k_{1,2,3,4}$ over ordinary frequency $f = \frac{\omega}{2\pi}$. Data is for air at standard conditions $T_0 = 298\,\mathrm{K}$, $p_0 = 1\,\mathrm{bar}$, where $R = 0.287\frac{\mathrm{kJ}}{\mathrm{kg\,K}}$, $c_\mathsf{v} = 0.716\frac{\mathrm{kJ}}{\mathrm{kg\,K}}$, $\lambda = 0.9\times 10^{-5}\frac{\mathrm{kg}}{\mathrm{m\,s}}$, $\eta = 1.81\times 10^{-5}\frac{\mathrm{kg}}{\mathrm{m\,s}}$, $\kappa = 0.025\frac{\mathrm{W}}{\mathrm{m\,K}}$.

The first pair of solutions, $k_{1,2}(\omega)$, shown in Fig. 6.3, describes wave-like behavior, that is sound waves with little dispersion, and weak damping. These

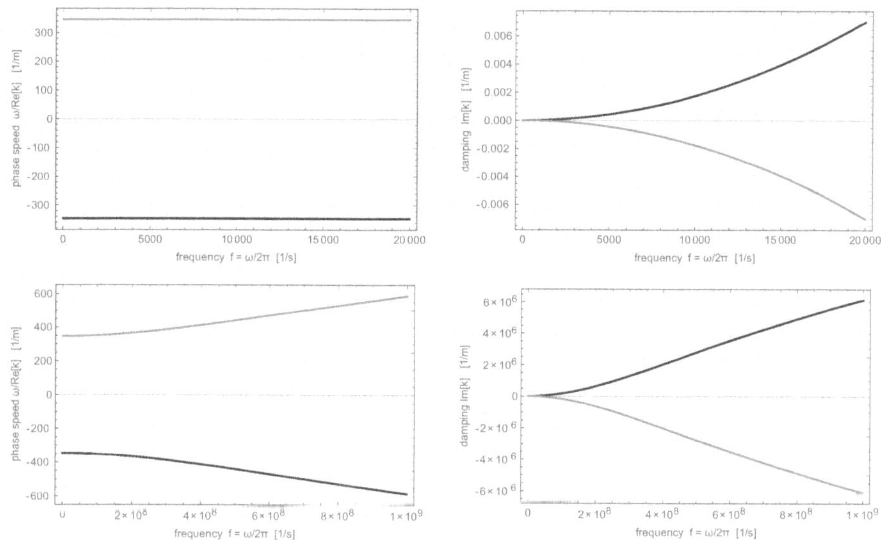

Fig. 6.3 Sound modes: Speed of sound and damping for air at standard conditions over ordinary frequency $f = \frac{\omega}{2\pi}$. The upper row shows results in the audible spectrum, with frequencies up to $20000\frac{1}{\mathrm{s}}$. The lower row shows results for a large range of frequencies up to extreme ultrasound.

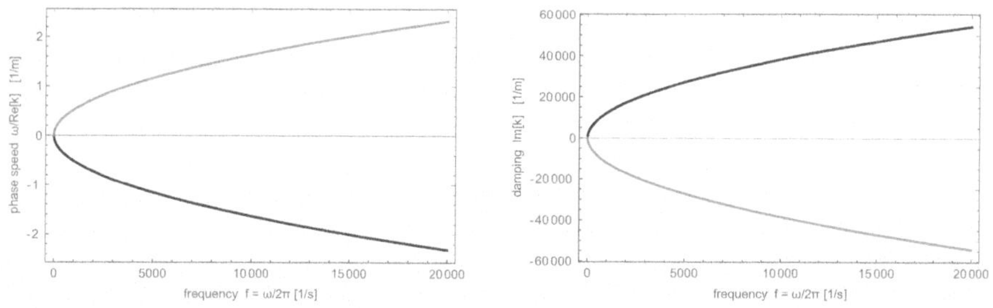

Fig. 6.4 Diffusion mode: Phase speed and damping for air at standard conditions over ordinary frequency.

are the irreversible counterparts of the waves that are obtained from the (reversible) Euler equations. The figure shows that in the audible spectrum ($f < 20000\frac{1}{\mathrm{s}}$) all waves move with the same speed, which is the speed of sound, and at extremely low damping. For instance, a wave of frequency $f = 2400\frac{1}{\mathrm{s}}$ has a damping coefficient of $k^{(i)} = 10^{-4}\frac{1}{\mathrm{m}}$, hence its amplitude falls to $1/e$ over a distance of about 10 km! For ultrasound waves, with f up to $10^9\frac{1}{\mathrm{s}}$, we observe marked dispersion, that is the phase speed increases significantly with frequency, while damping is extremely strong.

The second pair of solutions, $k_{3,4}(\omega)$, shown in Fig. 6.4, describes diffusive behavior, with strong dispersion and damping. These wave modes have no reversible counterpart in the Euler equations, which give only two sound modes. The diffusive modes describe the diffusion of temperature disturbances, that is they correspond to the heat equation. The diffusion mode describes the transmission of periodic temperature disturbances, with marked damping and dispersion. Due to the strong damping, these modes can be ignored in the discussion of acoustic phenomena.

Acoustic waves are studied further in several end-of-chapter problems.

6.6 Wave Equation in Gravity

As another exercise on manipulation of equations and linearization, we study the Euler equations for an ideal gas in the gravitational field, i.e., we consider (6.1) with external body force $f_i \neq 0$,

$$\begin{aligned} \frac{\partial \rho}{\partial t} + \frac{\partial \rho v_k}{\partial x_k} &= 0 \\ \rho \left(\frac{\partial v_i}{\partial t} + v_k \frac{\partial v_i}{\partial x_k} \right) + \frac{\partial p}{\partial x_i} &= \rho f_i \\ \frac{\partial s}{\partial t} + v_k \frac{\partial s}{\partial x_k} &= 0 \end{aligned} \tag{6.88}$$

The undisturbed ideal gas assumes a time independent equilibrium ground state $T_G = const.$, $\rho_G = \frac{p_G}{RT_G}$, $v_k^G = 0$ with entropy $s_G = s(T_G, \rho_G)$ and barometric pressure according to

$$\frac{\partial p_G}{\partial x_i} = \rho_G f_i \ . \tag{6.89}$$

Again, we linearize for small deviations from the ground state, and write the variables as

$$\rho = \rho_G + \tilde{\rho} \ , \quad p = p_G + \tilde{p} \ , \quad s = s_G + \tilde{s} \ , \quad v_k = \tilde{v}_k \ , \tag{6.90}$$

where the properties with the overbar are small disturbances of the ground state, i.e., $\tilde{\rho} \ll \rho_G$, etc. Inserting these into the transport equations, and deleting all contributions which are non-linear in the deviations, we find the equations for the deviations as

$$\begin{aligned} \frac{\partial \tilde{\rho}}{\partial t} + \frac{\partial \rho_G \tilde{v}_k}{\partial x_k} &= 0 \\ \frac{\partial \rho_G \tilde{v}_i}{\partial t} + \frac{\partial \tilde{p}}{\partial x_i} &= \tilde{\rho} f_i \\ \frac{\partial \tilde{s}}{\partial t} + \tilde{v}_k \frac{\partial s_G}{\partial x_k} &= 0 \ . \end{aligned} \tag{6.91}$$

The ground state density $\rho_G(x_i)$ varies in space, but is time independent, hence it can be moved under the time derivative.

From the Gibbs equation, we find for the ground state that

$$\frac{\partial s_G}{\partial x_k} = \frac{1}{T_G}\left(\frac{\partial h_G}{\partial x_k} - \frac{1}{\rho_G}\frac{\partial p_G}{\partial x_k}\right) = -\frac{1}{T_G \rho_G}\frac{\partial p_G}{\partial x_k} = -\frac{f_k}{T_G} \,. \tag{6.92}$$

The enthalpy term vanishes, since the ground state has homogeneous temperature, and for the ideal gas enthalpy depends only on temperature, $h(T)$.

Since changes are small, we can assume constant specific heats. Writing entropy of the gas as a function of pressure and density, we have

$$\tilde{s} = s - s_G = c_v \ln \frac{p}{p_G} - c_p \ln \frac{\rho}{\rho_G} \,, \tag{6.93}$$

or, with $\gamma = \frac{c_p}{c_v}$, and linearization in small deviations[10]

$$\frac{\tilde{s}}{c_v} = \ln \frac{p_G + \tilde{p}}{p_G} - \gamma \ln \frac{\rho_G + \tilde{\rho}}{\rho_G} = \frac{\tilde{p}}{p_G} - \left(\gamma \frac{\tilde{\rho}}{\rho_G}\right) \tag{6.94}$$

hence

$$\tilde{p} = \gamma \frac{p_G}{\rho_G}\tilde{\rho} + \frac{p_G}{c_v}\tilde{s} = \gamma R T_G \tilde{\rho} + \frac{p_G}{c_v}\tilde{s} = a^2 \tilde{\rho} + \frac{p_G}{c_v}\tilde{s} \,, \tag{6.95}$$

Here, $a = \sqrt{\gamma R T_G}$ is the speed of sound of the ideal gas in the ground state. Inserting the above, and introducing the abbreviation $J_k = \rho_G v_k$, we find

$$\begin{aligned} \frac{\partial \tilde{\rho}}{\partial t} + \frac{\partial J_k}{\partial x_k} &= 0 \\ \frac{\partial J_i}{\partial t} + a^2 \frac{\partial \tilde{\rho}}{\partial x_i} + \frac{p_G}{c_v}\frac{\partial \tilde{s}}{\partial x_i} + \frac{\tilde{s}}{c_v}\rho_G f_i &= \tilde{\rho} f_i \\ \frac{\partial \tilde{s}}{\partial t} - \frac{J_k}{\rho_G}\frac{f_k}{T_G} &= 0 \,. \end{aligned} \tag{6.96}$$

We proceed by combining the above into a single equation for J_k. For this, we take time and space derivatives of the equations above, so that

$$\begin{aligned} \frac{\partial^2 \tilde{\rho}}{\partial x_i \partial t} + \frac{\partial^2 J_k}{\partial x_i \partial x_k} &= 0 \\ \frac{\partial^2 J_i}{\partial t^2} + a^2 \frac{\partial^2 \tilde{\rho}}{\partial t \partial x_i} + \frac{p_G}{c_v}\frac{\partial^2 \tilde{s}}{\partial t \partial x_i} + \frac{\rho_G}{c_v} f_i \frac{\partial \tilde{s}}{\partial t} &= \frac{\partial \tilde{\rho}}{\partial t} f_i \\ \frac{\partial^2 \tilde{s}}{\partial x_i \partial t} - \frac{\partial J_k}{\partial x_i}\frac{1}{\rho_G}\frac{f_k}{T_G} + \frac{1}{\rho_G^2}\frac{\partial \rho_G}{\partial x_i}\frac{J_k f_k}{T_G} &= 0 \end{aligned} \tag{6.97}$$

[10] $\ln \frac{\rho_G + \tilde{\rho}}{\rho_G} = \ln\left(1 + \frac{\tilde{\rho}}{\rho_G}\right) \simeq \frac{\tilde{\rho}}{\rho_G}$.

Eliminating all terms that do not contain J_k in the middle equation, we finally find

$$\frac{\partial^2 J_i}{\partial t^2} - a^2 \frac{\partial^2 J_k}{\partial x_i \partial x_k} + \frac{R}{c_V} f_k \frac{\partial J_k}{\partial x_i} + f_i \frac{\partial J_k}{\partial x_k} = 0 \tag{6.98}$$

Compared to the standard wave equation from above, we notice additional contributions related to the body force.

To understand the influence of gravity, we consider wave solutions of the form

$$J_i = \hat{J}_i e^{i(\omega t - k n_k x_k)} \ . \tag{6.99}$$

for which we find the dispersion relation

$$-\omega^2 \hat{J}_i + a^2 k^2 n_i n_k \hat{J}_k - ik \frac{R}{c_V} f_k \hat{J}_k n_i - ik f_i n_k \hat{J}_k = 0 \ . \tag{6.100}$$

Again we distinguish between transversal and longitudinal waves. The dispersion relation for transversal waves results from scalar product with the direction of propagation n_i, and with $\hat{J}_i n_i = 0$, as

$$-ik \frac{R}{c_v} f_k \hat{J}_k = 0 \ , \tag{6.101}$$

hence $J_i = 0$—there are no transversal waves.

For longitudinal waves $\hat{J}_i = \hat{J} n_i$, hence multiplying with n_i yields the dispersion relation

$$\left[-\omega^2 + a^2 k^2 - ik \frac{c_p}{c_V} f_k n_k \right] \hat{J} = 0 \ . \tag{6.102}$$

As one would expect, horizontal waves, which travel perpendicular to the body force so that $n_k f_k = 0$, are not affected by gravity. For vertical waves, which travel in the direction of the body force, we find the dispersion relation

$$k^2 - ik \frac{n_k f_k}{R T_G} - \frac{\omega^2}{a^2} = 0 \ . \tag{6.103}$$

In Earth's atmosphere, the body force term is rather small, with a value of $\frac{n_k f_k}{R T_G} = \frac{g}{R T_G} \simeq 10^{-4} \frac{1}{\mathrm{m}}$. Treating this term as small, completion of the square yields a first order correction for the dispersion relation $k(\omega)$,

$$k = \pm \frac{\omega}{a} + i \frac{n_k f_k}{2 R T_G} \tag{6.104}$$

For a wave travelling in the direction of the body force, e.g., downwards in the atmosphere ($n_k f_k > 0$), this implies an extremely small increase in amplitude, while a wave travelling upwards will be slightly damped.

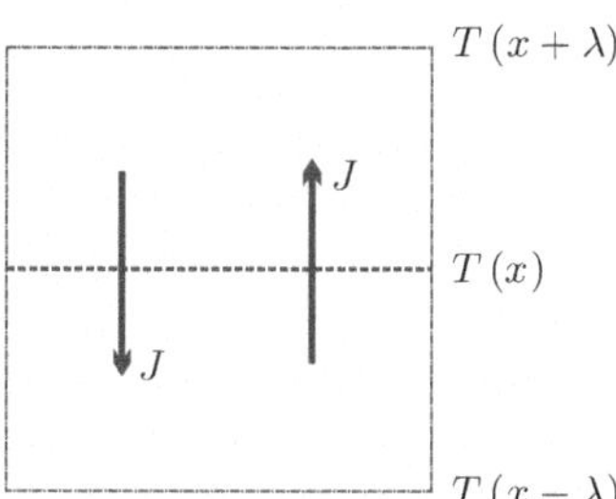

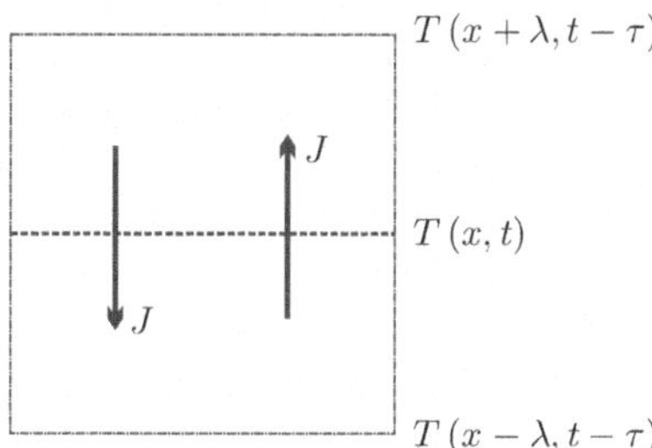

Fig. 6.5 Energy transfer by particles travelling upward and downward with directed fluxes J over a mean free path λ: a) Classical diffusion (time delay ignored), b) Cattaneo correction with time delay τ.

6.7 Cattaneo Equation

An interesting observation on the diffusion equation is that it describes infinitely fast transfer. Indeed, the solution (6.71) for the heat pulse is non-zero infinitely far from the origin immediately, this corresponds to infinite phase speeds for infinite frequency in (6.44). For engineering applications, typically this is not a problem, since the values are extremely small, and overall the agreement with experiments is very good.

We will now consider a simple model to derive the Fourier law for heat transfer, as well as a modification that guarantees finite speed of propagation.

6.7.1 Classical Diffusion: Fourier Heat Transfer

To proceed, we assume that heat transfer is effected by microscopic thermal motion of particles that carry energy, and exchange energy by collisions, where we assume a system at rest, with constant density and constant coefficients. In a simple one-dimensional setting, we have particles travelling upwards and downwards with the directed mass flow rate J, as depicted in Fig. 6.5a. The particles transfer energy over the distance of a mean free path λ, and then collide distributing their energy.

Particles passing the position x picked up energy at $x \pm \lambda$, and carry the energy $c_\mathsf{v} T(x \pm \lambda)$, where $T(x \pm \lambda)$ is the temperature at the location of their last collision. The net amount of energy transferred at the position x results from transfer in both directions as (mass flux times energy)

$$q = J c_\mathsf{v} T(x-\lambda) + J c_\mathsf{v} T(x+\lambda) \ . \tag{6.105}$$

With the help of the Taylor series

$$T(x \pm \lambda) = T(x) \pm \frac{\partial T}{\partial x}\lambda + \frac{1}{2}\frac{\partial^2 T}{\partial x^2}\lambda^2 \pm \mathcal{O}(\lambda^3) \tag{6.106}$$

the Fourier law for the heat flux is found as

$$q = -2c_v J\lambda\frac{\partial T}{\partial x} + \mathcal{O}(\lambda^3) = -\kappa\frac{\partial T}{\partial x}\ , \tag{6.107}$$

where $\kappa = 2c_v J\lambda$ is the heat conductivity. It is assumed that the mean free path λ is sufficiently small, so that higher order contributions to the Taylor series can be ignored.

With this argument, we found a somewhat intuitive underpinning for Fourier's law.

6.7.2 Diffusion with Delay: Unstable Model

A similar argument can be used to discuss a modification of Fourier's law. Carlo Cattaneo (1911-1979) assumed that transport takes time, that is the particles that arrive at x at the time t require the time interval τ to travel over the distance of the mean free path, that is their last collision occurred at time $t-\tau$ as indicated in Fig. 6.5b. If this delay is accounted for, the energy transfer is given by

$$q = J\left[c_v T(x-\lambda, t-\tau) - c_v T(x+\lambda, t-\tau)\right]\ . \tag{6.108}$$

Now, the Taylor series must consider space and time, i.e.,

$$T(x \pm \lambda, t-\tau) = T(x) \pm \frac{\partial T}{\partial x}\lambda - \frac{\partial T}{\partial t}\tau + \frac{1}{2}\frac{\partial^2 T}{\partial x^2}\lambda^2 \mp \frac{\partial^2 T}{\partial t \partial x}\lambda\tau \pm \mathcal{O}(\lambda^3) \tag{6.109}$$

so that

$$q = -2c_v J\left[\lambda\frac{\partial T}{\partial x} - \lambda\tau\frac{\partial^2 T}{\partial t \partial x}\right] = -\kappa\left[\frac{\partial T}{\partial x} - \tau\frac{\partial^2 T}{\partial t \partial x}\right] \tag{6.110}$$

Compared to Fourier's law there is the additional term $\kappa\tau\frac{\partial^2 T}{\partial t \partial x}$ which accounts for the variation of temperature in time and space.

This argument provides a simple example for using a microscopic model to find a macroscopic transport law, which here is the modified Fourier law. While before we have used the second law to find our phenomenological force-flux relations, and found stable equations, the present relation is not tied to the second law. Hence, it is a good idea to test its soundness by evaluation of its dispersion relation for stability. Inserting the heat flux (6.110) into the first law for a substance at rest,

$$\rho c_{\mathsf{V}} \frac{\partial T}{\partial t} + \frac{\partial q}{\partial x} = 0 \ , \tag{6.111}$$

gives, assuming all coefficients to be constant,

$$\rho c_{\mathsf{V}} \frac{\partial T}{\partial t} - \kappa \frac{\partial^2 T}{\partial x^2} + \kappa \tau \frac{\partial^3 T}{\partial t \partial x^2} = 0 \ , \tag{6.112}$$

with the dispersion relation

$$\rho c_{\mathsf{V}} i\omega + k^2 \kappa - i\omega k^2 \kappa \tau = 0 \ . \tag{6.113}$$

Considering initial value problems, frequency is purely complex,

$$\omega = i\omega^{(i)} = i \frac{k^2 \frac{\kappa}{\rho c_{\mathsf{V}}}}{1 - k^2 \frac{\kappa \tau}{\rho c_{\mathsf{V}}}} \ . \tag{6.114}$$

A good model should give positive temporal damping $\omega^{(i)} > 0$ for all wave numbers to guarantee stability. However, the present model gives unstable waves, with $\omega^{(i)} < 0$, for larger wave numbers, for which

$$k^2 > \frac{\rho c_{\mathsf{V}}}{\kappa \tau} \ . \tag{6.115}$$

Hence, waves of small wavelength (or large wave number) are unstable! In solutions of these equations, arbitrary small initial disturbances will exponentially grow over time. Note that $\kappa > 0$ to ensure validity of Fourier's law in time independent problems, and $\tau > 0$ is implied by the derivation.

One might be inclined to restrict use of the model to waves with larger wavelengths, but this cannot be recommended. Indeed, analytical solutions can only be found for a small number of problems, while in general one will seek to solve a set of transport equations numerically. Then, derivatives are approximated by finite differences (say) on a grid. Accurate solutions require sufficiently small grids, and one tests results by grid refinement, i.e., reducing grid spacing to check whether the results are already converged. Due to unavoidable numerical errors, one will have small disturbances in the simulations, which can be described as superpositions of waves of many wavelengths. The finer the grid, the smaller the wavelengths in the error. When the grid becomes so fine that the stability threshold is overcome, numerical errors will blow up, and no meaningful results can be obtained.

Hence, we put some effort into deriving an unstable, and hence unusable, model.

6.7.3 Diffusion with Delay: Corrected Model

But the case is not lost, since a small modification leads to a stable alternative. For this, the modified Fourier law (6.110) is written as

$$q=-\kappa\left[1-\tau\frac{\partial}{\partial t}\right]\frac{\partial T}{\partial x} \tag{6.116}$$

and the differential operator is inverted by using a formal Taylor series in τ,

$$\left[1-\tau\frac{\partial}{\partial t}\right]^{-1}\simeq\left[1+\tau\frac{\partial}{\partial t}\right] \tag{6.117}$$

so that

$$q+\tau\frac{\partial q}{\partial t}=-\kappa\frac{\partial T}{\partial x}\ . \tag{6.118}$$

Together with the first law, this yields a system of two balance laws for the variables (T,q), the Cattaneo equations

$$\rho c_{\mathsf{v}}\frac{\partial T}{\partial t}+\frac{\partial q}{\partial x}=0\quad,\quad\frac{\partial q}{\partial t}+\frac{\kappa}{\tau}\frac{\partial T}{\partial x}=-\frac{q}{\tau}\ , \tag{6.119}$$

where the second equation is a balance law for heat flux, with the flux[11] $\frac{\kappa}{\tau}T$, and the production $-\frac{q}{\tau}$.

With this, heat flux q, which was a constitutive function before, has now become another thermodynamic variable. This is the earliest, and simplest example for a theory of *Extended Thermodynamics*, where higher degrees of nonequilibrium are accounted for by extending [sic!] the number of variables.

Eliminating q from the two equations gives the telegraphers equation for temperature,

$$\tau\frac{\partial^2 T}{\partial t^2}+\frac{\partial T}{\partial t}-\frac{\kappa}{\rho c_{\mathsf{v}}}\frac{\partial^2 T}{\partial x^2}=0\ . \tag{6.120}$$

This equation is well known to be stable. The telegraphers equation exhibits all terms occurring in the diffusion and wave equations. For $\tau\to 0$ it reduces to the diffusion equation, but for $\tau\to\infty$ and finite $\frac{\kappa}{\rho c_{\mathsf{v}}\tau}$ it describes wave-like transport of energy.[12]

---

[11] The (one-dimensional) flux of the heat flux...

[12] In small systems at low temperatures one might indeed observe a wave-like transport mechansim for heat through phonons that describe the vibrations in a solid lattice, see W. Dreyer, H. Struchtrup, Cont. Mech. Thermodyn. **5**, 3-50 (1993); doi:10.1007/BF01135371.

6.7.4 The Second Law for the Cattaneo Equation

Of course, a proper thermodynamic theory is accompanied by a formulation of the second law of thermodynamics. In theories of *Extended Thermodynamics*, one must leave the assumption of Local Thermal Equilibrium behind, and construct nonequilibrium entropies. The simplest nonequilibrium entropy for the Cattaneo model has the form

$$\tilde{s}\left(T,q\right)=\tilde{s}_0-\frac{\alpha}{2}c_v\tilde{T}^2-\frac{\beta}{2\rho}q^2 \ , \tag{6.121}$$

where $\alpha > 0$ and $\beta > 0$ are constant coefficients that will be determined below. This quadratic entropy is suitable for the linear Cattaneo model that describes small deviations from an equilibrium state, which here is given by T_0, $\tilde{s}_0$, and vanishing heat flux $q_0 = 0$. Since entropy in equilibrium is maximum, deviations must be such that disturbed entropy is less than equilibrium entropy, and the above quadratic form is the simplest to fulfill this requirement ($\tilde{T} = T - T_0$).

The balance law for entropy is constructed in several steps. With the quadratic form for η we have the time derivative

$$\rho\frac{\partial\tilde{s}}{\partial t}=-\alpha\rho c_v\tilde{T}\frac{\partial\tilde{T}}{\partial t}-\beta q\frac{\partial q}{\partial t} \tag{6.122}$$

and insertion of energy and heat flux balances (6.119) gives

$$\rho\frac{\partial\tilde{s}}{\partial t}=\alpha\tilde{T}\frac{\partial q}{\partial x}-\beta q\left(-\frac{\kappa}{\tau}\frac{\partial\tilde{T}}{\partial x}-\frac{q}{\tau}\right) . \tag{6.123}$$

By means of the product rule, this can be rewritten as

$$\rho\frac{\partial\tilde{s}}{\partial t}-\frac{\partial\alpha\tilde{T}q}{\partial x}=\left(-\alpha+\beta\frac{\kappa}{\tau}\right)q\frac{\partial\tilde{T}}{\partial x}+\frac{\beta}{\tau}q^2 . \tag{6.124}$$

From this, the (one-dimensional) entropy flux for this simple model is $\varphi = -\alpha\tilde{T}q$, and the right hand side is the entropy production rate which must be non-negative.

To proceed, we recall that in the Cattaneo model q and $\tilde{T}$ are independent variables, which implies that the product $q\frac{\partial\tilde{T}}{\partial x}$ has no definite sign, while the factor $\left(-\alpha+\beta\frac{\kappa}{\tau}\right)$ is a constant. To ensure non-negative entropy generation for all values of the variables, this factor must vanish,

$$\alpha=\beta\frac{\kappa}{\tau} . \tag{6.125}$$

To ensure the proper sign for the q^2 terms in the entropy generation rate (6.121), the coefficient β must be positive. To ensure agreement with the non-linear Gibbs equation, we chose $\alpha = \beta\frac{\kappa}{\tau} = 1/T_0^2$ so that the entropy and its balance become, finally,

$$\tilde{s}\left(T,q\right) = \tilde{s}_0 - \frac{c_v}{2}\frac{\tilde{T}^2}{T_0^2} - \frac{\tau}{2T_0^2}\frac{q^2}{\rho\kappa} \quad , \quad \rho\frac{\partial\tilde{s}}{\partial t} - \frac{\partial}{\partial x}\left(\frac{\tilde{T}q}{T_0^2}\right) = \frac{q^2}{\kappa T_0^2} \, . \tag{6.126}$$

The entropy balance for these linear equations looks somewhat unusual, which is a result of the simplifying assumptions. To relate to our previous work, we divide the energy balance by T_0 and add it to the entropy balance to find at first

$$\rho\frac{\partial\left(\tilde{s} + c_\mathsf{v}\frac{T}{T_0}\right)}{\partial t} + \frac{\partial\frac{1}{T_0}\left(1 - \frac{\tilde{T}}{T_0}\right)q}{\partial x} = \frac{1}{T_0^2}\frac{q^2}{\kappa} \, . \tag{6.127}$$

Here, we identify the alternative nonequilibrium entropy

$$s = \tilde{s} + c_\mathsf{v}\frac{T}{T_0} = \tilde{s}_0 + c_\mathsf{v}\left(\frac{\tilde{T}}{T_0} - \frac{1}{2}\frac{\tilde{T}^2}{T_0^2}\right) - \frac{\tau}{2T_0^2}\frac{q^2}{\rho\kappa} \, . \tag{6.128}$$

A closer look at the temperature term shows that this is, within second order in $\tilde{T}$, the classical contribution to entropy $c_\mathsf{v}\left(\frac{\tilde{T}}{T_0} - \frac{1}{2}\frac{\tilde{T}^2}{T_0^2}\right) \simeq c_\mathsf{v}\ln\frac{T}{T_0}$. As well, within the linearization we have $\frac{1}{T_0}\left(1 - \frac{\tilde{T}}{T_0}\right) \simeq \frac{1}{T_0}\frac{1}{1+\frac{\tilde{T}}{T_0}} = \frac{1}{T_0+\tilde{T}} = \frac{1}{T}$, hence we find the familiar entropy flux q/T, and the entropy balance assumes the usual form

$$\rho\frac{\partial s}{\partial t} + \frac{\partial}{\partial x}\left(\frac{q}{T}\right) = \frac{1}{T_0^2}\frac{q^2}{\kappa} \, . \tag{6.129}$$

6.7.5 The Limit for Small τ

The Fourier law is recovered from the Cattaneo heat flux when the process time scale is large, so that the relaxation time τ is relatively small. For a proper argument, we rescale observational time as $t = \hat{t}/\omega_0$, where ω_0 is a typical frequency of the processes considered. With this, the Cattaneo equation $(6.119)_2$ assumes the form

$$\omega_0\tau\frac{\partial q}{\partial\hat{t}} + \kappa\frac{\partial T}{\partial x} = -q \tag{6.130}$$

For $\omega_0\tau \ll 1$ the first term can be ignored.

A more formal argument relies on a so-called Chapman-Enskog expansion, where the heat flux is written as a series in the in the dimensionless smallness parameter $\omega_0\tau$,

$$q = q_0 + \omega_0 \tau q_1 + \cdots \tag{6.131}$$

Inserting this into the transport equation gives

$$(\omega_0 \tau) \frac{\partial q_0}{\partial \hat{t}} + (\omega_0 \tau)^2 \frac{\partial q_1}{\partial \hat{t}} + \kappa \frac{\partial T}{\partial x} = -q_0 - (\omega_0 \tau) q_1 \tag{6.132}$$

and matching of powers retrieves the Fourier law

$$q_0 = -\kappa \frac{\partial T}{\partial x} \tag{6.133}$$

as the leading term, as well as the first order correction

$$q_1 = -\frac{\partial q_0}{\partial \hat{t}} = \frac{\kappa}{\omega_0} \frac{\partial}{\partial t} \frac{\partial T}{\partial x} . \tag{6.134}$$

Somewhat interestingly, the first two terms of the series recover the original unstable expression (6.110),

$$q = -\kappa \left[\frac{\partial T}{\partial x} - \tau \frac{\partial^2 T}{\partial t \partial x} \right] . \tag{6.135}$$

Indeed, it is well known that leading order Chapman-Enskog expansion gives stable equations, while higher order contributions are often unstable.[13]

Problems

6.1. Speed of Sound in Ideal Gas

We found speed of sound as change of pressure with density in isentropic process,

$$a = \sqrt{\left(\frac{\partial p}{\partial \rho}\right)_s} .$$

Show that for an ideal gas $a = \sqrt{\gamma R T}$, where $\gamma = \frac{c_p}{c_v}$. Do *not* assume constant specific heats! Recall, that for an ideal gas: $p = \rho R T$, $du = c_v dT$, $c_p = c_v + R$, and start with the Gibbs equation $T ds = du - \frac{p}{\rho^2} d\rho$.

6.2. Speed of Sound

Estimate the speed of sound for water vapor at 10 MPa, 550 °C

1. From the general formula $a = \sqrt{\left(\frac{\partial p}{\partial \rho}\right)_s}$ and steam tables.
2. From assumption that vapor behaves as an ideal gas.

[13] H. Struchtrup: Macroscopic Transport Equations for Rarefied Gas Flows, Springer 2005

6.3. Speed of Sound

Use property tables to determine the speed of sound

1. In CO_2 at $300\,\mathrm{K}, 1500\,\mathrm{K}$.
2. In R134a vapor at $70\,°\mathrm{C}$, $1.4\,\mathrm{MPa}$.

6.4. Wave and Diffusion Equation

Write the 3-dimensional wave and diffusion equations in direct tensor notation and in index notation.

6.5. Diffusion Equations

Simplify the balance equations for the following time dependent problems in simple geometries:

1. One dimensional heat transfer in an incompressible fluid at rest.
2. Shear flow in an incompressible fluid at constant temperature.
3. Diffusion in a binary mixture at rest.

Show that all problems lead to the diffusion equation

$$\frac{\partial w}{\partial t} - \mathsf{a}\frac{\partial^2 w}{\partial x^2} = 0 \,.$$

Identify variable w and diffusion coefficients a for the three problems.

6.6. Baking

You plan to bake bread in a convection oven, which ensures constant temperature on the surface of the bread. From previous experience you remember that a bread of 25 cm diameter needs a baking time of 30 min, but the bread you want to bake now has a diameter of 32 cm.

To estimate the baking time for the larger bread, assume that the heat transfer into the bread obeys the heat equation

$$\frac{\partial T}{\partial t} - \mathsf{a}\frac{\partial^2 T}{\partial x_k \partial x_k} = 0$$

where a is the temperature conductivity. Find the required baking time without solving the equation.

Hint: Introduce dimensionless time and space variables.

6.7. Delta Distribution

The delta distribution $\delta\left(x_i - x_i^*\right)$ is defined such that

$$\int_V f\left(x_i\right)\delta\left(x_i - x_i^*\right) dV = \begin{cases} f\left(x_i^*\right) & \text{if } x_i^* \in V \\ 0 & \text{if } x_i^* \notin V \end{cases}$$

Show that the expression

$$\frac{\partial}{\partial x_i}\frac{x_i - x_i^*}{|\mathbf{x} - \mathbf{x}^*|^3}$$

is a delta distribution in this sense. Hint: Use partial integrations, and variable transforms to $z_i = x_i - x_i^*$ and spherical coordinates centered at $z_i = 0$.

6.8. Green's Function

We found the solution of a heat pulse as Green's function

$$u(x,t) = \frac{U_0}{\sqrt{4\pi a t}} e^{-\frac{x^2}{4at}}$$

1. Show that overall property is conserved, that is $\int_{-\infty}^{\infty} u(x,t)\, dx = U_0$.
2. The initial condition for this problem was $u(x, t = 0) = U_0 \delta(x)$. Comparing with the solution this indicates that in the limit $t \to 0$ the Green function is a delta distribution,

$$\lim_{t\to 0}\frac{e^{-\frac{x^2}{4at}}}{\sqrt{4\pi a t}} = \delta(x)$$

Show that this is the case, i.e., show that

$$\lim_{t\to 0}\int_{-\infty}^{\infty} f(x)\frac{e^{-\frac{x^2}{4at}}}{\sqrt{4\pi a t}}dx = f(0)$$

Hint: Try a similar strategy to the solution of the integral in the footnote on page 117.

6.9. Plane Waves in Euler Fluid

Consider the linearized Euler equations for an ideal gas. Set

$$\left\{\tilde{\rho}, \tilde{v}_i, \tilde{T}\right\} = \left\{\hat{\rho}, \hat{v}_i, \hat{T}\right\} \exp\left[i\left(\omega t - k\, n_k x_k\right)\right]$$

and write the equations as a linear system for the complex amplitudes $\tilde{\rho}, \tilde{v}_i, \tilde{T}$.

1. Longitudinal waves: Assume that $\hat{v}_i = \hat{v} n_i$, and simplify the equations accordingly. Compute phase velocity and damping.
2. Transversal waves: Assume that $\hat{v}_i n_i = 0$, and simplify the equations accordingly. Compute phase velocity and damping.
3. Sound is a longitudinal wave. In a noisy concert, the pressure amplitude is $\bar{p} = 20Pa$. Compute the corresponding amplitudes $\hat{\rho}, \hat{v}_i, \hat{T}$.

6.10. Wave Equation: D'Alembert Solution

Consider the wave equation

$$\frac{\partial^2 u}{\partial t^2} - a^2\frac{\partial^2 u}{\partial x^2} = 0$$

on an infinite interval (no boundaries) with the initial conditions

$$u(x, t = 0) = \varphi(x) \quad , \quad \frac{\partial u}{\partial t}(x, t = 0) = \vartheta(x) \ .$$

Show that the ansatz $u = f(x - at) + g(x + at)$ fulfills the equation. Use this and the initial conditions to show that the solution must have the form

$$u(x, t) = \frac{1}{2}\left[\varphi(x - at) + \varphi(x + at)\right] + \frac{1}{2a}\int_{x-at}^{x+at} \vartheta(x')\, dx' \ .$$

Use a computer with appropriate software (Mathematica, Maple, ...) to plot the solution for the following initial conditions defined on the infinite interval

$$\varphi(x) = \begin{cases} 1 + x \ , & -1 < x < 0 \\ 1 - x \ , & 0 < x, 1 \\ 0 & \text{else} \end{cases} \quad , \quad \vartheta(x) = \begin{cases} -\alpha \ , & -1 < x < 0 \\ \alpha \ , & 0 < x, 1 \\ 0 & \text{else} \end{cases}$$

Set $a = 1$ and then consider the solutions for $\alpha = 0$ and for $\alpha = 1$.

6.11. Initial Conditions for Wave Equation

The figure below shows a solution $u(x, t)$ of the wave equation at times $t = 0$ and $t = 3$, with a signal running only to the left.

Use the data in the figure to determine the speed of sound a and the initial conditions $\varphi(x) = u(x, t = 0)$, $\vartheta(x) = \frac{\partial u}{\partial t}(x, t = 0)$ that yield the depicted solution.

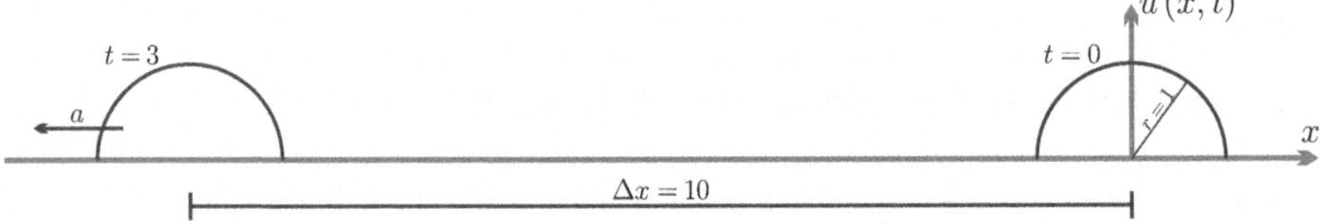

6.12. Sound Waves

Sound pressure level is given in dB (decibel), defined as

$$L_p = 20 \log_{10} \frac{p_{\text{rms}}}{p_{\text{ref}}}$$

where $p_{\text{ref}} = 20\mu\text{Pa}$ is the threshold of human hearing, and p_{rms} is the root mean square of the wave pressure.

1. In a sinusoidal plane wave, the pressure is $p = p_0 + p_{\text{amp}} \sin(kx - \omega t)$, with frequency ω and wave number $k = \frac{\omega}{a}$. Show that pressure amplitude p_{amp} and root mean square are related as $p_{\text{rms}} = \frac{p_{\text{amp}}}{\sqrt{2}}$.

2. A jet engine in 1 m distance produces sound at 150dB and the threshold for long-term hearing damage is 85dB. Determine the corresponding pressure amplitudes.
3. Since the sound wave is isentropic, pressure, temperature and density are related by the isentropic relations. Show first that for an ideal gas with constant specific heats one has, with $\gamma = \frac{c_p}{c_v}$,

$$\frac{p}{p_0} = \left(\frac{\rho}{\rho_0}\right)^{\gamma} \quad , \quad \frac{T}{T_0} = \left(\frac{\rho}{\rho_0}\right)^{\gamma-1} \quad , \quad \frac{T}{T_0} = \left(\frac{p}{p_0}\right)^{\frac{\gamma-1}{\gamma}}$$

Now consider small deviations from equilibrium, so that $p = p_0 + \tilde{p}$, $\rho = \rho_0 + \tilde{\rho}$, $T = T_0 + \tilde{T}$. Keep only linear terms in the above, and find relations between the the deviations. Show that all three properties have no phase shift. Determine pressure, temperature and density amplitudes for the jet engine and at the damage threshold.
4. Use the mass balance or the momentum balance for a plane sound wave to relate density variation and velocity. Determine the velocity amplitudes for the above cases.

6.13. Plane Waves in a Navier-Stokes-Fourier Fluid

Consider the balances of mass, momentum, and energy with the laws of Navier-Stokes and Fourier, for an ideal gas with constant specific heat $c_v = \frac{3}{2}R$. Heat conductivity and viscosity are related as $\kappa = \frac{15}{4}R\mu$.

1. Linearize the equations with respect to an equilibrium state ρ_0, T_0 and $v_i^0 = 0$. For this introduce dimensionless quantities as

$$\rho = \rho_0 (1 + \tilde{\rho}) \quad , \quad v_i = \sqrt{RT_0}\tilde{v}_i \quad , \quad T = T_0 \left(1 + \tilde{T}\right)$$

and ignore all products of $\tilde{\rho}, \tilde{v}_i, \tilde{T}$ (first order terms stay).
2. Introduce dimensionless space and time coordinates as

$$x_k = LX_k \quad , \quad t = \frac{L}{\sqrt{RT_0}}\tau$$

where L is a relevant reference length. Also introduce the Knudsen number $\mathrm{Kn} = \frac{\mu\sqrt{RT_0}}{p_0 L}$ and write the linear equations in dimensionless form.
3. Assume that the solutions are plane harmonic waves

$$\left\{\tilde{\rho}, \tilde{v}_i, \tilde{T}\right\} = \left\{\hat{\rho}, \hat{v}_i, \hat{T}\right\} \exp\left[i\left(\omega t - k\, n_k x_k\right)\right]$$

Perform all necessary derivations in the Navier-Stokes-Fourier equations and write the equations as an homogeneous system for the (dimensionless) amplitudes $\hat{\rho}, \hat{v}_i, \hat{T}$.
4. Transversal waves: Assume that $\hat{v}_i n_i = 0$, and simplify the equations accordingly. Compute phase velocity and damping.

5. Longitudinal waves: Assume that $\hat{v}_i = \hat{v} n_i$, and simplify the equations accordingly. Compute phase velocity and damping (use a computer!), and plot both as function of inverse frequency for Kn = 1. Consider the limit of small frequency, $\omega \to 0$, and discuss.

6.14. Wave Equation in Spherical Coordinates

1. In spherical coordinates, the Laplacian of a scalar reads

$$\triangle u = \frac{\partial^2 u}{\partial r^2} + \frac{2}{r}\frac{\partial u}{\partial r} + \frac{1}{r^2}\frac{\partial^2 u}{\partial \theta^2} + \frac{\cot\theta}{r^2}\frac{\partial u}{\partial \theta} + \frac{1}{r^2 \sin\theta}\frac{\partial^2 u}{\partial \phi^2}$$

where r is radius, ϕ represents the azimuthal angle and θ the zenith angle. For a problem that is independent of the angles, show that the above reduces to

$$\triangle u = \frac{1}{r}\frac{\partial^2 (ru)}{\partial r^2}$$

2. Show from the wave equation $\frac{\partial^2 u}{\partial t^2} - a^2 \triangle u = 0$ that density and pressure amplitude for a spherical wave are inversely proportional to radius.
3. How far do you have to step away from the above jet engine so that the sound is at 85dB (threshold for hearing damage in long term exposure), and how far to just not hear it anymore?

6.15. Temperature Profile in the Ground

Solve the heat equation

$$\frac{\partial T}{\partial t} - \mathsf{a}\frac{\partial^2 T}{\partial x^2} = 0$$

with the boundary condition

$$T(t,0) = T_0 + \varDelta T \cos\frac{2\pi t}{\tau} \; .$$

You my chose one of the following methods for solution:

a) Assume that the solution is a periodic function of the type $T = T_0 + A(x)\cos\omega t + B(x)\sin\omega t$ where $\omega = 2\pi/\tau$, and find the functions $A(x), B(x)$. Give arguments why the temperature is expected in this form.
b) Assume a plane wave solution for the T-disturbance in complex form, determine the dispersion relation, and take it from there.

Consider $T_0 = 273\,\mathrm{K}$, $\mathsf{a} = 2\times 10^{-7}\frac{\mathrm{m}^2}{\mathrm{s}}$, and

$$\tau = 1\text{year} \quad \text{and} \quad \varDelta T = 20K$$

or

$$\tau = 1\text{day} \quad \text{and} \quad \varDelta T = 5K \; .$$

For each case, determine the depth at which the temperature amplitude changes by just $1K$. At which depth is it coldest in Summer or at noon, respectively?

6.16. Telegrapher's Equation

We found the Cattaneo model comprised of the energy balance, and a balance law for heat flux,

$$\rho c_{\mathsf{v}} \frac{\partial T}{\partial t} + \frac{\partial q}{\partial x} = 0 \quad , \quad \frac{\partial q}{\partial t} + \frac{\kappa}{\tau} \frac{\partial T}{\partial x} = -\frac{q}{\tau}$$

In the linear model considered, density ρ, specific heat c_v, heat conductivity κ and relaxation time τ are positive constants. Show that the above can be combined into the telegrapher's equation

$$\alpha \frac{\partial^2 T}{\partial t^2} + \beta \frac{\partial T}{\partial t} - \gamma \frac{\partial^2 T}{\partial x^2} = 0$$

and identify the coefficients.

6.17. Dispersion Relation

Consider the telegrapher's equation

$$\tau \frac{\partial^2 u}{\partial t^2} + \frac{\partial u}{\partial t} - \mathsf{a} \frac{\partial^2 u}{\partial x^2} = 0$$

and assume plane wave solutions of the form $u = \bar{u} \exp\left[i\left(\omega t - kx\right)\right]$.

1. Consider boundary value problems where ω is given, and compute the complex wave number k as function of ω. Identify damping and phase velocity, and plot both as function of ω.
2. Consider initial value problems where k is given, and compute the complex frequency ω as function of k. Identify damping and phase velocity, and plot both as function of k.

For plotting use the following values: (i) $\tau = \mathsf{a} = 1$, (ii) $\tau = 0$, $\mathsf{a} = 1$.

6.18. Dispersion Relations $\omega(k)$

Determine the dispersion relations $\omega(k)$ for:

wave equation

$$\frac{\partial^2 u}{\partial t^2} - a^2 \frac{\partial^2 u}{\partial x^2} = 0$$

heat equation

$$\frac{\partial u}{\partial t} - \mathsf{a} \frac{\partial^2 u}{\partial x^2} = 0$$

telegrapher's equation

$$\tau \frac{\partial^2 u}{\partial t^2} + \frac{\partial u}{\partial t} - \mathsf{a} \frac{\partial^2 u}{\partial x^2} = 0$$

and the equations

$$\frac{\partial u}{\partial t} - \frac{\partial^2 u}{\partial x^2} - \frac{\partial^4 u}{\partial x^4} = 0 ,$$

$$\frac{\partial u}{\partial t} - \frac{\partial^2 u}{\partial x^2} + \frac{\partial^4 u}{\partial x^4} = 0 .$$

For each equation, determine phase speed and damping and plot both over k (set $a = \mathsf{a} = 1$). Discuss the stability of the equations.

6.19. Dispersion Relations $k(\omega)$

Determine the dispersion relations $k(\omega)$ for the equations of the previous problem.

For each equation, determine phase speed and damping and plot both over ω (set $a = \mathsf{a} = 1$). Discuss the stability of the equations.

6.20. More Dispersion Relations

Determine dispersion relations and discuss stability for the equations

$$\frac{\partial u}{\partial t} \pm \frac{\partial u}{\partial x} - \frac{\partial^2 u}{\partial x^2} = 0 ,$$

$$\frac{\partial^2 u}{\partial t^2} - \frac{\partial^3 u}{\partial x^3} = 0 .$$

6.21. Dispersion Relation

In an old book you found an interesting equation, where due to faded print you cannot recognize the signs:

$$\frac{\partial u}{\partial t} (\pm) (\frac{\partial^2 u}{\partial x^2} (\pm) \frac{\partial^3 u}{\partial t \partial x^2} = 0$$

For the equation to be stable in time for all wavelengths, which signs are required at the second and third term?

6.22. Spectrum of a Function

The spectrum of a function is given by the amplitudes of its Fourier transform, i.e.

$$F(\omega) = \frac{1}{2\pi} \int_{-\infty}^{\infty} f(t)\, e^{-i\omega t} dt .$$

Since the inverse transform is given by

$$f(t) = \int_{-\infty}^{\infty} F(\omega)\, e^{i\omega t} d\omega ,$$

i.e., as superposition of oscillations $e^{i\omega t}$, the amplitude $F(\omega)$ is related to the contribution of the frequency ω to the original function.

Consider the function

$$f(t) = \begin{cases} 1\ , 0 \le t \le t_0 \\ 0\ ,\ \text{elsewhere} \end{cases}$$

and show that higher frequencies become important when the signal becomes shorter (smaller signal duration t_0).

6.23. Dispersion and Stability

Consider dispersion relations for macroscopic equations derived within kinetic theory. The dimensionless, linearized, one-dimensional equations can be written as:

Conservation laws for mass, momentum, energy:

$$\frac{\partial \tilde{\rho}}{\partial \tilde{t}} + \frac{\partial \tilde{v}}{\partial \tilde{x}} = 0\ ,\quad \frac{\partial \tilde{v}}{\partial \tilde{t}} + \frac{\partial \tilde{\theta}}{\partial \tilde{x}} + \frac{\partial \tilde{\rho}}{\partial \tilde{x}} + \frac{\partial \tilde{\sigma}}{\partial \tilde{x}} = 0\ ,\quad \frac{3}{2}\frac{\partial \tilde{\theta}}{\partial \tilde{t}} + \frac{\partial \tilde{v}}{\partial \tilde{x}} + \frac{\partial \tilde{q}}{\partial \tilde{x}} = 0$$

Navier-Stokes-Fourier:

$$\tilde{\sigma} = -\frac{4}{3}\frac{\partial \tilde{v}}{\partial \tilde{x}} \qquad \tilde{q} = -\frac{15}{4}\frac{\partial \tilde{\theta}}{\partial \tilde{x}}$$

Grad 13:

$$\frac{\partial \tilde{\sigma}}{\partial \tilde{t}} + \frac{8}{15}\frac{\partial \tilde{q}}{\partial \tilde{x}} + \frac{4}{3}\frac{\partial \tilde{v}}{\partial \tilde{x}} = -\tilde{\sigma}\ ,\quad \frac{\partial \tilde{q}}{\partial \tilde{t}} + \frac{\partial \tilde{\sigma}}{\partial \tilde{x}} + \frac{5}{2}\frac{\partial \tilde{\theta}}{\partial \tilde{x}} = -\frac{2}{3}\tilde{q}$$

Burnett equations:

$$\tilde{\sigma} = -\frac{4}{3}\frac{\partial \tilde{v}}{\partial \tilde{x}} - \frac{4}{3}\frac{\partial^2 \tilde{\rho}}{\partial \tilde{x}^2} + \frac{2}{3}\frac{\partial^2 \tilde{\theta}}{\partial \tilde{x}^2}\ ,\quad \tilde{q} = -\frac{15}{4}\frac{\partial \tilde{\theta}}{\partial \tilde{x}} - \frac{7}{4}\frac{\partial^2 \tilde{v}}{\partial \tilde{x}^2}$$

Here, the dimensionless quantities are related to the dimensional quantities as (note that $\theta = \frac{k}{m}T$ is temperature in energy units, and $\lambda = \frac{\mu_0\sqrt{\theta_0}}{p_0}$ is a measure for the mean free path)

$$\tilde{\rho} = \frac{\rho}{\rho_0}\ ,\ \tilde{v} = \frac{v}{\sqrt{\theta_0}}\ ,\ \tilde{\theta} = \frac{\theta}{\theta_0}\ ,\ \tilde{\sigma} = \frac{\sigma}{p_0}\ ,\ \tilde{q} = \frac{q}{p_0\sqrt{\theta_0}}\ ,\ \tilde{t} = \frac{p_0}{\mu_0}t\ ,\ \tilde{x} = \frac{p_0}{\mu_0\sqrt{\theta_0}}x\ .$$

For the above sets of dimensionless equations, i.e. Navier-Stokes-Fourier, Grad 13, and Burnett, assume plane wave solutions for the variables of the form

$$u_A = \hat{u}_A \exp\left[i\left(\omega\tilde{t} - k\tilde{x}\right)\right]$$

where $\tilde{u}_A$ is the complex amplitude, ω is the frequency of the wave, and k is the complex wave number. Note: For NSF and Burnett $u_A = \{\rho, v, \theta\}$ and for Grad 13 $u_A = \{\rho, v, \theta, \sigma, q\}$.

1. ω and k are dimensionless. Discuss their dimensional form, and relate the dimensionless values to Knudsen number.

2. With the help of a computer algebra system (Maple, Mathematica, etc.), compute the dispersion relation $k(\omega) = k_r(\omega) + ik_i(\omega)$ and plot the inverse phase velocity $\frac{1}{v_{ph}} = \frac{k_r}{\omega}$ and the reduced damping coefficient $\frac{\alpha}{\omega} = -\frac{k_i}{\omega}$ as functions of the inverse dimensionless frequency $z = \frac{1}{\omega}$. Note: plot the phase velocities of all three sets of equations in one single plot, likewise for the damping coefficients.
3. Discuss the results, in particular you should see the following:

   a. Navier-Stokes-Fourier: For $\omega \to \infty$, the phase velocity goes to infinity (infinite speed of propagation), while for $\omega \to 0$, the phase speed is the speed of sound (for one solution).
   b. Grad 13: the phase velocity remains finite for all ω - give the limits of v_{ph} for $\omega \to 0, \omega \to \infty$.
   c. Show that the Burnett equations might be unstable, that is they violate the condition $k_r/k_i < 0$. Tip: Plot k_r/k_i as function of z.

4. For the three models determine the solutions $\omega(k)$ and check stability of all modes (use computer).

6.24. Initial Value Problem

Consider the equation

$$\tau\frac{\partial^2 u}{\partial t^2} + b\frac{\partial u}{\partial t} - a\frac{\partial^2 u}{\partial x^2} = 0$$

on an infinite interval, with the initial conditions

$$u(0,x) = \begin{cases} x & ,0 < x < 1 \\ 2-x & ,1 < x < 2 \\ 0 & else \end{cases} \quad , \quad \frac{\partial u}{\partial t}(0,x) = \begin{cases} A & ,0 < x < 2 \\ 0 & else \end{cases} .$$

1. Solve the equation formally as a superposition of plane harmonic waves, and determine the amplitudes from the initial condition as function of the wave number (explicit!).
2. Use Mathematica, or any other software, to plot the solution as a function of time. Consider the following sets of values for the parameters:

$$\begin{aligned} \tau = 1, b = 0, a = 1 \quad & \text{(wave equation)} \\ \tau = 0, b = 1, a = 1 \quad & \text{(heat equation)} \\ \tau = 1, b = 1, a = 1 \quad & \text{(telegraph equation)} \end{aligned}$$

and $A = 0$ or $A = 1$.

Chapter 7
Boundaries and Interfaces

So far we have studied the partial differential equations that describe the behavior of fluids in bulk regions without discontinuities. These bulk regions are connected by sharp transitions—described as singular surfaces—such as wall-fluid boundaries, or phase interfaces. In this chapter we develop boundary and interface conditions, where once more the second law serves as the guiding principle. Topics discussed include velocity slip and temperature jump at wall-fluid boundaries, as well as resistivities in evaporation and condensation processes.

7.1 Initial and Boundary Conditions

Full solutions of the transport equations do not only require the transport equations themselves, with full knowledge of all transport coefficients from suitable measurements or models, but also a proper set of initial and boundary conditions. We study this in detail for the case of a simple substance, with the conservation laws

$$\begin{aligned} \frac{\partial \rho}{\partial t}+v_k\frac{\partial \rho}{\partial x_k}+\rho\frac{\partial v_k}{\partial x_k}&=0\ ,\\ \rho\frac{\partial v_i}{\partial t}+\rho v_k\frac{\partial v_i}{\partial x_k}-\frac{\partial t_{ij}}{\partial x_j}&=\rho f_i\ ,\\ \rho\frac{\partial u}{\partial t}+\rho v_k\frac{\partial u}{\partial x_k}+\frac{\partial q_k}{\partial x_k}&=t_{ij}\frac{\partial v_i}{\partial x_j}\ , \end{aligned} \tag{7.1}$$

and the NSF constitutive relations

$$t_{ij}=-p\delta_{ij}+\lambda\frac{\partial v_k}{\partial x_k}\delta_{ij}+2\eta\frac{\partial v_{\langle i}}{\partial x_{j\rangle}}\quad ,\quad q_k=-\kappa\frac{\partial T}{\partial x_k}\ . \tag{7.2}$$

H. Struchtrup, *A Thermodynamic Introduction to Transport Phenomena*,
https://doi.org/10.1007/978-3-031-61868-0_7

Boundary and initial conditions come into the solutions of these equations by integration. To determine the constants of integration, each integration in time requires an initial condition, and each integration in space requires a boundary condition.

The above system requires five integrations in time, hence fives initial conditions. A standard set of initial conditions would be the values of the primary variables $\{\rho, v_i, T\}$ in all points of the system, at the starting time t_0 of the process,

$$\begin{aligned} \rho\left(\mathbf{x}, t=t_0\right) &= \rho_0\left(\mathbf{x}\right) \;, \\ v_i\left(\mathbf{x}, t=t_0\right) &= v_i^0\left(\mathbf{x}\right) \qquad (i=1,2,3) \;, \\ T\left(\mathbf{x}, t=t_0\right) &= T_0\left(\mathbf{x}\right) \;. \end{aligned} \tag{7.3}$$

Other initial conditions are possible as well, e.g., initial values of pressure, or gradients of the variables.

Space integrations are more difficult to count. To proceed, we interpret the above set of equations (7.1, 7.2) as a system of first order partial differential equations (pde's) for the variables $u_A = \{\rho, v_i, u, t_{ij}, q_k\}_A$. In rather compact notation, the equations can be written as

$$\mathcal{B}_{AB}\frac{\partial u_B}{\partial t} + \mathcal{A}_{AB}^k\frac{\partial u_B}{\partial x_k} = \mathcal{C}_{AB}u_B \tag{7.4}$$

where the matrices $\mathcal{A}_{AB}^k$, $\mathcal{B}_{AB}$, $\mathcal{C}_{AB}$ depend on the u_A. For instance, the matrix $\mathcal{B}_{AB}$ reads

$$\mathcal{B}_{AB} = \begin{bmatrix} 1 & 0 & \cdots & & & & \cdots & 0 \\ 0 & \rho & \ddots & & & & & \vdots \\ \vdots & \ddots & \rho & & & & & \\ & & & \rho & & & & \\ & & & & \rho & & & \\ & & & & & 0 & \ddots & \vdots \\ \vdots & & & & & \ddots & \ddots & 0 \\ 0 & \cdots & & & & \cdots & 0 & 0 \end{bmatrix} \tag{7.5}$$

This matrix is diagonal, and has five non-zero eigenvalues, while all other entries vanish. The required number of integrations equals the number of non-zero eigenvalues. With $\mathcal{B}_{AB}$ in front of the time derivatives, we confirm the need for five initial conditions.

At a boundary with normal $\mathbf{n}$, we decompose the terms with spatial derivatives as

$$\mathcal{A}_{AB}^k\frac{\partial u_B}{\partial x_k} = \mathcal{A}_{AB}^n\frac{\partial u_B}{\partial n} + \mathcal{A}_{AB}^{\tau_\alpha}\frac{\partial u_B}{\partial \tau_\alpha} \tag{7.6}$$

where n indicates the normal coordinate (i.e., $\frac{\partial}{\partial n} = n_k \frac{\partial}{\partial x_k}$), and τ_α, $\alpha = 1,2$ are the tangential coordinates (summation over α is implied). Boundary conditions are required for integration from the boundary into the system, that is in direction n. Hence, the number of boundary conditions required is equal to the number of non-zero eigenvalues of the matrix $\mathcal{A}^n_{AB} = \mathcal{A}^k_{AB} n_k$.

While in analytical calculations the number of required boundary conditions typically becomes evident, this is not so in numerical solutions of pde's. A computer does what it is told to do, and it can happen that one prescribes too many, or not enough boundary conditions. In the first case—too many—the numerical results will most likely show unexpected features, such as oscillations close to the boundary, where the code tries—and fails—to find solutions of the equations that match the boundary conditions. In the second case—too few—one might find solutions, but with boundary values that the code finds somehow "by itself", which might be different for different runs. For instance, a change of parameters for the problem might result in changes that are affected by missing boundary conditions.

7.1.1 Example: Heat Transfer

As an example, we consider one-dimensional heat transfer, with the transport equations

$$\rho c_v \frac{\partial T}{\partial t} + \frac{\partial q}{\partial x} = 0 \quad , \quad \frac{\partial T}{\partial x} = -\frac{q}{\kappa} \, . \tag{7.7}$$

For the variables $u_A = \{T, q\}_A$, this has the matrix form (7.4) with

$$\mathcal{A}_{AB} = \begin{bmatrix} 0 & 1 \\ 1 & 0 \end{bmatrix} \quad , \quad \mathcal{B}_{AB} = \begin{bmatrix} \rho c_v & 0 \\ 0 & 0 \end{bmatrix} \quad , \quad \mathcal{C}_{AB} = -\begin{bmatrix} 0 & 0 \\ 0 & \frac{1}{\kappa} \end{bmatrix} \tag{7.8}$$

The matrix $\mathcal{B}_{AB}$ is diagonal with one non-zero eigenvalue, hence one initial condition is required. The matrix $\mathcal{A}_{AB}$ has the eigenvalues[1] $\lambda_{1,2} = \pm 1$, hence two boundary conditions must be prescribed.

There is a wide variety of initial and boundary conditions that can be prescribed, for instance

$$T(x, t = t_0) \ ; \ T(x = 0, t) \ , \ T(x = L, t)$$

or

$$T(x, t = t_0) \ ; \ q(x = 0, t) \ , \ q(x = L, t)$$

or

$$T(x, t = t_0) \ ; \ T(x = 0, t) \ , \ q(x = 0, t)$$

[1] The eigenvalue problem reads $\det \begin{bmatrix} -\lambda & 1 \\ 1 & -\lambda \end{bmatrix} = \lambda^2 - 1 = 0$

and so on. All these lead to proper mathematical solutions.

However, one should also think about what boundary conditions can be realized experimentally. To control temperature at a boundary, e.g., $T(x=0,t)$, one uses heating and cooling with a feedback system that adjusts $q(x=0,t)$ such that the boundary temperature has the desired values. Then, obviously, it is impossible to independently control the heat at the same wall—that is the last set of conditions above cannot be realized in an experiment. Similarly, when prescribing a defined heat flux at a boundary, e.g., $q(x=L,t)$, the temperature at that boundary cannot be controlled. In other words, a set of mathematically meaningful initial and boundary conditions might not be realizable in an actual experimental set-up.

7.1.2 Example: Euler Equations

As a second example, we have a look at the non-linear Euler equations in two dimensions, with the variables[2] $u_A = \{\rho, v_x, v_y\}_A$ and the equations

$$\begin{aligned}
\frac{\partial \rho}{\partial t} + v_x \frac{\partial \rho}{\partial x} + v_y \frac{\partial \rho}{\partial y} + \rho \left(\frac{\partial v_x}{\partial x} + \frac{\partial v_y}{\partial y} \right) &= 0 \\
\frac{\partial v_x}{\partial t} + v_x \frac{\partial v_x}{\partial x} + v_y \frac{\partial v_x}{\partial y} + \frac{a^2}{\rho} \frac{\partial \rho}{\partial x} &= 0 \\
\frac{\partial v_y}{\partial t} + v_x \frac{\partial v_y}{\partial x} + v_y \frac{\partial v_y}{\partial y} + \frac{a^2}{\rho} \frac{\partial \rho}{\partial y} &= 0
\end{aligned} \tag{7.9}$$

Hence, we have the matrices,

$$\mathcal{A}^x_{AB} = \begin{bmatrix} v_x & \rho & 0 \\ \frac{a^2}{\rho} & v_x & 0 \\ 0 & 0 & v_x \end{bmatrix}_{AB} , \quad \mathcal{A}^y_{AB} = \begin{bmatrix} v_y & 0 & \rho \\ 0 & v_y & 0 \\ \frac{a^2}{\rho} & 0 & v_y \end{bmatrix}_{AB} , \quad \mathcal{B}_{AB} = \begin{bmatrix} 1 & 0 & 0 \\ 0 & 1 & 0 \\ 0 & 0 & 1 \end{bmatrix}_{AB} \tag{7.10}$$

and $\mathcal{C}_{AB} = 0$. For a boundary with normal in x-direction, we have to determine the eigenvalues of $\mathcal{A}^x_{AB}$, which are

$$\lambda_1 = v_x \quad , \quad \lambda_{2,3} = v_x \pm a \ . \tag{7.11}$$

The number of non-zero eigenvalues depend on the flow state (v_x is zero, or not), and this gives us a hint that the setting of proper boundary conditions for the non-linear Euler equations might be a non-trivial task. Indeed, these equations form a set of hyperbolic pde's, and are widely studied in mathematics. Here is not the room to discuss this further, but the interested reader

[2] The Euler equations are isentropic, which introduces the speed of sound a, see Sec. 6.1.

is pointed at the method of characteristics, which shows that only boundary conditions can be prescribed for characteristics that travel into the domain.

7.2 Wall-Fluid Boundaries

7.2.1 The Second Law at the Boundary

At wall-fluid boundaries we typically manipulate the wall, e.g., by moving it, and by heating or cooling. Boundary conditions relate the controlled properties of the wall to the fluid properties directly in front of the wall, which also depend on the state of the fluid further inside.

How can we formulate appropriate boundary conditions? Once more we employ the second law of thermodynamics.

A wall-fluid boundary can be described as a singular surface, hence we consider the jump conditions for mass, momentum and energy (3.27, 3.45, 3.71), without surface supply term,

$$[[\rho\left(v_k - v_k^W\right)]]\, n_k = 0\ , \tag{7.12}$$

$$[[\rho v_i\left(v_k - v_k^W\right) - t_{ik}]]\, n_k = 0\ , \tag{7.13}$$

$$\left[\left[\rho\left(u + \frac{1}{2}v^2\right)\left(v_k - v_k^W\right) + q_k - t_{ik}v_i\right]\right] n_k = 0\ , \tag{7.14}$$

together with the second law (4.20),

$$\left[\left[\rho s\left(v_k - v_k^W\right) + \frac{q_k}{T}\right]\right] n_k = \Sigma_{\mathcal{S}} \geq 0\ . \tag{7.15}$$

Here, v_k is the fluid velocity, v_k^W is the wall velocity, and n_k is the wall normal, pointing towards the fluid. For an impermeable wall

$$\left(v_k - v_k^W\right) n_k = V_k n_k = 0\ , \tag{7.16}$$

hence the slip velocity

$$V_i = v_i - v_i^W \tag{7.17}$$

is tangential to the wall. With this, the mass balance is identically fulfilled, and the other balances reduce to

$$-t_{ik}n_k = -t_{ik}^W n_k\ , \tag{7.18}$$

$$q_k n_k - t_{ik}v_i n_k = q_k^W n_k - t_{ik}^W v_i^W n_k\ , \tag{7.19}$$

$$\frac{q_k n_k}{T} - \frac{q_k^W n_k}{T_W} = \Sigma_{\mathcal{S}} \geq 0\ , \tag{7.20}$$

where the index W denotes properties of the wall at the interface, while properties without subscript refer to the fluid directly at the interface. The first condition is the force balance for the interface, which gives the total stress $t_{ik}^W n_k$ on the wall, i.e., the forces required for the process under consideration.

We are interested in boundary conditions for the fluid, hence use momentum and energy conservation to eliminate stress t_{ik}^W and heat flux q_k^W in the wall, to find the second law for the interface as

$$q_k n_k \left(\frac{1}{T} - \frac{1}{T_W}\right) + \frac{1}{T_W} t_{ik} n_k V_i = \Sigma_S \geq 0 \ . \tag{7.21}$$

Again, we interpret the entropy generation as a sum of products of thermodynamic forces and fluxes, i.e., $\Sigma_S = \sum_A \mathcal{F}_A \mathcal{J}_A$, where

Force $\mathcal{F}_A$	**Flux** $\mathcal{J}_A$	**Description**
$\left(\frac{1}{T} - \frac{1}{T_W}\right)$	$q_k n_k$	temperature jump
$V_i = v_i - v_i^W$	$\frac{1}{T_W} t_{ik}^\tau n_k$	velocity slip

Since the slip velocity V_i is tangential to the wall, only the tangential component of the stress vector $t_{ik}^\tau n_k$ plays a role, where

$$t_{ik}^\tau = t_{ik} - t_{lm} n_l n_m n_i n_k \quad \text{with} \quad t_{ik}^\tau n_i n_k = 0 \ . \tag{7.22}$$

Entropy generation and forces vanish in equilibrium, where temperature and velocity of wall and fluid are equal, $T = T_W$, $v_i = v_i^W$.

Just as in the bulk, we use the principles of LIT to relate fluxes and forces by linear relations. Thus, boundary conditions arise as phenomenological equations, where the flux is proportional to the force to guarantee positive entropy generation. We have one scalar and one vectorial force-flux pair, which according to Curie's principle are independent.

7.2.2 Temperature Jump

The temperature jump boundary condition arises from the scalar force-flux pair as

$$q_k n_k = \hat{\xi} \left(\frac{1}{T} - \frac{1}{T_W}\right) = \xi \left(T_W - T\right) \tag{7.23}$$

where $\xi > 0$ and $\hat{\xi} = \xi T_W T > 0$ are positive coefficients that must be determined from measurements; $1/\xi$ is known as the Kapitza resistance (Pyotr Kapitza, 1894-1984).

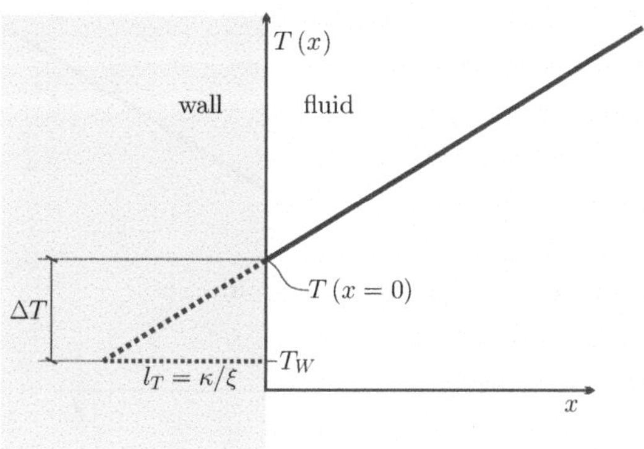

Fig. 7.1 On the interpretation of temperature jump ΔT and jump length l_T.

This boundary condition can be read in two ways: Looking at temperatures, due to the difference between T_W and T, there will be a normal heat flux $q_k n_k$ from warm to cold; this is Newton's law of cooling between wall and fluid. Inversely, looking at heat flux, the temperature jump condition states that a heat transfer through the wall is accompanied by a temperature jump. As always, forces and fluxes vanish in equilibrium, where $T_W = T$ and $q_k n_k = 0$.

With the Fourier law $q_k = -\kappa \frac{\partial T}{\partial x_k}$, the jump condition can be re-written as

$$\frac{\partial T}{\partial n} = \frac{T - T_W}{\kappa/\xi} = \frac{T - T_W}{l_T} \,, \tag{7.24}$$

where $\frac{\partial}{\partial n} = n_k \frac{\partial}{\partial x_k}$ is the space derivative in normal direction, and $l_T = \kappa/\xi$ is the jump length, which results from extrapolation of the fluid temperature $T(x)$ into the wall to the point where it reaches the wall temperature T_W, as shown in Fig. 7.1.

Under conditions where temperature jumps occur, these will also affect temperature measurements, since jumps might occur at thermometer boundaries. Hence, temperature measurements in small nonequilibrium systems are particularly difficult.

7.2.3 Velocity Slip

The vectorial force-flux pair gives the velocity slip relation

$$t^{\tau}_{ik} n_k = \zeta V_i \tag{7.25}$$

with positive slip coefficient $\zeta > 0$. For the Navier-Stokes stress tensor (5.3), we find

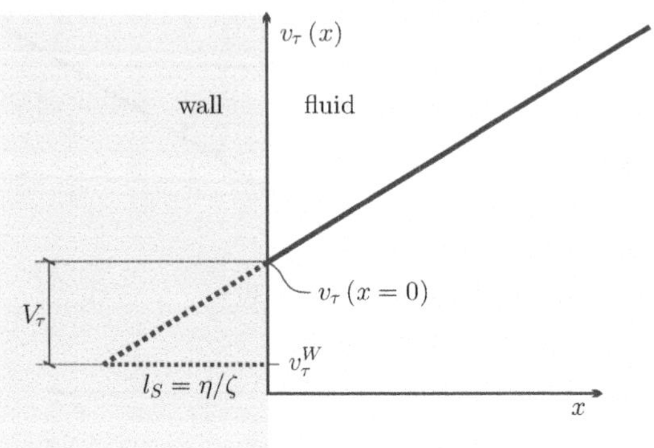

Fig. 7.2 On the interpretation of slip velocity V_τ and slip length l_T.

$$
\begin{aligned}
t^{\tau}_{ik} n_k &= t_{ik} n_k - t_{lm} n_l n_m n_i \\
&= \eta \left(\frac{\partial v_i}{\partial x_k} n_k + \frac{\partial v_k}{\partial x_i} n_k - \frac{\partial v_l}{\partial x_m} n_l n_m n_i - \frac{\partial v_m}{\partial x_l} n_l n_m n_i \right) \\
&= \eta \left(\frac{\partial}{\partial n} (v_i - v_n n_i) + \left(\frac{\partial}{\partial x_i} - n_i \frac{\partial}{\partial n} \right) v_n \right) \\
t^{\tau}_{ik} n_k &= \eta \left(\frac{\partial v_\tau}{\partial n} + \frac{\partial v_n}{\partial \tau} \right)
\end{aligned}
\tag{7.26}
$$

where $v_n = v_k n_k$ is the normal velocity at the wall, $v_\tau = (v_i - v_n n_i)$ is the tangential velocity, $\frac{\partial}{\partial n} = n_k \frac{\partial}{\partial x_k}$ is the derivative in normal direction, and $\frac{\partial}{\partial \tau} = \left(\frac{\partial}{\partial x_i} - n_i \frac{\partial}{\partial n} \right)$ is the derivative in tangential direction. Here, we have considered locally flat walls, so that the normal vector n_i can be pulled under the derivatives without compensation.

For a rigid wall, according to (7.16) normal velocity of the fluid equals the normal velocity of the wall, $v_n = v_n^W$, and there is no variation of the wall velocity in tangential direction, hence $\frac{\partial v_n}{\partial \tau} = \frac{\partial v_n^W}{\partial \tau} = 0$. With this, the slip condition reduces to a relation between shear stress and (tangential!) slip velocity,

$$
t^{\tau}_{ik} n_k = \eta \frac{\partial v_\tau}{\partial n} = \zeta V_\tau = \zeta \left(v_\tau - v_\tau^W \right) . \tag{7.27}
$$

Introducing the slip length $l_S = \eta/\zeta$, the slip condition assumes the intuitive form

$$
\frac{\partial v_\tau}{\partial n} = \frac{v_\tau - v_\tau^W}{\eta/\zeta} = \frac{v_\tau - v_\tau^W}{l_S} , \tag{7.28}
$$

which is interpreted along the same lines as the jump length, see Fig. 7.2

Note that the normal part of the momentum balance gives $[[t_{ik}]] n_k n_i = 0$, which relates the normal force in the fluid at the wall to the force on the wall. Thus, the normal stress acting on the wall is obtained from the interface

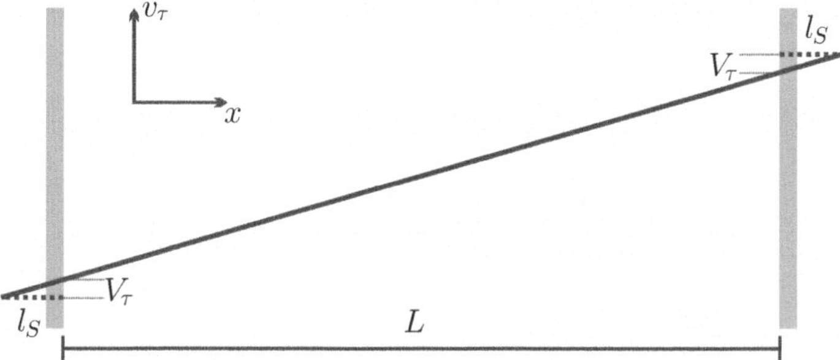

Fig. 7.3 Slip length l_S compared to system size L. For $L \gg l_S$ slip velocity V_τ can be ignored.

condition for momentum as

$$t^W_{ik} n_i n_k = t_{ik} n_i n_k = -p + \left(\lambda - \frac{2}{3}\eta\right)\frac{\partial v_{\tau_\alpha}}{\partial \tau_\alpha} + \left(\lambda + \frac{4}{3}\eta\right)\frac{\partial v_n}{\partial x_n} \tag{7.29}$$

7.2.4 When Do We Need Slip and Jump Conditions?

Typically, in engineering problems jump and slip are ignored, that is the jump and slip lengths are set to zero, so that temperature and (tangential *and* normal) velocity at the wall-fluid boundary are continuous,

$$T = T_W \quad , \quad v_i = v_i^W \ . \tag{7.30}$$

To discuss the relevance of slip and jump lengths, Fig. 7.3 shows the velocity for a simple linear shear flow between two walls at distance L, with slip length l_S indicated at both walls. From the figure it becomes clear that slip—and jump—phenomena matter when L and l_S are not too different, or for rather large gradients. For many applications $L \gg l_S$, and setting $l_S = 0$ leads to no noticeable error. Thus, jump and slip boundary conditions are important for micro- and nanoflows, where L is relatively small.

In gases, slip and jump lengths l_S and l_T are of the order of the mean free path of the molecules, which at standard conditions is of the order of 10^{-7} m, but for vacuum flows can be of the order of centimeters, or even meters. Jump and slip result from interaction between gas particles and wall, where particles collide with the wall and are reflected back into the gas. Here, the structure of the wall plays a role as well, smooth walls might have a larger slip length than rough walls.

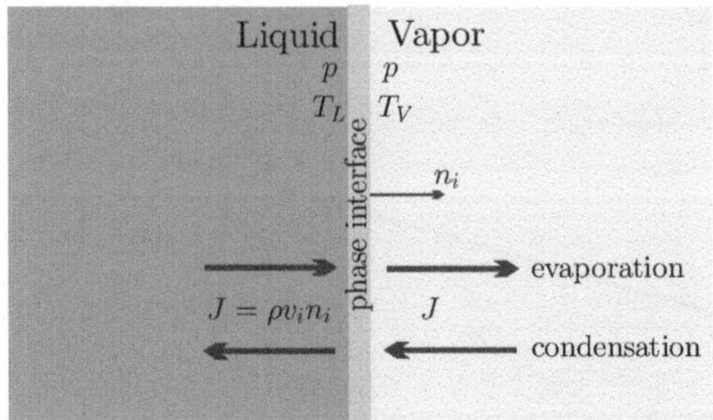

Fig. 7.4 Evaporation and condensation in nonequilibrium, with temperature jump $T_V - T_L$ across the phase interface.

7.3 Evaporation and Condensation at Phase Interfaces

7.3.1 Interface Conditions

Figure 7.4 shows a phase interface between vapor and liquid of the same substance with mass flux J indicated for evaporation of the liquid as well as condensation of the vapor. We consider jump conditions across the phase interface, where now $[[f]] = f_V - f_L$ is the difference between vapor (V) and liquid (L) properties at the interface; the normal $\mathbf{n}$ points into the vapor. For convenience we denote the mass flux normal to the interface as $J = \rho v_i n_i$, and chose a viewpoint resting with the interface, that is velocities are measured relative to the interface. Also, viscous stresses at the interface are ignored.

With this, the jump condition for mass simply states that the mass flux leaving one phase enters the other,

$$[[\rho v_k]]\, n_k = 0 \quad \Rightarrow \quad J_L = J_V = J \tag{7.31}$$

The momentum balance reduces to

$$[[\rho v_i v_k + p\delta_{ik}]]\, n_k = 0 \quad \Rightarrow \quad J\,[[v_i]] + [[p]]\, n_i = 0 \tag{7.32}$$

For most processes, evaporation or condensation velocities v_i are very small, hence the quadratic term $Jv_i = \rho v_i v_k n_k$ is extremely small and can be ignored. Then, the balance of momentum gives the simple result that both phases are at the same pressure, as indicated in the figure,

$$[[p]]\, n_i = 0 \quad \Rightarrow \quad p_L = p_V = p\,. \tag{7.33}$$

Note that this is valid for plane interfaces since surface tension is not considered.

Again ignoring quadratic terms in velocity, the energy balance reduces to $(h = u + p/\rho)$

$$[[\rho h v_k + q_k]]\, n_k = 0 \quad \Rightarrow \quad J h_L + q_L = J h_V + q_V \ , \tag{7.34}$$

where $q_L = -\kappa_L \frac{\partial T}{\partial n}$ and $q_V = -\kappa_V \frac{\partial T}{\partial n}$ are the normal heat fluxes in the two phases.

The entropy balance (4.20) for singular surfaces gives the entropy generation at the phase interface,

$$\left[\left[\rho s v_k + \frac{q_k}{T}\right]\right] n_k = \Sigma_S \quad \Rightarrow \quad \Sigma_S = J\left(s_V - s_L\right) + \frac{q_V}{T_V} - \frac{q_L}{T_L} \ . \tag{7.35}$$

As always, we combine the conditions to find a useful expression for the entropy generation rate. Eliminating the heat flux in the liquid, q_L, and introducing Gibbs free energies $g_L\left(T_L, p\right) = h_L - T_L s_L$, $g_V\left(T_V, p\right) = h_V - T_V s_V$, we find

$$\Sigma_S = J\left[\frac{g_L}{T_L} - \frac{g_V}{T_V} + h_V\left(\frac{1}{T_V} - \frac{1}{T_L}\right)\right] + q_V\left(\frac{1}{T_V} - \frac{1}{T_L}\right) \ . \tag{7.36}$$

Once more, the entropy generation rate is a sum of products of forces and fluxes, i.e., $\Sigma_S = \sum_A \mathcal{F}_A \mathcal{J}_A$, where

Force $\mathcal{F}_A$	**Flux** $\mathcal{J}_A$	**Description**
$\left[\frac{g_L}{T_L} - \frac{g_V}{T_V} + h_V\left(\frac{1}{T_V} - \frac{1}{T_L}\right)\right]$	J	evaporation flux
$\frac{1}{T_V} - \frac{1}{T_L}$	q_V	heat flux

7.3.2 Phase Equilibrium

In phase equilibrium, the forces must vanish, hence both phases have the same temperature $T_V = T_L = T$, and, since they have the same pressure as well, their Gibbs free energies fulfill

$$g_V\left(p, T\right) = g_L\left(p, T\right) \ . \tag{7.37}$$

This is the well known equilibrium condition for coexistence of liquid and vapor, which states that for given temperature T phase equilibrium is possible only at the saturation pressure $p_{\mathrm{sat}}\left(T\right)$, which obeys

$$g_V\left(p_{\mathrm{sat}}\left(T\right), T\right) = g_L\left(p_{\mathrm{sat}}\left(T\right), T\right) \ . \tag{7.38}$$

7.3.3 Close to Equilibrium

The physical interpretation of the forces $\mathcal{F}_A$ becomes clearer for small deviations from equilibrium, where the pressure is close to saturation, and the temperature jump is small, hence

$$\Delta p = p - p_{\rm sat}(T_L) \quad , \quad \Delta T = T_V - T_L \tag{7.39}$$

are small. Taylor expansion of the thermodynamic forces to leading order yields (see Prob. 7.7)

$$\mathcal{F}_A = \left\{ -\left(\frac{1}{\rho_{V,\rm sat}(T_L)} - \frac{1}{\rho_{L,\rm sat}(T_L)} \right) \frac{\Delta p}{T_L}, \; -\frac{\Delta T}{T_L^2} \right\}_A \, . \tag{7.40}$$

We recognize that the force for the evaporation flux is associated with the deviation of pressure from saturation pressure, while the force for heat transfer is given by the temperature jump.

Further simplification is possible for states well below the critical temperature, where the liquid density is much larger than the vapor density, such that $\frac{1}{\rho_V} - \frac{1}{\rho_L} \simeq \frac{1}{\rho_V}$. Further assuming that the vapor behaves as an ideal gas, within first order of Δp and ΔT we can write

$$\mathcal{F}_A = \left\{ -\frac{R\Delta p}{p_{\rm sat}(T_L)}, \; -\frac{\Delta T}{T_L^2} \right\}_A \, . \tag{7.41}$$

We proceed with this simplified form of the forces to have more compact equations; the extension to the full forces is straightforward.

7.3.4 Force-Flux Relations

Since both forces, and both fluxes, are scalars, cross effects appear and the phenomenological equations $J_A = \sum L_{AB}\mathcal{F}_B$ read, with a symmetric non-negative definite matrix of phenomenological coefficients l_{AB},

$$\begin{bmatrix} J \\ \\ q_V \end{bmatrix} = \begin{bmatrix} L_{11} & L_{12} \\ \\ L_{12} & L_{22} \end{bmatrix} \begin{bmatrix} -\frac{R\Delta p}{p_{\rm sat}(T_L)} \\ \\ -\frac{\Delta T}{T_L^2} \end{bmatrix} \tag{7.42}$$

For $\Delta p = p - p_{\rm sat}(T_L) < 0$ the actual pressure is below the saturation pressure, and we expect positive evaporation flux J. We can say that evaporation occurs to increase the pressure in the vapor to the saturation pressure. For condensation, the pressure is above saturation, and condensation from the vapor occurs to lower the pressure.

The force-flux relations can be rewritten such that the matrix of phenomenological coefficients becomes dimensionless, as

$$\begin{bmatrix} J \\ \\ \frac{q_V}{RT_L} \end{bmatrix} = \begin{bmatrix} L_{11}\frac{R\sqrt{2\pi RT_L}}{p_{\text{sat}}(T_L)} & L_{12}\frac{\sqrt{2\pi RT_L}}{p_{\text{sat}}(T_L)T_L} \\ \\ L_{12}\frac{\sqrt{2\pi RT_L}}{p_{\text{sat}}(T_L)T_L} & L_{22}\frac{\sqrt{2\pi RT_L}}{p_{\text{sat}}(T_L)RT_L^2} \end{bmatrix} \begin{bmatrix} -\frac{\Delta p}{\sqrt{2\pi RT_L}} \\ \\ -\frac{p_{\text{sat}}(T_L)}{\sqrt{2\pi RT_L}}\frac{\Delta T}{T_L} \end{bmatrix} \tag{7.43}$$

where forces and fluxes are rescaled to have the unit of mass flux, and the factor $\sqrt{2\pi}$ is motivated by expressions found in kinetic theory of gases. Taking the inverse of the matrix, we can express the forces through the fluxes by means of a symmetric non-negative definite matrix of dimensionless interface resistivities $\hat{r}_{ab}$,

$$\begin{bmatrix} -\frac{\Delta p}{\sqrt{2\pi RT_L}} \\ \\ -\frac{p_{\text{sat}}(T_L)}{\sqrt{2\pi RT_L}}\frac{\Delta T}{T_L} \end{bmatrix} = \begin{bmatrix} \hat{r}_{11} & \hat{r}_{12} \\ \\ \hat{r}_{12} & \hat{r}_{22} \end{bmatrix} \begin{bmatrix} J \\ \\ \frac{q_V}{RT_L} \end{bmatrix} \tag{7.44}$$

This form of the flux-force-relations can be found in the literature of kinetic theory of ideal gases, where the resistivities are determined as

$$\hat{r}_{AB} = \begin{bmatrix} \frac{1}{\psi} - 0.40044 & 0.126 \\ \\ 0.126 & 0.294 \end{bmatrix} \tag{7.45}$$

Here, $\psi \in (0,1)$ is the condensation coefficient, defined as the probability that a vapor particle hitting the interface will condense, and not bounce back into the vapor.

As is the case for jump and slip at a boundary, the condensation/evaporation conditions become relevant for sufficiently small systems. In macroscopic systems, the jumps can be ignored, so that the interface conditions reduce to

$$p = p_{\text{sat}}(T_L) \quad , \quad T_V = T_L \tag{7.46}$$

A closer look at the phase interface reveals that it is not a sharp jump between densities of liquid and vapor, but rather a steep continuous transition over a thickness of a few atomic diameters. The attracting forces between atoms in this layer lead to surface tension effects.

7.3.5 Hertz-Knudsen Theory

Considering the vapor as an ideal gas, the mass flux from vapor particles with thermal velocity towards the interface is $\rho\sqrt{\frac{RT_V}{2\pi}} = \frac{p}{\sqrt{2\pi RT_V}}$, see Prob. 7.8. Accordingly, the rate of condensation at the interface can be written as $J_c =$

$\sigma_c \frac{p}{\sqrt{2\pi RT_V}}$, with the condensation coefficient σ_c defined as the probability for a vapor particle hitting the interface to condense. Similarly, assuming that evaporated particles are in the saturated state at T_L, the rate of evaporation can be written as $J_e = \sigma_e \frac{p_{\rm sat}(T_L)}{\sqrt{2\pi RT_L}}$ with the evaporation coefficient σ_e.

The net mass flux across the interface is the difference between evaporation and condensation rates, which gives the classical Hertz-Knudsen model for evaporation as[3]

$$J_{HK} = J_e - J_c = \sigma_e \frac{p_{\rm sat}(T_L)}{\sqrt{2\pi RT_L}} - \sigma_c \frac{p}{\sqrt{2\pi RT_V}} . \tag{7.47}$$

The effective evaporation and condensation coefficients, σ_e and σ_c, must be obtained from modeling or measurement. Obviously, in equilibrium both coefficients must agree, $\sigma_{e|E} = \sigma_{c|E}$, and this is often assumed also in nonequilibrium processes. Experimental determination of these coefficients is quite inconclusive, with different authors suggesting values that differ by three orders of magnitude.[4]

If one is interested in interfaces with temperature jump and pressure deviation, the Hertz-Knudsen mass flux relation must be complemented by a corresponding expression for heat flux. In particular, the Hertz-Knudsen model alone can not predict temperature jumps.

A simplified kinetic model yields the fluxes for the Hertz-Knudsen-Schrage model as[5]

$$J = \frac{2\psi}{2-\psi}\left[\frac{p_{\rm sat}(T_L)}{\sqrt{2\pi RT_L}} - \frac{p}{\sqrt{2\pi RT_V}}\right] \tag{7.48}$$

$$\begin{aligned}\frac{q_V}{RT_L} = \frac{2\psi}{2-\psi}\left(2 - \frac{5}{2}\frac{T_V}{T_L}\right)\left[\frac{p_{\rm sat}(T_L)}{\sqrt{2\pi RT_L}} - \frac{p}{\sqrt{2\pi RT_V}}\right] \\ + \frac{4\left[\psi(1-\vartheta)+\vartheta\right]}{2-\psi(1-\vartheta)-\vartheta}\frac{p}{\sqrt{2\pi RT_V}}\left[\frac{T_L - T_V}{T_L}\right] ,\end{aligned} \tag{7.49}$$

Here, ψ is the evaporation coefficient of the kinetic theory evaporation model and ϑ is the accommodation coefficient; the effective evaporation and condensation coefficients are $\sigma_e = \sigma_c = \frac{2\psi}{2-\psi}$. Linearization of these expressions in Δp and ΔT leads to the thermodynamic model (7.44) with a symmetric resistivity matrix, see Prob. 7.9.

---

[3] H. Hertz, Ann. Physik **253**, 177-193 (1882); doi: 10.1002/andp.18822531002
M. Knudsen, Ann. Physik **352**, 697-708 (1915); doi: 10.1002/andp.19153521306

[4] R. Marek, J. Straub, Int. J. Heat Mass Transf. **44**, 39-53 (2001); doi: 10.1016/S0017-9310(00)00086-7

[5] J.P. Caputa, H. Struchtrup, Physica A **390**, 31–42 (2011); doi:10.1016/j.physa.2010.09.019

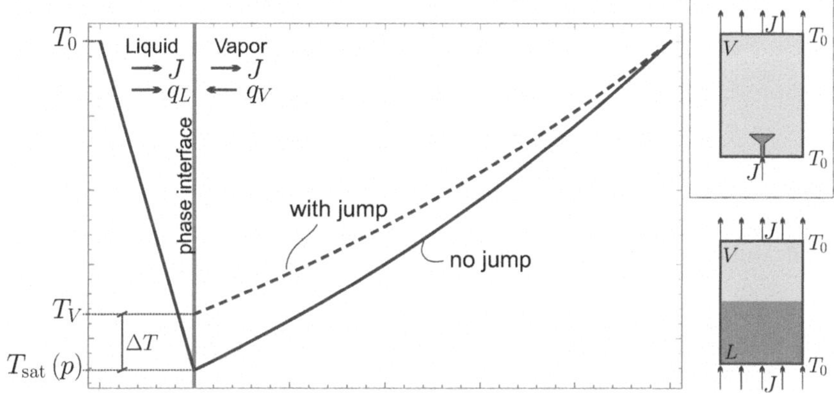

Fig. 7.5 Temperature curve in one-dimensional evaporative cooling process with and withiut jump. The inserts shows the phyiscal set-up in the experiments by Ward and co-workers.

7.3.6 Example: Evaporative Cooling

A simple application of phase boundaries without and with jump behavior is given by evaporative cooling where a vapor mass flux J is removed, hence lowering the pressure below saturation. This triggers evaporation, where the heat of evaporation is taken mainly from the liquid, which therefore cools down. A simple setting where the evaporated liquid is replaced such that the interface is stationary is shown in Fig. 7.5. The temperature curves shown are for the case that the boundary temperatures for liquid and vapor are equal; due to the evaporative cooling at the interface, heat is transferred from both boundaries towards the interface. Problems 7.10 and 7.11 lead through the necessary computations, where the combination of convective and conductive heat transfer leads to the curvature of temperature shown in the figure.

Just as other jump and slip phenomena, temperature jump and deviation of pressure from saturation are only expected to play a role in micro- and nanosystems, they can be ignored in macroscopic technical applications. Nevertheless, in a series of experiments, Ward and co-workers observed temperature jumps of about 3 °C in an apparatus as sketched in the insert of Fig. 7.5, where evaporation takes place in a small funnel in the center of a bigger chamber.[6] Due to geometric effects, the heat flux at the phase interface is much larger than in the one-dimensional setting, which from (7.44) enhances the temperature jump.[7] Experimental error, however, is significant, and up to this day there is rather limited understanding of the interfacial resistivities.

---

[6] G. Fang and C.A. Ward, Phys. Rev. E **59**, 417 (1999); doi: 10.1103/PhysRevE.59.417

[7] M. Bond and H. Struchtrup, Phys. Rev. E **70**, 061605 (2004); doi: 10.1103/PhysRevE.70.061605

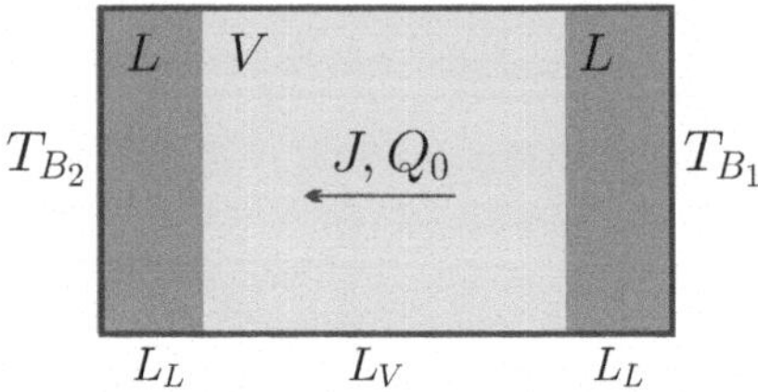

Fig. 7.6 Vapor between liquid layers in temperature gradient.

7.3.7 Example: Vapor between Liquid Layers

We consider a one-dimensional setting with vapor (thickness L_V) between two liquids layers (thickness L_L) that are kept at different boundary temperatures T_{B_1}, T_{B_2}, see Fig. 7.6. Assuming low flow velocities so that quadratic terms and viscosity contributions can be ignored, mass, momentum and energy conservation give at each point of the system

$$\rho v = J = const. \quad , \quad p = P = const \quad , \quad Jh + q = Q = const \tag{7.50}$$

where mass flux J, normal stress P, and energy flux Q arise as constants of integration. Assuming the simple model of Prob. 7.10 for liquid and vapor states, the energy balances for liquid and vapor read

$$Jc_L (T - T_0) - \kappa_L \frac{dT}{dx} = Q \tag{7.51}$$

$$J [c_p (T - T_0) + h_{LV}] - \kappa_V \frac{dT}{dx} = Q \tag{7.52}$$

These must be solved with boundary conditions $T(x = 0) = T_{B_1}$, $T(x = L) = T_{B_2}$, where $L = 2L_L + L_V$, and the interface conditions (7.44) evaluated at $x = L_L$ and $x = L_L + L_V$.

Considering a macroscopic system only, we can ignore temperature jump and pressure deviation. Then, for the pressures we find, with $\Delta p = p - p_{\text{sat}}(T_L) = 0$ and $p = P$,

$$p_{\text{sat}}(T(L_L)) = p_{\text{sat}}(T(L_L + L_V)) = P \ , \tag{7.53}$$

which implies that both interfaces are at the same temperature T_I, hence

$$T_L(L_L) = T_V(L_L) = T_V(L_L + L_V) = T_L(L_L + L_V) = T_I \tag{7.54}$$

Accordingly, the temperature gradient in the vapor, and the heat flux q_V vanish, which gives the energy flux in terms of the interface temperature,

$$Q = J [c_p (T_I - T_0) + h_{LV}] \tag{7.55}$$

The temperatures in the liquid layers are obtained by integration as

$$0 < x < L_L : T = \left(T_0 + \frac{Q}{Jc_L}\right) + \left(T_{B_1} - T_0 - \frac{Q}{Jc_L}\right) e^{\frac{J_0 c_L}{\kappa_L} x} \tag{7.56}$$

$$L_L + L_V < x < L : T = \left(T_0 + \frac{Q}{Jc_L}\right) + \left(T_{B_2} - T_0 - \frac{Q}{Jc_L}\right) e^{\frac{J_0 c_L}{\kappa_L}(x-L)} \tag{7.57}$$

where the outer boundary conditions where used. Evaluating the liquid temperatures at the phase boundaries ($x = L_L$ and $x = L_L + L_V$) gives two coupled equations for the common interface temperature T_I, which is hidden in Q, and the mass flow J,

$$T_I = \left[T_0\left(1 - \frac{c_p}{c_L}\right) + \frac{h_{LV}}{c_L} + \frac{c_p}{c_L}T_I\right]\left(1 - \exp\left[\frac{Jc_L}{\kappa_L}L_L\right]\right) + \\ + T_{B_1}\exp\left[\frac{Jc_L}{\kappa_L}L_L\right] \tag{7.58}$$

$$T_I = \left[T_0\left(1 - \frac{c_p}{c_L}\right) + \frac{h_{LV}}{c_L} + \frac{c_p}{c_L}T_I\right]\left(1 - \exp\left[-\frac{Jc_L}{\kappa_L}L_L\right]\right) + \\ + T_{B_2}\exp\left[-\frac{Jc_L}{\kappa_L}L_L\right] \tag{7.59}$$

These equations must be solved numerically. Considering the data for water of Prob. 7.10 with $L_L = 5\,\text{cm}$, $L_V = 15\,\text{cm}$ and boundary temperatures $T_{B_1} = 15\,°\text{C}$, $T_{B_2} = 30\,°\text{C}$, we find $T_I = 22.55\,°\text{C}$, and $J_0 = -3.37\times 10^{-5}\frac{\text{kg}}{\text{m}^2\,\text{s}}$. The pressure in the system is $P_0 = p_{\text{sat}}(T_I) = 2.73\,\text{kPa}$.

Figure 7.7 shows the temperature curve for this case without jumps (black, continuous). When jumps are considered, depending on choice of parameters, one might encounter a curve with an inverted temperature profile (gray, dashed), where in agreement with the second law the overall energy transfer $Q = Jh + q$ is from warm to cold (right to left), but the conductive heat flux in the vapor, $q_V = -\kappa_V\frac{dT}{dx}$, is in the opposite direction.[8] The computations are left to the interested reader.

7.4 Non-Uniqueness of Fluxes and Forces

There is no general rule on how to split the entropy generation into force-flux pairings. Evaporation offers an accessible case to study the rules for transfor-

[8] H. Struchtrup, S. Kjelstrup, D. Bedeaux, Phys. Rev. E **85**, 061201 (2012), doi: 10.1103/PhysRevE.85.061201

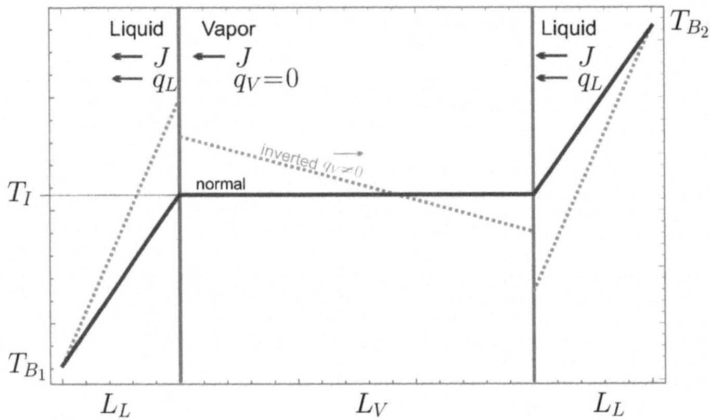

Fig. 7.7 Heat and mass transfer in liquid-vapor-liquid. Without jumps both interfaces are at the same temperature T_I (black, continuous). With jumps an inverted temperature profile is possible (gray, dashed).

mations between different choices of fluxes and forces. Indeed, eliminating q_V instead of q_L in (7.35), we find the entropy generation rate in the alternative form

$$\Sigma_S = J\left[\frac{g_L}{T_L} - \frac{g_V}{T_V} + h_L\left(\frac{1}{T_V} - \frac{1}{T_L}\right)\right] + q_L\left(\frac{1}{T_V} - \frac{1}{T_L}\right) . \tag{7.60}$$

which suggests alternative forces and fluxes as

$$\bar{\mathcal{F}}_A = \left\{\frac{g_L}{T_L} - \frac{g_V}{T_V} + h_L\left(\frac{1}{T_V} - \frac{1}{T_L}\right), \frac{1}{T_V} - \frac{1}{T_L}\right\}_A , \quad \bar{\mathcal{J}}_A = \{J, q_L\}_A . \tag{7.61}$$

The transformation between the new forces and fluxes, $(\bar{\mathcal{F}}_A, \bar{\mathcal{J}}_A)$, and the original ones $(\mathcal{F}_A, \mathcal{J}_A)$ is linear, with a matrix A_{AB},

$$\bar{\mathcal{J}}_A = \begin{bmatrix} 1 & 0 \\ h_{LV} & 1 \end{bmatrix}_{AB} \begin{bmatrix} J \\ q_V \end{bmatrix}_B = A_{AB}\mathcal{J}_B \tag{7.62}$$

$$\bar{\mathcal{F}}_A = \begin{bmatrix} 1 & -h_{LV} \\ 0 & 1 \end{bmatrix}_{AB} \begin{bmatrix} \frac{g_L}{T_L} - \frac{g_V}{T_V} + h_V\left(\frac{1}{T_V} - \frac{1}{T_L}\right) \\ \frac{1}{T_V} - \frac{1}{T_L} \end{bmatrix}_B = A_{AB}^{-T}\mathcal{F}_B \tag{7.63}$$

where A_{AB}^{-T} is the transposed inverse matrix, and $h_{LV} = h_V - h_L$ is the difference between the enthalpies of liquid and vapor in nonequilibrium, which has values in the order of the heat of vaporization $h_{LV}^{\text{sat}}(T_L)$.

By inversion, we have as well

$$\mathcal{J}_A = A_{AB}^{-1}\bar{\mathcal{J}}_B \quad , \quad \mathcal{F}_A = A_{AB}^T\bar{\mathcal{F}}_B \tag{7.64}$$

The entropy generation is invariant against the transformation,

$$\Sigma_S = \bar{\mathcal{J}}_A\bar{\mathcal{F}}_A = A_{AB}\mathcal{J}_B A_{AC}^{-T}\mathcal{F}_C = A_{CA}^{-1}A_{AB}\mathcal{J}_B\mathcal{F}_C = \mathcal{J}_B\mathcal{F}_B \tag{7.65}$$

Also the matrix of phenomenological coefficients transforms with the same matrix A_{AB}. In the original formulation we have $\mathcal{J}_A = L_{AB}\mathcal{F}_B$ with symmetric matrix $L_{AB} = L_{BA}$. From this,

$$\begin{aligned}\mathcal{J}_A = L_{AB}\mathcal{F}_B \quad &\Rightarrow \quad A_{AB}^{-1}\bar{\mathcal{J}}_B = L_{AB}A_{BC}^T\bar{\mathcal{F}}_C \\ &\Rightarrow \quad \bar{\mathcal{J}}_D = A_{DA}L_{AB}A_{BC}^T\bar{\mathcal{F}}_C = \bar{L}_{DC}\bar{\mathcal{F}}_B\end{aligned} \tag{7.66}$$

with the transformed phenomenological matrix $\bar{L}_{DC}$, which inherits the symmetry of L_{AB}, since

$$\bar{L}_{DC} = A_{DA}L_{AB}A_{BC}^T = A_{BC}^T L_{BA}A_{DA} = A_{CB}L_{BA}A_{AD}^T = \bar{L}_{CD} \ . \tag{7.67}$$

Here we see for a simple example that forces and fluxes are not unique. Linear transformations, as with the matrix A_{AB} above, leave the structure of forces, fluxes, and phenomenological force-flux relations unchanged.

One might be tempted to declare diagonal elements in the Onsager matrix L_{AB} to be 'more important', since they link the force and its flux as paired in the entropy generation. This, however, cannot be justified, since forces and fluxes are not unique.

While one has the freedom to chose force-flux pairs, some choices will be more meaningful than others, as can be seen from studying the dimensionless resistivities.

The dimensionless resistivities $\bar{r}_{AB}$ for the alternative force-flux pair $(\bar{\mathcal{F}}_A, \bar{\mathcal{J}}_A)$ are obtained from the inverse of $\bar{L}_{AB}$, specifically we find

$$\begin{bmatrix} \frac{p_{\text{sat}}(T_L)}{\sqrt{2\pi RT_L}}\left[\frac{g_L}{RT_L} - \frac{g_V}{RT_V} + h_L\left(\frac{1}{RT_V} - \frac{1}{RT_L}\right)\right] \\ \\ \frac{p_{\text{sat}}(T_L)}{\sqrt{2\pi RT_L}}T_L\left[\frac{1}{T_V} - \frac{1}{T_L}\right]\end{bmatrix} = \begin{bmatrix}\bar{r}_{11} & \bar{r}_{12} \\ \\ \bar{r}_{21} & \bar{r}_{22}\end{bmatrix}\begin{bmatrix} J \\ \\ \frac{q_L}{RT_L}\end{bmatrix} \tag{7.68}$$

with the resistivities $\bar{r}_{AB}$ related to our original resistivities $\hat{r}_{\alpha\beta}$ as

$$\begin{bmatrix}\bar{r}_{11} & \bar{r}_{12} \\ \\ \bar{r}_{12} & \bar{r}_{22}\end{bmatrix} = \begin{bmatrix}\left[\hat{r}_{11} - \frac{h_{LV}}{RT_L}\left(\hat{r}_{12} + \hat{r}_{21}\right) + \left(\frac{h_{LV}}{RT_L}\right)^2\hat{r}_{22}\right] & \left[\hat{r}_{12} - \frac{h_{LV}}{RT_L}\hat{r}_{22}\right] \\ \\ \left[\hat{r}_{21} - \frac{h_{LV}}{RT_L}\hat{r}_{22}\right] & \left[\hat{r}_{22}\right]\end{bmatrix} . \tag{7.69}$$

The resistivities in (7.68) are strongly affected by the relatively large values of the dimensionless heat of evaporation, $\frac{h_{LV}}{RT_L}$, which far from the critical point is of order 10. Considering (7.69) with the kinetic theory values for

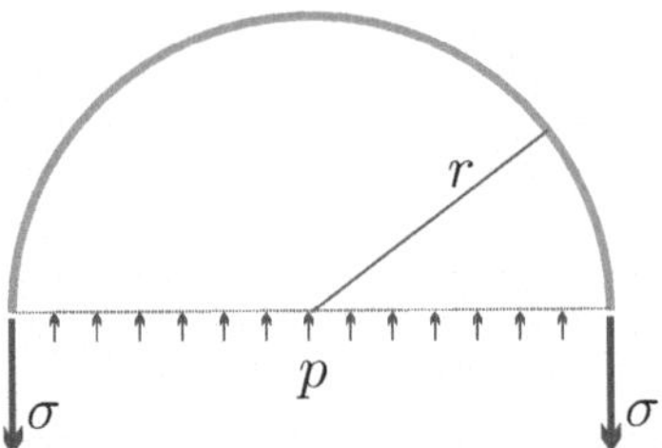

Fig. 7.8 Droplet in vacuum: Pressure p in the droplet must balance surface tension σ.

$\hat{r}_{\alpha\beta}$ (7.45) in mind, it is evident that $\bar{r}_{11}$, $\bar{r}_{12}$ and $\bar{r}_{21}$ are dominated by the contributions $\left(\frac{h_{LV}}{RT_L}\right)^2 \hat{r}_{22}$ and $\left(\frac{h_{LV}}{RT_L}\right) \hat{r}_{22}$. That is, evaluating experimental data based on the flux-force pair $\left(\bar{\mathcal{F}}_A, \bar{\mathcal{J}}_A\right)$, one would effectively measure the temperature jump coefficient $\hat{r}_{22}$ and the heat of evaporation. Notably, this would make it difficult to settle the question of Onsager symmetry from experimental results, where a significant asymmetry between $\hat{r}_{12}$ and $\hat{r}_{21}$ would be invisible as an asymmetry between $\bar{r}_{12}$ and $\bar{r}_{21}$. One might say that the $\hat{r}_{\alpha\beta}$, that is the resistivities for the fluxes $\mathcal{J}_A = \{J, q_V\}$, give the pure data for interface behavior, while those for the the fluxes $\bar{\mathcal{J}}_A = \{J, q_L\}$ are overwhelmingly affected by the heat of evaporation.

7.5 Surface Tension and Vapor Pressure in Droplets

7.5.1 Surface Tension and Energy

Due to attractive intermolecular forces molecules in a liquid stay together. While a molecule in the bulk is interacting with other molecules all around, a molecule at the liquid-vapor interface has strong interactions only with the molecules on the liquid side. Effectively, this causes surface tension.

To determine surface energy, we consider a droplet of radius r in vacuum. Due to surface tension, the liquid inside is at a pressure p. Figure 7.8 shows one half of a spherical droplet, for which pressures and surface tension must balance. Since droplet pressure p acts on the surface πr^2 and surface tension σ acts on the circumference $2\pi r$ we have

$$p\pi r^2 = \sigma 2\pi r \quad \Rightarrow \quad p = \frac{2\sigma}{r} \,, \tag{7.70}$$

just as for membrane stress, see Sec. 3.8.5.

$\frac{T}{°C}$	$\frac{\sigma}{10^{-3}\frac{N}{m}}$	$\frac{T}{°C}$	$\frac{\sigma}{10^{-3}\frac{N}{m}}$	$\frac{T}{°C}$	$\frac{\sigma}{10^{-3}\frac{N}{m}}$	$\frac{T}{°C}$	$\frac{\sigma}{10^{-3}\frac{N}{m}}$
0.01	75.64	60	66.24	120	54.96	250	26.06
10	74.23	70	64.47	140	50.85	275	20.11
20	72.75	80	62.67	160	46.58	300	14.30
30	71.20	90	60.82	180	42.19	325	8.73
40	69.60	100	58.91	200	37.69	350	3.65
50	67.94	110	56.96	225	31.93	374	0.00

Table 7.1 Surface tension of water as function of temperature [Vargaftik, Volkov, and Voljak, J. Phys. Ref. Data **12**, 817-820 (1983)]

The work to increase the droplet volume by dV through injection of liquid is[9]

$$\delta W = -pdV = -\frac{2\sigma}{r} d\left(\frac{4\pi}{3}r^3\right) = -\sigma d\left(4\pi r^2\right) = -\sigma dA \;, \tag{7.71}$$

where $V = \frac{4\pi}{3}r^3$ and $A = 4\pi r^2$ are droplet volume and surface. The corresponding change of energy is

$$dE_S = -\delta W = \sigma dA \;, \tag{7.72}$$

where E_S is the surface energy, which we identify through integration as

$$E_S = \sigma A = 4\pi\sigma r^2 \;. \tag{7.73}$$

Surface tension results from attractive forces between liquid particles. As temperature increases volume expands, hence average particle distances increase, which leads to weaker forces, hence weaker surface tension. Table 7.1 shows surface tension of water as function of temperature. Starting from the triple point (0.01 °C) surface tension decreases until it vanishes at the critical point (374.14 °C) where the differences between liquid and vapor vanish.

7.5.2 Droplet in Vapor

Next, we study a droplet in equilibrium with its vapor. Specifically, we consider a system consisting of a single droplet in vapor, where temperature T, volume V, and mass m of the total system are kept constant, as depicted in Fig. 7.9. With obvious notation we have

$$m = m_V + m_D = const. \quad , \quad V = V_V + V_D = const \;. \tag{7.74}$$

First and second law (1.8, 1.10) for this closed system read

[9] Injection of liquid is work done on the droplet, hence must be negative.

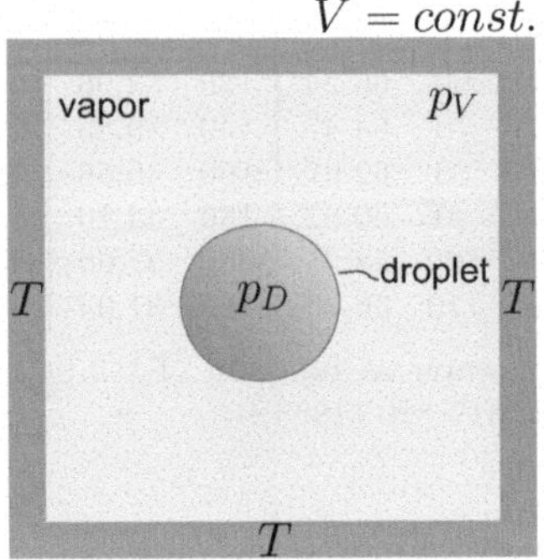

Fig. 7.9 A single droplet in equilibrium with its vapor. The system is thermalized at temperature T and its volume V is fixed.

$$\frac{d}{dt}\left(U+E_s\right)=\dot{Q}-\dot{W} \quad , \quad \frac{dS}{dt}-\frac{\dot{Q}}{T}=\dot{S}_{gen}\geq 0 \ , \tag{7.75}$$

where $U = U_V + U_D$ and $S = S_V + S_D$ are total internal energy and entropy within the system boundaries, and T is the constant temperature at the boundary. Note that we have to account for the surface energy E_S explicitly. Also, since the system boundary is rigid, there is no work, $\dot{W} = p\frac{dV}{dt} = 0$. Eliminating heat, the two equations combine to

$$\frac{d}{dt}\left(U+E_s-TS\right)=-T\dot{S}_{gen}\leq 0 \ . \tag{7.76}$$

Since the right hand side of this equation is negative or zero, the equilibrium condition for this system is [TEC, Chap. 17]

$$\Phi = U - TS + E_s = F + E_s \rightarrow \min \tag{7.77}$$

where $F = F_V + F_D$ is the total Helmholtz free energy of vapor and droplet.

As variables of the system we chose droplet mass m_D and volume V_D. Hence, we have to minimize

$$\Phi=\left(m-m_D\right)f_V\left(T,\mathsf{v}_V=\frac{V-V_D}{m-m_D}\right)+m_D f_D\left(T,\mathsf{v}_D=\frac{V_D}{m_D}\right)+E_s\left(V_D\right) \ , \tag{7.78}$$

where f_V, f_D are the specific free energies as functions of the specific volumes of vapor, v_V, and droplet, v_D, respectively.[10]

Equilibrium derivatives of Φ with respect to m_D and V_D must vanish. Evaluation requires product and chain rule and use of the relation $p = -\left(\frac{\partial f}{\partial \mathsf{v}}\right)_T$

[10] This implies that the liquid phase is compressible. Incompressibility will be used only further below.

which follows from the Gibbs equation. Specifically one finds

$$0 = \frac{\partial \Phi}{\partial m_D} = -f_V - p_V \mathrm{v}_V + f_D + p_D \mathrm{v}_D = -g_V + g_D \tag{7.79}$$

$$0 = \frac{\partial \Phi}{\partial V_D} = p_V - p_D + \frac{2\sigma}{r} \tag{7.80}$$

Thus, when the vapor is at pressure p_V, due to surface tension the droplet pressure is elevated to

$$p_D = p_V + \frac{2\sigma}{r} \,. \tag{7.81}$$

The equilibrium vapor pressure p_V results from evaluation of the above requirement of equal Gibbs free energies, which explicitly reads

$$g_D\left(T, p_D\right) = g_V\left(T, p_V\right) \,. \tag{7.82}$$

Note that droplet and vapor are at the same temperature, but, due to surface tension, at different pressures.

Since in good approximation the liquid is incompressible, we can write[11]

$$\begin{aligned} g_D\left(T, p_D\right) &= g_D\left(T, p_{\mathrm{sat}}\right) + \frac{1}{\rho_D}\left(p_D - p_{\mathrm{sat}}\right) \\ &= g_D\left(T, p_{\mathrm{sat}}\right) + \frac{1}{\rho_D}\left(p_V + \frac{2\sigma}{r} - p_{\mathrm{sat}}\right) , \end{aligned} \tag{7.83}$$

where $p_{\mathrm{sat}} = p_{\mathrm{sat}}(T)$ is the saturation pressure at the given temperature, that is the equilibrium pressure for a flat interface.

Considering the vapor as an ideal gas, its Gibbs free energy is

$$g_V\left(T, p_V\right) = g_V\left(T, p_{\mathrm{sat}}\right) + RT \ln \frac{p_V}{p_{\mathrm{sat}}} \tag{7.84}$$

Thus, the equilibrium condition becomes

$$g_D\left(T, p_{\mathrm{sat}}\right) + \frac{1}{\rho_D}\left(p_V + \frac{2\sigma}{r} - p_{\mathrm{sat}}\right) = g_V\left(T, p_{\mathrm{sat}}\right) + RT \ln \frac{p_V}{p_{\mathrm{sat}}} \tag{7.85}$$

where due to the definition of saturation pressure $g_D\left(T, p_{\mathrm{sat}}\right) = g_V\left(T, p_{\mathrm{sat}}\right)$. The resulting equation for the equilibrium vapor pressure p_V reads

$$\ln \frac{p_V}{p_{\mathrm{sat}}} - \frac{p_V - p_{\mathrm{sat}}}{\rho_D RT} = \frac{2\sigma}{\rho_D RT} \frac{1}{r} \tag{7.86}$$

---

[11] ... another Taylor series, with $\left(\frac{\partial g_L}{\partial p}\right)_T = \frac{1}{\rho_L}$:
$g_L\left(T, p_D\right) = g_L\left(T, p_{\mathrm{sat}} + \left(p_D - p_{\mathrm{sat}}\right)\right) = g_L\left(T, p_{\mathrm{sat}}\right) + \left(\frac{\partial g_L}{\partial p}\right)_T \left(p_D - p_{\mathrm{sat}}\right)$

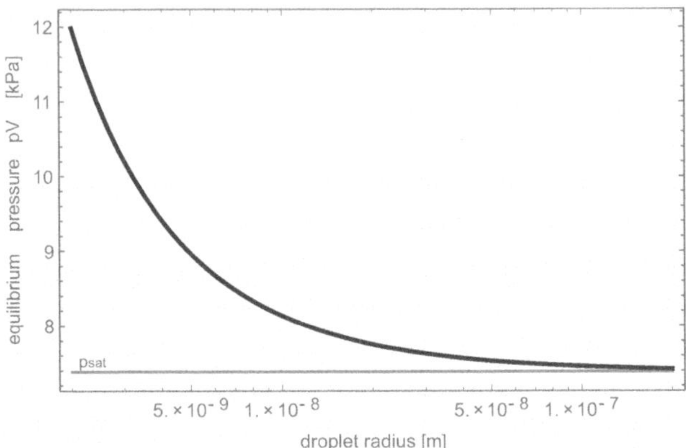

Fig. 7.10 Equilibrium vapor pressure p_V over droplet radius r (logarithmic scale) for water at 313 K ($p_{\rm sat} = 7.384\,{\rm kPa}$, $\rho_L = 1000\frac{\rm kg}{\rm m^3}$, $\sigma = 0.07\,{\rm Pa\,m}$). A water molecule has a size of $\sim 2.75 \times 10^{-10}\,{\rm m}$, hence the figure considers droplets with as little as 400 molecules onward.

Since the mass density of the liquid is significantly larger than the density of the vapor, we have $\frac{p_V - p_{\rm sat}}{\rho_D RT} \ll 1$, and can ignore this contribution. Solving for the equilibrium pressure in the vapor we find

$$p_V = p_{\rm sat} \exp\left[\frac{2\sigma}{\rho_D RT}\frac{1}{r}\right] . \tag{7.87}$$

Figure 7.10 shows the resulting equilibrium pressure over droplet radius for water at 40 °C. Already for droplets with radius of $10^{-7}\,{\rm m}$, the deviation from saturation pressure is negligible. Effectively, the large vapor pressure is required to keep the droplet at size.

7.5.3 Formation of Droplets

The derivation above ignores that the function Φ that is minimized not only exhibits the local minimum $r_{\rm min}$ that results in the above equilibrium curve, but as well a boundary minimum for vanishing radius $r = 0$ and a local maximum at smaller radius $r_{\rm max} < r_{\rm min}$. Details of the calculation are left to the reader as formulated in Prob. 7.16. Figure 7.11 shows, for the data of Prob. 7.16, the rescaled function $\hat{\Phi}(T, r)$ in different intervals so that the boundary minimum, the maximum, and the bulk minimum are visible.

Droplet growth, such as formation of droplets in a supersaturated vapor at $p_V > p_{\rm sat}(T)$, is hampered by the local maximum. If a droplet forms, say as a result of a density fluctuation, with a radius below $r_{\rm max}$, this droplet will

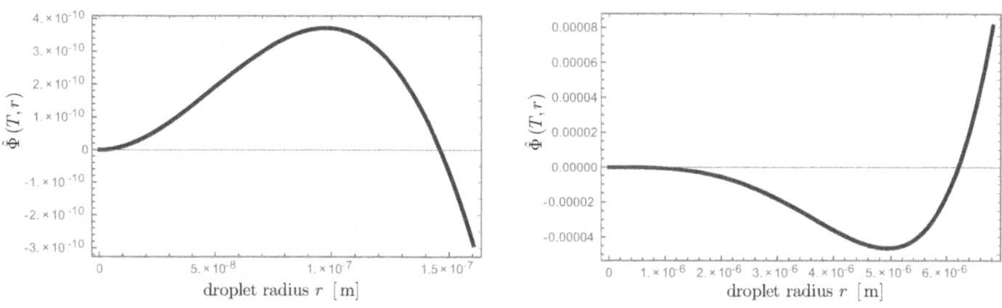

Fig. 7.11 Minima of the function $\hat{\Phi}(T, r)$ determine the stable states of the system of droplet and vapor shown in Fig. 7.10. When plotting the curve, it is easy to miss the maximum at $r_{\min} = 10^{-7}$ m as seen on the left, wheras in the plot on the right only the minimum at $r_{\max} = 5 \times 10^{-6}$ m is recognized. See Prob. 7.16 for details and data.

shrink to approach the boundary minimum at $r = 0$, that is the droplet will vanish. Only if the initial droplet radius is above $r_{\max}$ it can be expected to grow towards the bulk minimum value, that is to size $r_{\min}$. The value $r_{\max}$ can be considered as a critical radius for nucleation, in Prob. 7.16 it is found as

$$r_{\max} = \frac{2\sigma}{\rho_D RT \ln \frac{p_V^0}{p_{\text{sat}}}} . \tag{7.88}$$

where $p_V^0 = \frac{mRT}{V} > p_{\text{sat}}(T)$ is the pressure in the supersaturated vapor before a droplet is formed.

Since condensation releases the heat of evaporation, there must be a mechanism in place to remove heat from the growing droplet. This is difficult in the bulk vapor, where heat conduction in the bulk is rather weak, and much easier on a surface which can quickly disperse the heat of condensation.

There is a particle of dust in every rain droplet which helps to overcome the critical radius, and condensation is more likely to occur on suitable surfaces—such as a window pane—than in the bulk. In careful experiments, no droplets form in vapors even at three times the saturation pressure.

Problems

7.1. Slip Length

Measurement of a plane Poiseuille flow between two resting parallel plates in distance $2L$ gives the velocity field between the plates as

$$v(x) = v_0 \left[1 - a \left(\frac{x}{L} \right)^2 \right] ,$$

1. Determine the slip length for the (identical) plates at $x_W = \pm L$.

2. What is the range of acceptable values for the constant a?

7.2. Temperature Curves and Jump

A fluid is enclosed between walls at $x_1 = -0.1\,\text{mm}$ and $x_2 = 0.2\,\text{mm}$. In a particular steady state process, the temperature in the fluid is found as $T(x) = ax^2 + cx^4$, with $T(x_2) = 20\,°\text{C}$ and $T(x_1) = 30\,°\text{C}$

1. Determine the constants a and c.
2. The wall temperature at x_2 is $T_{W,2} = 25\,°\text{C}$. Determine the jump length.

Show all units, and present all calculations clearly, with comments!!

7.3. Heat Flux with Temperature Jump

Assume constant heat conductivity and jump length of a fluid, so that heat flux and jump condition read

$$q_i = -\kappa \frac{\partial T}{\partial x_i} \quad , \quad n_i \frac{\partial T}{\partial x_i} = \frac{T - T_W}{l_T}$$

Consider the fluid enclosed between two walls in distance L at temperatures $T_W = T_0 \pm \Delta T$ in a one-dimensional setting at steady state. Determine the equation for fluid temperature $T(x)$, and plot for a variety of values of l_T/L. Discuss the results.

7.4. Poiseuille Flow with Jump and Slip

Consider pressure driven Poiseuille flow in an ideal gas. The gas is confined between two parallel plates at rest, with temperatures T_0, the wall normals point into y-direction. Due to the pressure gradient $P = \frac{dp}{dx} = const.$, the gas moves with a velocity $\mathbf{v} = \{v(y), 0, 0\}$.

1. Simplify the balance equations, together with the Navier-Stokes and Fourier laws, and the jump and slip boundary conditions at the plates, for this one-dimensional flow. Assume that viscosity η and heat conductivity κ are constants.
2. Introduce dimensionless coordinates based on the plate distance L, the plate temperature T_0, reference pressure $p_0 = \rho_0 R T_0$. In the dimensionless equations, introduce the dimensionless parameters Knudsen number $\text{Kn} = \frac{\lambda_{\text{mfp}}}{L} = \frac{\eta\sqrt{RT_0}}{p_0 L}$ and Prandtl number $\text{Pr} = \frac{5}{2}\frac{\eta R}{\kappa}$, and set slip and jump length to $l_S = \alpha_S \lambda_{\text{mfp}}$, $l_T = \alpha_T \lambda_{\text{mfp}}$, for plotting use $\alpha_S = \alpha_T = 1$; λ_{mfp} is the mean free path of the gas.
3. Solve the equations for velocity, temperature, shear stress, and heat flux, all as functions of the space variable y. Plot the results for a variety of Knudsen numbers between 0.001 and 0.1. Use the measured value of the Prandtl number, $\text{Pr} = 2/3$.

7.5. Non-Linear Heat Transfer with Temperature Jump

In a simplified dimensionless formulation, kinetic theory of gases expresses the Fourier heat flux as

$$q_i = -\frac{15}{4}\mathrm{Kn}\theta^{\varpi}\frac{\partial\theta}{\partial x_i}$$

where $\theta = T/T_0$ is dimensionless temperature, $\mathrm{Kn} = \frac{\lambda_{\mathrm{mfp}}}{L}$ denotes the Knudsen number in the reference state, that is the ratio between mean free path λ_{mfp} and macroscopic length L, and the exponent ϖ is a constant that for most gases is between 0.5 and 1.

Moreover, kinetic theory yields the corresponding temperature jump condition in dimensionless form as

$$\theta - \theta_W = \frac{2-\vartheta}{2\vartheta}\frac{15}{4}\sqrt{\frac{\pi}{2}}\mathrm{Kn}\theta^{\varpi+\frac{1}{2}}\frac{\partial\theta}{\partial x_i}n_i$$

where $\vartheta \in [0,1]$ is the accommodation coefficient, which vanishes for a specularly reflecting surface, and is unity for an ideal rough surface

An ideal gas is confined between two rigid walls of temperatures $\theta_{W_1}, \theta_{W_2}$, the dimensionless distance of the walls is $L = 1$, and the gas is at rest.

1. Simplify the balances for mass, momentum, energy for this problem. Show that the pressure is constant.
2. Solve the problem for $\theta_{W_1} = 1, \theta_{W_2} = 1.01,\ 1.1,\ 1.5$ and Knudsen numbers $\mathrm{Kn} = 0.001,\ 0.01,\ 0.1$ for $\varpi = 1,\ \vartheta = 1$ and other values
   Hint: solve first symbolically for the bulk, with the gas temperatures at the walls denoted as θ_0, θ_L. Then use a computer to find the values for θ_0, θ_L from the jump conditions for $\theta_0 - \theta_{W_1}$, $\theta_L - \theta_{W_2}$.
3. Study and discuss the influence of the accommodation coefficient by varying its value.

7.6. Approximate Equation for Saturation Pressure

In order to find an approximate equation for the saturation pressure of a substance, assume that the liquid can be considered as an incompressible liquid, and the vapor as an ideal gas.

1. For the liquid (L) show first that incompressibility implies that the specific heats at constant pressure and constant volume are the same. Next, assume constant specific heat c_L and incompressibility and find internal energy, enthalpy and entropy by integration. Show that

$$u_L = c_L(T - T_0) \quad , \quad h_L = c_L(T - T_0) \quad , \quad s_L = c_L \ln\frac{T}{T_0} .$$

   The assumption of incompressibility is not sufficient to obtain the relation for h_L. What contribution is missing, and why (or when) can it be ignored?
2. Consider vapor (V) as an ideal gas with constant specific heats, and show that specific enthalpy and entropy are given by

$$h_V = c_p(T - T_0) + h_{LV}(T_0) \quad , \quad s_V = c_p \ln\frac{T}{T_0} - R\ln\frac{p}{p_{\mathrm{sat}}(T_0)} + \frac{h_{LV}(T_0)}{T_0} ;$$

c_p is the specific heat of the vapor, $h_{LV}(T_0)$ is the specific heat of evaporation at reference temperature T_0, and $p_{\text{sat}}(T_0)$ is the saturation pressure at T_0. Discuss the choice of integrating constants and give clear arguments why $h_{LV}(T_0)$ appears in both relations.
3. Determine the heat of evaporation $h_{LV}(T)$.
4. Use the condition for phase equilibrium $g_L(T, p_{\text{sat}}(T)) = g_V(T, p_{\text{sat}}(T))$ to find an equation for the saturation pressure $p_{\text{sat}}(T)$.
5. Use data for water and chose $T_0 = 273.15\,\text{K}$ to make a table with values of saturation pressure and heat of evaporation for several temperatures. Compare to tabled data: When an error of 5% is acceptable, what is the maximum temperature for which the approximation can be used?
6. An equation that is regularly used, and gives a better fit, is the Antoine equation

$$\log p_{\text{sat}}(T) = A - \frac{B}{C+T}\,.$$

Use data for water at 0.01 °C, 50 °C, and 100 °C to obtain the constants A, B, C. You may use a computer program to find the constants. Plot the saturation pressure as function of T, and make a list of values for temperatures up to the critical temperature 374.14 °C. Compare with tabulated data.

7.7. Driving Forces in Evaporation

We have seen that the driving forces in nonequilibrium evaporation are

$$\mathcal{F}_A = \left\{\frac{g_L}{T_L} - \frac{g_V}{T_V} + h_V\left(\frac{1}{T_V} - \frac{1}{T_l}\right), \frac{1}{T_V} - \frac{1}{T_L}\right\}\,.$$

The second force, $\frac{1}{T_V} - \frac{1}{T_L}$, simply is the (non-linear) difference between the temperatures of liquid and vapor, but the first force is, on the first glance, not easy to interpret. In order to understand it better, assume that all differences from equilibrium are small. In particular introduce the temperature difference between vapor and liquid at the interface, $\Delta T = T_V - T_L$, and the difference between actual and saturation pressure at the interface, $\Delta p = p_{\text{sat}}(T_L) - p$. Expand the forces X_A to first order in ΔT and Δp. Here $p = p_L = p_V$ is the prescribed pressure (why is there no difference between p_L and p_V?). Discuss the result.

Hints: Note that in the above, of course, $g_L = g_L(T_L, p)$, $g_V = g_V(T_V, p)$. The saturation pressure is defined through $g_L(T, p_{\text{sat}}(T)) = g_V(T, p_{\text{sat}}(T))$. Use the Gibbs equation to relate derivatives of g to other properties.

7.8. Half Mass Flux

The velocity distribution function f of an monatomic ideal gas is defined such that $fd\mathbf{C}$ gives the local mass density of particles with velocities $\mathbf{C}$ in the interval $d\mathbf{C} = dC_x dC_y dC_z$. For a gas in equilibrium with mass density ρ and temperature T, microscopic velocities follow the Maxwellian distribution $(C^2 = C_x^2 + C_y^2 + C_z^2)$

$$f_M = \frac{\rho}{\sqrt{2\pi RT}^3} \exp\left[-\frac{C^2}{2RT}\right] .$$

1. Show that the integral of f_M over the full velocity space yields the mass density,
$$\rho = \int_{-\infty}^{\infty}\int_{-\infty}^{\infty}\int_{-\infty}^{\infty} f_M d\mathbf{C} .$$
2. For an ideal gas, the internal energy is the kinetic energy of its particles. Show that
$$\rho u = \int_{-\infty}^{\infty}\int_{-\infty}^{\infty}\int_{-\infty}^{\infty} \frac{C^2}{2} f_M d\mathbf{C} = \frac{3}{2}\rho RT .$$
3. Show that the full mass flux vanishes
$$J_i = \int_{-\infty}^{\infty}\int_{-\infty}^{\infty}\int_{-\infty}^{\infty} C_i f_M d\mathbf{C} = 0$$
and the half mass flux in x-direction is
$$J_{x>0} = \int_{0}^{\infty}\int_{-\infty}^{\infty}\int_{-\infty}^{\infty} C_x f_M d\mathbf{C} = \rho\sqrt{\frac{RT}{2\pi}} .$$

7.9. Hertz-Knudsen-Schrage Model

Use Taylor series for small deviation from saturation state $p_{\text{sat}}(T_L)$ to determine the matrix of dimensionless resistivities $\hat{r}_{AB}$ for the simplified Hertz-Knudsen-Schrage model (7.48, 7.49).

7.10. Evaporation of Water

For the following problem, we consider liquid water as an incompressible ideal fluid ($\rho_L = 1000\,\text{kg/m}^3$), with enthalpy and entropy given as

$$h_L = c_L (T - T_0) \quad , \quad s_L = c_L \ln \frac{T}{T_0}$$

where $c_L = 4.18\,\text{kJ/kg K}$ is the specific heat, and $T_0 = 273.15\,\text{K}$ is a reference temperature; the heat conductivity is $\kappa_L = 55 \times 10^{-2} \frac{\text{J}}{\text{m s K}}$.

The water vapor is considered as an ideal gas, with

$$\begin{aligned} p_V &= \rho_V RT , \\ h_V &= c_p (T - T_0) + h_{LV}(T_0) , \\ s_V &= c_p \ln \frac{T}{T_0} - R \ln \frac{p}{p_{\text{sat}}(T_0)} + \frac{h_{LV}(T_0)}{T_0} \end{aligned}$$

where $R = 0.462\,\text{kJ/kg K}$, $c_p = 1.85\,\text{kJ/kg K}$ and $h_{LV}(T_0) = 2500\,\text{kJ/kg}$ is the heat of evaporation at T_0; the heat conductivity is $\kappa_V = 1.4 \times 10^{-2}\,\text{J/m s K}$.

The problem is as follows: A vessel contains liquid water at the bottom and vapor above. The temperature at bottom and top of the container is $T_W = 298\,\mathrm{K}$ and remains constant, the container height is 15 cm, and the liquid layer has a thickness of 5 cm. A mass flux J of vapor is pumped out at the top of the container, and the same mass flow of liquid water at T_W is pumped in at the bottom to keep the liquid surface at constant level. In this process, liquid is forced to evaporate at the surface, and—since evaporation needs energy—the temperature at the surface will lie below T_W. The goal is to determine temperature and pressure as a function of height z in the container.

For the computation assume: i.) stationary process, i.e., all time derivatives vanish, ii.) one-dimensional process, i.e., all variables depend only on z—chose z such that $z = 0$ at the interface between liquid and vapor, iii.) the gravitational force can be ignored, iv.) the mass flow, and thus the flow velocity is so small that v^2 can be ignored in the momentum and energy balances, v.) the heat conductivities can be considered to be constant.

1. Make a sketch of the problem, and write down the balance laws for mass momentum and energy for the bulk vapor and liquid, and at the interface.
2. Show that the total energy flux

$$Q_0 = \rho v h + q$$

is a constant, i.e., does not depend on z. Then show that for both, vapor and liquid, the energy balance can be written as

$$T - a_\alpha \frac{dT}{dz} = Q_\alpha \,.$$

Identify the constants a_V, a_L and Q_V, Q_L for vapor and liquid, respectively.
3. Solve the above differential equation for T and show that

$$T_L = Q_L + (T_W - Q_L) \exp\left[\frac{z + L_L}{a_L}\right]$$

$$T_V = Q_V + (T_W - Q_V) \exp\left[\frac{z - L_V}{a_V}\right]$$

where L_L, L_V are the thickness of liquid and vapor layer. Remark: this solution guarantees the proper temperature at bottom and top, but no boundary condition for the interface was used so far.
4. Assume that there is no temperature jump at the interface, $T_L\,(z = 0) = T_V\,(z = 0)$ to find the final result for the temperature curve. Plot the curve for a variety of values for mass flux J. What mass flux is required to have an interface temperature of 273.15 K?

5. Use the interface condition for entropy to show that $h_V - Ts_V = h_L - Ts_L$ at a phase interface where no entropy is produced. Use this condition and the above relations to determine the saturation pressure at the interface.
6. Inverse the mass flow, so that vapor is pumped in, and liquid removed. Plot the temperature curve.

7.11. T-Jump in Evaporation

The evaporation conditions at a liquid-vapor interface were found as

$$\begin{bmatrix} -\frac{\Delta p}{\sqrt{2\pi R T_L}} \\ \\ -\frac{p_{sat}(T_L)}{\sqrt{2\pi R T_L}} \frac{\Delta T}{T_L} \end{bmatrix} = \begin{bmatrix} 0.625 & 0.125 \\ \\ 0.125 & 0.294 \end{bmatrix} \begin{bmatrix} J \\ \\ \frac{q_V}{R T_L} \end{bmatrix}$$

with $\Delta p = p - p_{\text{sat}}(T_L)$ and $\Delta T = T_V - T_L$.

Consider the case of evaporating water with pressure maintained at $p = 0.8718\,\text{kPa}$. The measured surface temperature is $T_L = 278.15\,\text{K}$, and the temperature jump is measured as $\Delta T = 0.5\,\text{K}$. Heat conductivities of liquid and vapor are $\kappa_L = 55 \times 10^{-2} \frac{\text{J}}{\text{m}\,\text{s}\,\text{K}}$, $\kappa_V = 1.4 \times 10^{-2} \frac{\text{J}}{\text{m}\,\text{s}\,\text{K}}$.

1. Determine the mass flux J and the heat flux in the vapor q_V.
2. Determine the corresponding temperature gradient in the vapor.
3. If this is a simple 1D setting, and the vapor layer is 5 cm thick, to what temperature would the upper boundary have to be heated?
4. Discuss the data and the results.

Show all units, and present all calculations clearly, with comments!!

7.12. Vapor between Liquid Layers

Consider the set-up discussed in Sec. 7.3.7, where a vapor is confined between two liquid layers.

1. Follow through the computations for no-jump interfaces as presented, and solve for interface temperature T_I, mass flux J and system pressure p for a number of outer boundary temperatures. Plot the temperature curves. Why is the interfce temperature not just the average of the boundary temperatures?
2. Consider jump interface conditions and develop the (partly numerical) solution for these. Plot temperature curves. Vary the values of resistivities and identify values for which the inverted temperature profile occurs (a criterion is given in Struchtrup, Kjelstrup, Bedeaux, Phys. Rev. E **85**, 061201 (2012)).

7.13. Alternative Evaporation Forces

Show that in the limit of small deviations, by expansion with respect to the vapor state, the forces $\bar{\mathcal{F}}_A$ (7.61) reduce to

$$\begin{bmatrix} \rho_V\sqrt{\frac{RT_L}{2\pi}}\left[\frac{g_L}{RT_L}-\frac{g_V}{RT_V}+h_L\left(\frac{1}{RT_V}-\frac{1}{RT_L}\right)\right] \\ \rho_V\sqrt{\frac{RT_L}{2\pi}}T_L\left[\frac{1}{T_V}-\frac{1}{T_L}\right] \end{bmatrix} \simeq \begin{bmatrix} -\left[\left(1-\frac{\rho_V^{\rm sat}(T_V)}{\rho_L^{\rm sat}(T_V)}\right)\right]\frac{p-p_{\rm sat}(T_V)}{\sqrt{2\pi RT_V}} \\ -\rho_V^{\rm sat}(T_V)\sqrt{\frac{RT_V}{2\pi}}\frac{\Delta T}{T_V} \end{bmatrix}$$

Simplify for an ideal gas, where $p_{\rm sat}(T)=\rho_V^{\rm sat}(T)\,RT$.

7.14. Transformation of Resistivities

We have discussed the dimensionless resistivities for the fluxes $\mathcal{J}_A=\left\{J,\frac{q_V}{RT_L}\right\}$ and $\bar{\mathcal{J}}_A=\left\{J,\frac{q_L}{RT_L}\right\}$. As another alternative, consider the fluxes $\tilde{\mathcal{J}}_A=\left\{J,\frac{Q}{RT_L}\right\}$ where $Q=Jh_L+q_L=Jh_V+q_V$ is the overall energy flux through the interface.

1. Identify the corresponding forces $\tilde{\mathcal{F}}_A$.
2. Express the corresponding dimensionless resistivities $\tilde{r}_{AB}$ in terms of the resistivities $\hat{r}_{AB}$ and the dimensionless vapor enthalpy $\frac{h_V}{RT_L}$.
3. Discuss the transformation of the resistivities with regards of the influence of the vapor enthalpy. In your discussion note that $\frac{h_V}{RT_L}$ depends on the chosen enthalpy scale and reference enthalpy, and explore the implications.

7.15. Droplet Pressure

Using the same data as in Fig. 7.10 compare the pressure curves resulting from (7.87) and (7.86) to confirm that the omitted term has no influence, indeed. Equation (7.86) cannot be solved for $p_V(r)$; instead, plot $r(p_V)$ for both equations.

7.16. Droplet in Vapor

Consider the equilibrium between a single droplet and vapor in a system of fixed volume V, mass m and temperature T as in Fig. 7.9 under the assumption that the vapor behaves as an ideal gas and the liquid is incompressible.

1. Introduce the Gibbs free energies and show that the function to be minimized can be written as

$$\Phi=(m-m_D)(g_V-RT)+m_D\left(g_D-\frac{p_D}{\rho_D}\right)+E_s(V_D)$$

2. Show that the Gibbs free energies of (ideal gas) vapor and (incompressible) droplet can be expressed as

$$g_V(T,p_V)=g_V(T,p_{\rm sat})+RT\ln\frac{p_V}{p_{\rm sat}}\quad,\quad g_D(T,p_{\rm sat})+\frac{p_D-p_{\rm sat}}{\rho_D}$$

where $p_{\rm sat}(T)$ solves $g_V(T,p_{\rm sat})=g_D(T,p_{\rm sat})$ and $p_V=\frac{m_VRT}{V_V}$.

3. Due to incompressibility droplet mass is related to droplet volume by $m_D = \rho_D V_D = \rho_D \frac{4\pi}{3} r^3$, where r is the radius of the droplet. Argue that Φ is minimum when the function $\hat{\Phi}$ is a minimum, where

$$\hat{\Phi} = 1 + \frac{\Phi}{mRT} - \frac{g_V(T, p_{\text{sat}})}{RT} - \ln \frac{mRT}{V p_{\text{sat}}}$$

Show that

$$\hat{\Phi} = \frac{\rho_D V_D}{m}\left(1 - \ln \frac{mRT}{V p_{\text{sat}}}\right) + \left(1 - \frac{\rho_D V_D}{m}\right) \ln \frac{1 - \frac{\rho_D V_D}{m}}{1 - \frac{V_D}{V}} - \frac{p_{\text{sat}}(T)}{\rho_D RT} \frac{\rho_D V_D}{m} + \frac{4\pi\sigma}{mRT}\left(\frac{V_D}{4\pi/3}\right)^{2/3}$$

Insert $V_D = \frac{4\pi}{3} r^3$and identify the function $\hat{\Phi}(T, r)$.

4. Use data for water as

$$T = 40\,°\text{C}\ ,\quad V = 10^{-9}\,\text{m}^3\ ,\quad m = 1.01\frac{p_{\text{sat}}(T)\,V}{RT}\ ,\quad R = 0.462\frac{\text{kJ}}{\text{kg K}}$$

$$p_{\text{sat}}(T) = 7.384\,\text{kPa}\ ,\quad \sigma = 69.6 \times 10^{-6}\frac{\text{kN}}{\text{m}}\ ,\quad \rho_D = 992\frac{\text{kg}}{\text{m}^3}$$

and plot $\hat{\Phi}(r)$ as function of radius. Carefully chose intervals for plotting, the curve has a maximum and a minimum which both should be seen.

5. The condition for extremum of $\hat{\Phi}$ is $\frac{d\hat{\Phi}}{dr} = 0$. Plot $\frac{d\hat{\Phi}}{dr}$ in the intervals identified above, and determine numerical values for the radius at minimum and the maximum. For the droplet in stable equilibrium determine vapor mass and volume. Compute pressure p_V from the ideal gas law and compare to the result of Eq. (7.87). Determine relative mass of the droplet, $\frac{m_D}{m}$.

6. A good approximation for the critical radius is

$$r_{\max} = \frac{2\sigma}{\rho_D RT \ln \frac{p_V^0}{p_{\text{sat}}}}$$

where $p_V^0 = \frac{mRT}{V}$ is the pressure in the supersaturated vapor before a droplet is formed. Compare this value to your result from above. To find this approximation from the full equation consider $\hat{\Phi}$ for small volumes: linearize $\hat{\Phi}$ to first order (Taylor series in $\frac{\rho_D V_D}{m} \ll 1$, $\frac{V_D}{V} \ll 1$), then find the maximum.

Chapter 8
Shocks and Flames

Shocks and flames are irreversible changes of flow properties in inert or reacting gas flows over short distance, which are described as singular surfaces. In this chapter, after reviewing sub- and supersonic nozzle flows, we study both phenomena for the accessible case of ideal gases with constant specific heats, and a simplified reaction model. Shock and flame solutions arise from the conservation of mass, momentum, and energy across the shock, with the second law determining which solutions are possible. Specifically, we study normal and oblique shocks as well as Chapman-Jouguet flames.

8.1 Introduction

Shocks are steep irreversible changes from supersonic to subsonic flow. Experiments at very low densities reveal that the shock thickness is of the order of the mean free path of molecules, which is confirmed in the Kinetic Theory of Gases. Within the shock structure the flow is in a strong nonequilibrium state, for which the conditions for local equilibrium breaks down. Hence, the Navier-Stokes-Fourier equations are not valid within the shock zone. While one can solve the NSF equations across the shock, the resulting smooth curves do not match experimental data, and more accurate kinetic theory models must be used to determine the conditions within a shock.

In classical continuum theory the mean free path, and therefore the shock structure, is not resolved, and shocks are described as singular surfaces. Shocks are a non-linear phenomenon, hence linearizations are not possible, and one must evaluate the full non-linear conservation laws across the singular shock surface.

To simplify the mathematical treatment, we limit the discussion to ideal gases with constant specific heats, which allows us to find analytical solutions for property changes across the shock. The extension to ideal gases with variable specific heats and vapors requires numerical solutions based

H. Struchtrup, *A Thermodynamic Introduction to Transport Phenomena*,
https://doi.org/10.1007/978-3-031-61868-0_8

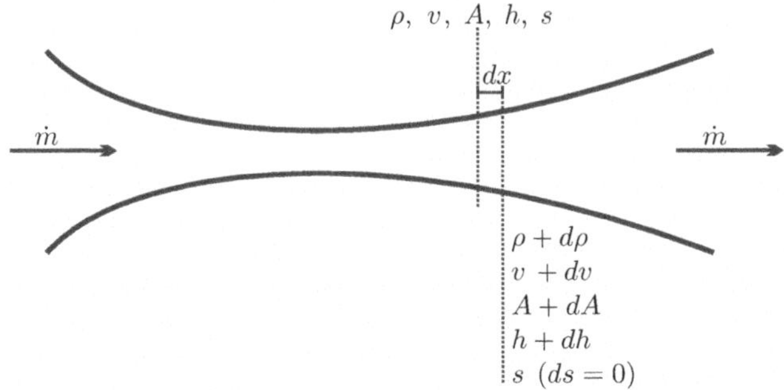

Fig. 8.1 Adiabatic reversible, i.e., isentropic, flow through a duct with changing cross section.

on computerized thermodynamic property tables. The overall behavior, however, will be as for the case of constant specific heats, and can be found in the specialized literature.

For context on sub- and supersonic flows we include the discussion of nozzle flows [TEC, Chap. 14]. Depending on nozzle geometry and the values of stagnation and back pressure, isentropic flow in a nozzle might not be possible, and normal or oblique shocks will be encountered. Supersonic aircraft induce shocks at their boundaries that travel with the aircraft. A sonic boom is heard when the shock—which exhibits a pressure jump—passes the observer.

8.2 Nozzle Flows

8.2.1 Area-Velocity Relation

The main design parameter for nozzles and diffusers is the variation of cross section $A(x)$ along their axis, and we ask how flow properties, in particular velocity and pressure, change with the cross section. Figure 8.1 shows an adiabatic and reversible, i.e., isentropic, flow through a duct with varying cross section. We consider a small slice of the duct of width dx, and apply the global balances of mass, energy and entropy.

The mass balance yields

$$\dot{m} = \rho v A = const. \tag{8.1}$$

and thus its logarithmic variation vanishes,

$$\frac{d\dot{m}}{\dot{m}} = \frac{dA}{A} + \frac{dv}{v} + \frac{d\rho}{\rho} = 0\,. \tag{8.2}$$

The first law gives

$$h + \frac{1}{2}v^2 = h_s = const. \quad , \quad dh + vdv = 0 \, , \tag{8.3}$$

where h_s is the stagnation enthalpy, defined as the enthalpy of the flow when brought to rest adiabatically.

Since the flow is isentropic, $ds = 0$, the Gibbs equation gives

$$Tds = dh - \frac{1}{\rho}dp = 0 \, . \tag{8.4}$$

Elimination of enthalpy between the last two equations yields the relation between pressure and velocity variations in isentropic nozzles and diffusers,

$$\frac{1}{\rho}dp = -vdv \, . \tag{8.5}$$

We use this to eliminate flow velocity v from the mass balance to find

$$\frac{dA}{A} = \frac{1}{v^2}\frac{1}{\rho}dp - \frac{d\rho}{\rho} = \frac{1}{\rho}dp\left(\frac{1}{v^2} - \frac{1}{a^2}\right) \, , \tag{8.6}$$

with the speed of sound $a = \sqrt{\left(\frac{\partial p}{\partial \rho}\right)_s}$.

To proceed, we introduce the Mach number

$$\mathrm{Ma} = \frac{v}{a} \, , \tag{8.7}$$

which compares flow velocity to speed of sound: $\mathrm{Ma} < 1$ for subsonic flows, $\mathrm{Ma} > 1$ for supersonic flows, and $\mathrm{Ma} = 1$ for sonic flows. Flows with $\mathrm{Ma} \gg 1$ are called hypersonic and flows with $\mathrm{Ma} \simeq 1$ are called transonic.

We thus can write the relation (8.6) as

$$\frac{dA}{A} = \frac{1}{\rho v^2}dp\left(1 - \mathrm{Ma}^2\right) \, , \tag{8.8}$$

or, by eliminating pressure by means of (8.5),

$$\frac{dA}{A} = -\frac{dv}{v}\left(1 - \mathrm{Ma}^2\right) \, . \tag{8.9}$$

Equations (8.8) and (8.9) are the area-pressure relation and the area-velocity relation for isentropic duct flows. Both relations carry the factor $\left(1 - \mathrm{Ma}^2\right)$ which has different sign for subsonic and supersonic flows. Accordingly, a change of cross section has different effect when applied to sub- and supersonic flows.

Subsonic flows (Ma<1): For a converging duct (8.8, 8.9) give

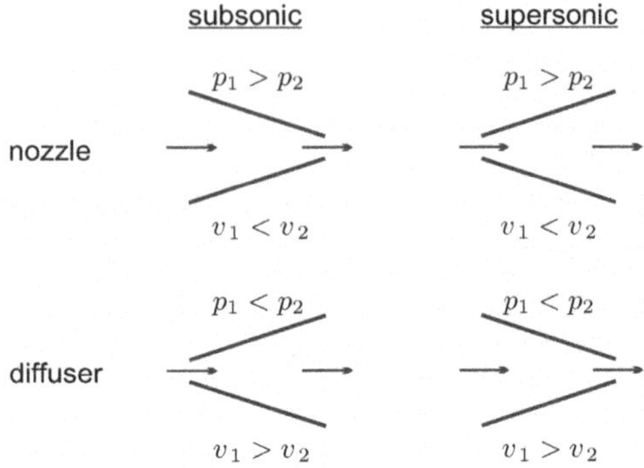

Fig. 8.2 Nozzles and diffuser cross section variation in sub- and supersonic flow.

$$dA < 0 \quad \Longrightarrow \quad dp < 0\,, \quad dv > 0\,; \tag{8.10}$$

the flow is accelerated while pressure drops; this is a nozzle.

For a diverging duct, on the other hand, we have the opposite signs,

$$dA > 0 \quad \Longrightarrow \quad dp > 0\,, \quad dv < 0\,; \tag{8.11}$$

pressure grows, and the flow decelerates; this is a diffuser.

Supersonic flows (Ma>1): For a converging duct (8.8, 8.9) give

$$dA < 0 \quad \Longrightarrow \quad dp > 0\,, \quad dv < 0\,; \tag{8.12}$$

pressure grows, and the flow decelerates; this is a diffuser.

For a diverging duct, on the other hand, we have the opposite signs,

$$dA > 0 \quad \Longrightarrow \quad dp < 0\,, \quad dv > 0\,; \tag{8.13}$$

the flow is accelerated while pressure drops; this is a nozzle.

In other words, a converging duct acts as a nozzle in subsonic flow, but as a diffuser in supersonic flow. A diverging duct acts as a diffuser in subsonic flow, but as a nozzle in supersonic flow. Figure 8.2 shows a summary.

8.2.2 Flow Function

Rocket motors and some jet engines expel supersonic flows for propulsion, and thus need appropriate nozzle geometries. To accelerate a subsonic flow to supersonic speed requires a converging-diverging nozzle, where the flow is accelerated to sonic speed in the converging part, and then to supersonic speed in the diverging part. After its inventor Gustaf de Laval (1845-1913), such a nozzle is called Laval nozzle.

Here we study isentropic flows, and, so that we can perform analytical calculations, we restrict the treatment to ideal gases with constant specific heats. We shall discuss flows through purely converging nozzles, and through converging-diverging nozzles. In both cases, the balances of mass, energy and entropy reduce to

$$\dot{m} = \rho v A = const. \,, \tag{8.14}$$

$$h + \frac{1}{2}v^2 = h_s = const. \,, \tag{8.15}$$

$$\frac{T}{p^{\frac{\gamma-1}{\gamma}}} = \frac{T_s}{p_s^{\frac{\gamma-1}{\gamma}}} \,, \quad \frac{p}{\rho^\gamma} = \frac{p_s}{\rho_s^\gamma} \,, \tag{8.16}$$

where ρ, v, T, h are the properties at a given cross section of the nozzle, and T_s, p_s, h_s are stagnation properties. The stagnation state is defined as the hypothetical state that is reached by bringing the flow to rest isentropically.

With $h - h_s = c_p (T - T_s)$ and $c_p = \frac{\gamma}{\gamma-1}R$ we find from the above the local velocity as

$$v = \sqrt{2(h_s - h)} = \sqrt{\frac{2\gamma R T_s}{\gamma - 1}\left(1 - \frac{T}{T_0}\right)} = \sqrt{\frac{2\gamma R T_s}{\gamma - 1}}\sqrt{1 - \left(\frac{p}{p_s}\right)^{\frac{\gamma-1}{\gamma}}} \,. \tag{8.17}$$

With the isentropic relation for density, the mass flow through the nozzle is

$$\dot{m} = \rho v A = \rho_s \sqrt{\frac{2\gamma R T_s}{\gamma - 1}} A \left[\left(\frac{p}{p_s}\right)^{\frac{1}{\gamma}} \sqrt{1 - \left(\frac{p}{p_s}\right)^{\frac{\gamma-1}{\gamma}}}\right] = const. \tag{8.18}$$

The mass flow is a product of three factors: The constant $\rho_s\sqrt{\frac{2\gamma RT_s}{\gamma-1}}$ which is fixed by the stagnation state (ρ_s, T_s), the cross section A, and a function of the pressure ratio, the flow function $\psi\left(\frac{p}{p_s}\right)$, defined as

$$\psi\left(\frac{p}{p_s}\right) = \left(\frac{p}{p_s}\right)^{\frac{1}{\gamma}} \sqrt{1 - \left(\frac{p}{p_s}\right)^{\frac{\gamma-1}{\gamma}}} \,. \tag{8.19}$$

Figure 8.3 shows the flow function for $\gamma = 1.33, 1.4, 1.67$ as function of pressure ratio p/p_s.

The argument $\left(\frac{p}{p_s}\right)$ of the flow function assumes values between 0 and 1, and the curve exhibits a maximum with the critical values

$$\frac{p^*}{p_s} = \left(\frac{2}{\gamma+1}\right)^{\frac{\gamma}{\gamma-1}} \,, \quad \psi^* = \left(\frac{2}{\gamma+1}\right)^{\frac{1}{\gamma-1}} \sqrt{\frac{\gamma-1}{\gamma+1}} \,. \tag{8.20}$$

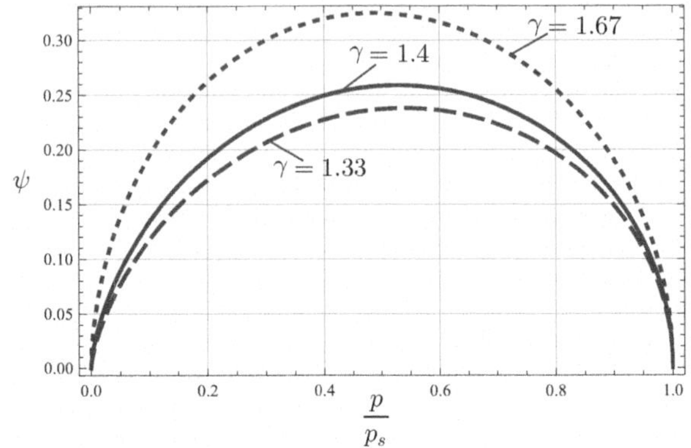

Fig. 8.3 Flow function $\psi\left(p/p_s\right)$ for $\gamma = 1.33, 1.4, 1.67$. The function has a maximum at $\frac{p^*}{p_s} = \left(\frac{2}{\gamma+1}\right)^{\frac{\gamma}{\gamma-1}} = 0.540, 0.528, 0.488$.

The condition of constant mass flow is equivalent to

$$A\,\psi\left(\frac{p}{p_s}\right) = const.\ , \tag{8.21}$$

which is central for the discussion of flows in converging and converging-diverging nozzles.

8.2.3 Converging Nozzle

We study the outflow from a large container into a nozzle, where the gas in the container is in the constant stagnation state (T_s, p_s). The flow is driven by the difference between the back pressure outside the nozzle, p_b, and the stagnation pressure p_s.

No flow occurs when $p_b = p_s$. As the back pressure p_b is lowered a bit, the flow develops. The cross section A is decreasing along the nozzle coordinate x, see Fig. 8.4. According to Fig. 8.3, the flow function is hill-shaped, and the nozzle feed state is on the right foot of that hill, at $\frac{p}{p_s} = 1$, $\psi = 0$. Since the product $A\psi\left(\frac{p}{p_s}\right)$ is constant, for decreasing cross section A the flow function must grow, i.e., go uphill. The flow function grows along x, and reaches its largest value in the smallest cross section A_{th}, the throat of the nozzle, which is the exit of the converging nozzle. As ψ grows along the nozzle coordinate, the pressure decreases, until it assumes the pressure $p_e = p_b$ in the exit. Further decrease of the back pressure leads to lower pressures along the nozzle, and larger values of the flow function.

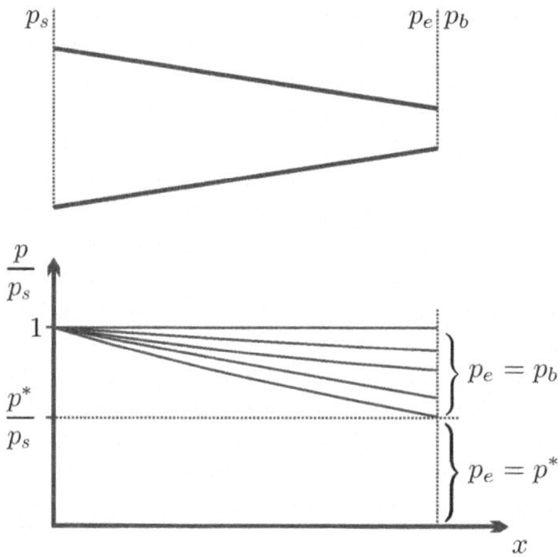

Fig. 8.4 Converging nozzle: Geometry and pressure profile.

When the back pressure p_b reaches the critical value p^*, the flow function in the exit is at its maximum ψ^*. No further growth of ψ is possible when the back pressure is lowered further. Thus, the exit state remains at $\psi_e = \psi^*$ and $p_e = p^*$, even when the back pressure is lowered below p^*. Figure 8.4 visualizes this behavior.

When the critical pressure p^* is reached in the nozzle throat, one speaks of choked flow. Indeed, under this condition, the mass flow obtains a maximum value,

$$\dot{m}^* = \rho_s \sqrt{\gamma R T_s} A_{th} \left(\frac{2}{\gamma + 1} \right)^{\frac{\gamma+1}{2\gamma-2}} , \tag{8.22}$$

where A_{th} is the cross section at the exit. No further increase of the mass flow through the nozzle is possible.

To understand this behavior, we determine temperature and velocity in the exit at choked conditions. From the adiabatic relation and (8.17, 8.20) we find

$$T^* = T_s \left(\frac{p^*}{p_s} \right)^{\frac{\gamma-1}{\gamma}} = \frac{2T_s}{\gamma + 1} \quad \text{and} \quad v^* = \sqrt{\frac{2\gamma R T_s}{\gamma + 1}} = \sqrt{\gamma R T^*} = a^* \, . \tag{8.23}$$

Thus, for choked flow, the exit speed is just the local speed of sound, a^*. We recall that the speed of sound is the velocity of a pressure disturbance. When the back pressure is lowered, the information on the pressure change travels with the speed of sound relative to the gas. Since the gas leaves with just the same speed, the information on pressure change is not transmitted into the nozzle, and no changes can occur inside, the exit velocity and the mass flow are limited.

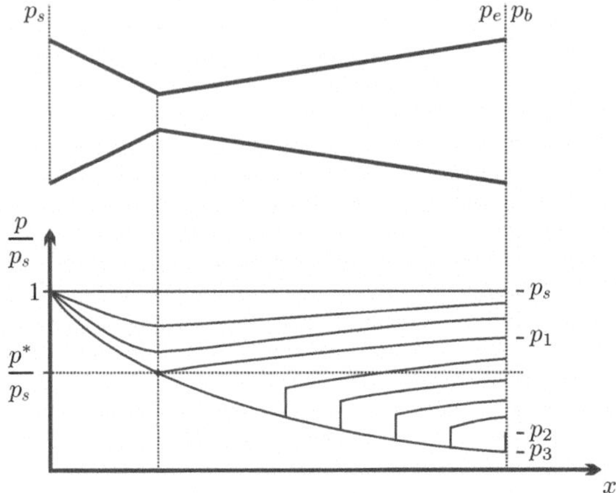

Fig. 8.5 Geometry and pressure profiles for a Laval nozzle. Pressure profiles depend on the back pressure as discussed in the text.

With $p_b < p^*$, a pressure discontinuity occurs at the nozzle exit, which contributes to thrust for the airplane or rocket. The exhaust expands outside the nozzle to the back pressure, and accelerates, and this expansion is somewhat irreversible, more so with bigger pressure differences. Most commercial airplanes have converging nozzles and subsonic outflow, to reduce noise.

8.2.4 Laval Nozzle

Exit velocities above the speed of sound are reached in converging-diverging nozzles. The cross section A is first decreasing until is reaches its smallest value A_{th} in the throat of the nozzle, and then it is increasing to the exit cross section A_e. Again we consider the pressure distribution in the nozzle which results from lowering the back pressure p_b, see Fig. 8.5.

When $p_b = p_s$, there is no flow and homogenous pressure p_s throughout the nozzle. When the back pressure is lowered somewhat, a flow occurs. According to Fig. 8.3, the flow function is hill-shaped, and the nozzle feed state is on the right foot of that hill, at $\frac{p}{p_s} = 1$, $\psi = 0$. Since $A\psi\left(\frac{p}{p_s}\right)$ is constant, the flow function must grow—go uphill—in the converging part of the nozzle, and it must decrease—go downhill—in the diverging part.

We first consider relatively large back pressures in the pressure range $p_1 < p_b < p_s$ in the figure. As the nozzle converges, the flow state climbs uphill until a value $p_{th} > p^*$, $\psi_{th} < \psi^*$ is reached in the throat. This point is to the right of the maximum. As the cross section grows in the diverging part, the flow function must decrease, and this is only possible by returning to higher pressures, that is by going back downhill towards the right. The flow is accelerated in the converging part of the nozzle, and decelerated in the diverging part, the outflow velocity is relatively low, and subsonic. The extreme case of this flow type is reached for $p_b = p_1$, when the air is in the critical state in the throat—on top of the hill with sonic speed—but then is decelerated again.

When the flow reaches the critical state in the throat, the flow function can decrease by going down the left side of the hill, towards lower pressures, and higher, i.e., supersonic, velocities. As indicated in the figure, this requires low back pressures $p_2 < p_b \leq 0$. In this range the flow is isentropic inside the nozzle. If the back pressure is just at p_3, the end pressure is equal to the back pressure, and no external irreversibilities occur. If the back pressure is in the range $p_2 < p_b < p_3$, the end pressure is below the back pressure and pressure is equilibrated through oblique shocks outside the nozzle. For lower back pressures $p_b < p_3$, the end pressure is above the back pressure and pressure is equilibrated through external expansion waves.

In the range $p_2 < p_b < p_1$, isentropic flow inside the nozzle is not possible. The flow will follow isentropic flow conditions for a while, accelerating to supersonic flow behind the throat, and then a irreversible normal shock will occur, that is a sudden jump from low pressure supersonic flow to high pressure subsonic flow. Behind the shock, the flow will be isentropic again, and the gas will expand to the back pressure.

In a shock, the flow changes from supersonic to subsonic, which means significant reduction of thrust. Nozzle geometry must be carefully designed so that the flow conditions are optimal. Some supersonic aircraft have nozzles with variable geometry, to adjust for the wide range of back pressures encountered between take-off and high altitude flight.

8.2.5 Rockets, Ramjet and Scramjet

Jet engines are air-breathing, that is they carry the fuel on board, and burn it with oxygen from the ambient air that passes through the engine. Rockets carry the oxygen on board, either as liquefied oxygen, or in the form of a compound. Thus, rockets are independent of ambient air, and can travel at extremely high altitude, and in space. In a rocket motor, fuel and oxidizer are burnt at high pressure in the combustion chamber and then expand through a Laval nozzle, which provides large supersonic exit velocities, and thus large thrust, see Fig. 8.6. The oxygen a rocket has to carry on board increases the take-off weight, and reduces the payload that can be carried along (Prob. 8.1).

Ramjet and scramjet are conceptually simple air-breathing engines for supersonic flight, hence they do not require to carry the heavy oxidizer on board. In both pressure build-up in the engine is affected only through diffusers, see Fig. 8.7 for sketches.

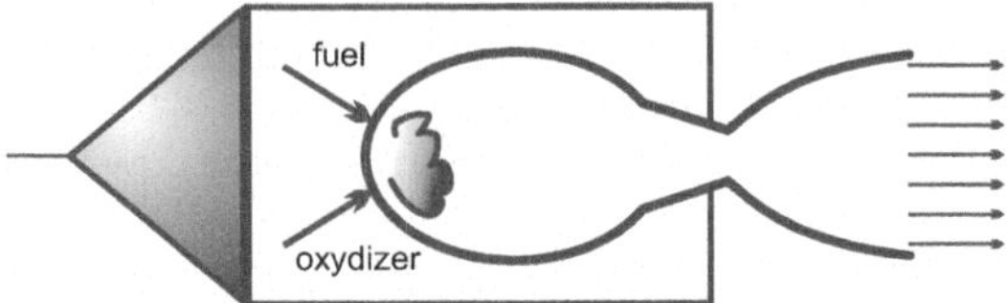

Fig. 8.6 A rocket.

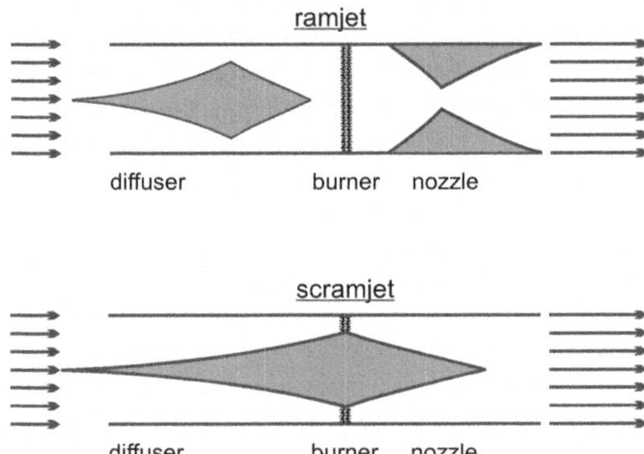

Fig. 8.7 Ramjet and scramjet.

In a ramjet, the incoming air is slowed down to subsonic speed and high pressure by means of a converging-diverging diffuser, which operates on the same principles as a Laval nozzle, only inversely. Fuel is injected into the compressed air and burned, the hot combustion product then expands through a Laval nozzle to high velocities. Ramjets can be used for supersonic flight with speeds up to $\mathrm{Ma} = 6$.

In a scramjet (supersonic combustion ramjet), the flow stays supersonic at all times. A supersonic diffuser decelerates the flow and pressure increases. Fuel is injected into the flow and burned, and the hot pressurized combustion product leaves through a supersonic nozzle. Flight speeds could be up to $\mathrm{Ma} = 15$ or so. Due to the high air velocity at the burner, it is quite difficult to maintain stable combustion of the fuel. To get the scramjet engine started it must be accelerated to supersonic speed first, so that the converging diffuser leads to pressure build-up.

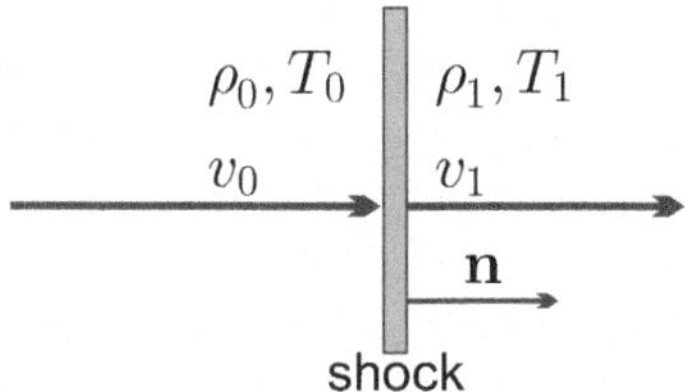

Fig. 8.8 Inflow and outflow for a normal shock.

8.3 Normal Shocks

8.3.1 Conservation Laws

We distinguish between normal shocks, where the flow velocity is perpendicular to the shock, and flow direction is unaltered, and oblique shocks, where inflow and outflow have different directions.

We first consider a normal shock as seen for an observer resting with the shock, see Fig. 8.8. Here, the state before the shock is denoted as state 0, state 1 is just behind the shock, and $v_0 = v_{0,i}n_i$, $v_1 = v_{1,i}n_i$ are the normal velocities relative to the shock.

To proceed, we assume that outside the shock itself irreversible effects, in particular viscous stress and heat transfer, can be ignored, so that $t_{ij} = -p\delta_{ij}$ and $q_i = 0$. Then, the jump conditions for mass (3.27), momentum (3.45), and energy (3.71) reduce to, for an observer resting with the shock, $w_i = 0$,

$$[[\rho v_k]]\, n_k = 0 \quad \Rightarrow \quad \rho_0 v_0 = \rho_1 v_1 \ , \tag{8.24}$$

$$[[\rho v_i v_k + p\delta_{ik}]]\, n_k = 0 \quad \Rightarrow \quad \rho_0 v_0^2 + p_0 = \rho_1 v_1^2 + p_1 \ , \tag{8.25}$$

$$\left[\left[\rho\left(h + \frac{1}{2}v^2\right) v_k\right]\right] n_k = 0 \quad \Rightarrow \quad h_0 + \frac{1}{2}v_0^2 = h_1 + \frac{1}{2}v_1^2 \ . \tag{8.26}$$

Together, these equations are known as the Rankine–Hugoniot conditions.

For given inflow state $\{\rho_0, v_0, T_0\}$ the above equations determine the outflow state $\{\rho_1, v_1, T_1\}$. While the equations can be analyzed in more detail for any substance, for accessible results we consider only ideal gases with constant specific heats, so that[1]

$$p = \rho RT \quad , \quad h = c_p T = \frac{c_p}{R}\frac{p}{\rho} = \frac{c_p}{c_p - c_v}\frac{p}{\rho} = \frac{\gamma}{\gamma - 1}\frac{p}{\rho} \ . \tag{8.27}$$

[1] Note that in general we must write $h = c_p\,(T - T_{\text{ref}}) + h_{\text{ref}}$. For the problem at hand the reference values T_{ref} and h_{ref} can be freely chosen since they cancel from the equations, and setting $h_{\text{ref}} = c_p T_{\text{ref}}$ is the most convenient choice.

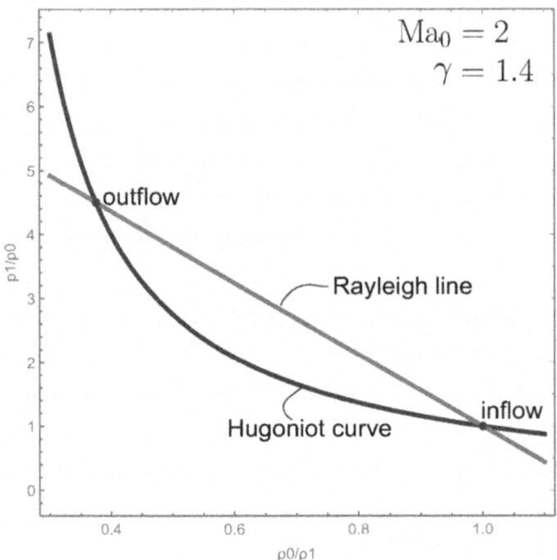

Fig. 8.9 Shock solutions are found as intersections of Rayleigh line and Hugoniot curve.

We also introduce the Mach numbers as the ratio of flow velocity (relative to the shock) and local speed of sound,

$$\mathrm{Ma} = \frac{v}{a} = \frac{v}{\sqrt{\gamma R T}}\,, \quad \mathrm{Ma}_0 = \frac{v_0}{a_0} = \frac{v_0}{\sqrt{\gamma R T_0}}\,, \quad \mathrm{Ma}_1 = \frac{v_1}{a_1} = \frac{v_1}{\sqrt{\gamma R T_1}}\,. \tag{8.28}$$

8.3.2 Rayleigh Line and Hugoniot Curve

To proceed, we combine the conservation laws for the shock, where we give some detail on the required manipulations of equations. Squaring the mass balance and using the momentum balance we obtain

$$\rho_0^2 v_0^2 = \rho_1^2 v_1^2 = \rho_1 \left(\rho_0 v_0^2 + p_0 - p_1\right) \tag{8.29}$$

with $v_0^2 = \gamma R T_0 \mathrm{Ma}_0^2$ and $\rho_0 R T_0 = p_0$ this gives

$$\gamma \rho_0 p_0 \mathrm{Ma}_0^2 = \rho_1 \left(\gamma p_0 \mathrm{Ma}_0^2 + p_0 - p_1\right) \tag{8.30}$$

hence

$$\left.\frac{p_1}{p_0}\right|_R = 1 + \gamma \mathrm{Ma}_0^2 \left(1 - \frac{\rho_0}{\rho_1}\right) \tag{8.31}$$

This combination of mass and momentum balance gives a straight line in a $\left[\frac{p_1}{p_0}, \frac{\rho_0}{\rho_1}\right]$-diagram, which is known as the Rayleigh line, see Fig. 8.9. With pre-shock properties ρ_0, p_0 known, the post-shock properties ρ_1, p_1 must lie on the Rayleigh line.

Another relation between pressure and density ratios follows from combining energy, momentum, and mass balances. With $h = \frac{\gamma}{\gamma-1}\frac{p}{\rho}$, the energy balance reads

$$\frac{\gamma}{\gamma-1}\frac{p_0}{\rho_0} + \frac{1}{2}v_0^2 = \frac{\gamma}{\gamma-1}\frac{p_1}{\rho_1} + \frac{1}{2}v_1^2 \tag{8.32}$$

From the momentum balance (8.25) we have

$$v_0^2 = \frac{p_1 - p_0}{\rho_0 - \frac{\rho_0^2}{\rho_1}} \quad , \quad v_1^2 = \frac{p_0 - p_1}{\rho_1 - \frac{\rho_1^2}{\rho_0}} \, , \tag{8.33}$$

where the second relation follows from exchanging $(0 \leftrightarrow 1)$.

After some reshuffling, the energy balance gives

$$\frac{p_0}{\rho_0}\left(\frac{2\gamma}{\gamma-1} - \frac{1}{1-\frac{\rho_0}{\rho_1}} - \frac{1}{\frac{\rho_1}{\rho_0}\left(1-\frac{\rho_1}{\rho_0}\right)}\right) = \frac{p_1}{\rho_1}\left(\frac{2\gamma}{\gamma-1} - \frac{1}{1-\frac{\rho_1}{\rho_0}} - \frac{1}{\frac{\rho_0}{\rho_1}\left(1-\frac{\rho_0}{\rho_1}\right)}\right) \tag{8.34}$$

To simplify this, we use

$$\frac{1}{1-x} + \frac{1}{\frac{1}{x}\left(1-\frac{1}{x}\right)} = \frac{1}{1-x} + \frac{x^2}{(x-1)} = \frac{1-x^2}{1-x} = 1+x \tag{8.35}$$

hence with $x = \frac{\rho_0}{\rho_1}$ and $x = \frac{\rho_1}{\rho_0}$, respectively,

$$\frac{p_0}{\rho_0}\left[\frac{\gamma}{\gamma-1} - \frac{1}{2}\left(1+\frac{\rho_0}{\rho_1}\right)\right] = \frac{p_1}{\rho_1}\left[\frac{\gamma}{\gamma-1} - \frac{1}{2}\left(1+\frac{\rho_1}{\rho_0}\right)\right] \tag{8.36}$$

and finally, we find the Hugoniot curve, on which pressure and density ratios across the shock obey

$$\left.\frac{p_1}{p_0}\right|_H = \frac{\frac{\gamma+1}{\gamma-1} - \frac{\rho_0}{\rho_1}}{\frac{\gamma+1}{\gamma-1}\frac{\rho_0}{\rho_1} - 1} \quad \text{or} \quad \left.\frac{\rho_1}{\rho_0}\right|_H = \frac{1 + \frac{\gamma+1}{\gamma-1}\frac{p_1}{p_0}}{\frac{\gamma+1}{\gamma-1} + \frac{p_1}{p_0}} \tag{8.37}$$

With pre-shock properties ρ_0, p_0 known, the post-shock properties ρ_1, p_1 must lie on the Hugoniot curve.

8.3.3 Shock Solution

Figure 8.9 shows Rayleigh line and Hugoniot curve in a $\left[\frac{p_1}{p_0}, \frac{\rho_0}{\rho_1}\right]$ diagram. Since shock solutions must lie on both of these, the actual solutions are the intersections of the two curves. As the figure shows, this is either the trivial no-shock solution, where outflow equals inflow, $\rho_1 = \rho_0$, $p_1 = p_0$, or the actual shock solution of interest. While the Hugoniot curve is independent of Mach number, the post-shock state is determined through the pre-shock Mach number Ma_0 which determines the slope of the Rayleigh line. From the figure it is evident that faster inflow will lead to stronger increase in density and pressure, that is stronger shocks.

To determine the shock solution in dependence of inflow Mach number Ma_0, we set $\frac{p_1}{p_0}\big|_R = \frac{p_1}{p_0}\big|_H$ to find a quadratic equation for the density ratio,

$$\left(\frac{\rho_1}{\rho_0}\right)^2 - \frac{1+\gamma\mathrm{Ma}_0^2}{1+\frac{\gamma-1}{2}\mathrm{Ma}_0^2}\frac{\rho_1}{\rho_0} + \frac{\frac{\gamma+1}{2}\mathrm{Ma}_0^2}{1+\frac{\gamma-1}{2}\mathrm{Ma}_0^2} = 0\ , \tag{8.38}$$

with the non-trivial root

$$\frac{\rho_1}{\rho_0} = \frac{v_0}{v_1} = \frac{\frac{\gamma+1}{2}\mathrm{Ma}_0^2}{1+\frac{\gamma-1}{2}\mathrm{Ma}_0^2}\ . \tag{8.39}$$

Inserting this into the equation for the Rayleigh line, we find the pressure ratio for the shock as

$$\frac{p_1}{p_0} = \frac{2\gamma}{\gamma+1}\mathrm{Ma}_0^2 - \frac{\gamma-1}{\gamma+1}\ . \tag{8.40}$$

The temperature ratio across the shock then follows as

$$\frac{T_1}{T_0} = \frac{p_1}{p_0}\frac{\rho_0}{\rho_1} = \frac{4\gamma}{(\gamma+1)^2}\left(1+\frac{\gamma-1}{2}\mathrm{Ma}_0^2\right)\left(1-\frac{\gamma-1}{2\gamma}\frac{1}{\mathrm{Ma}_0^2}\right)\ . \tag{8.41}$$

For $\mathrm{Ma}_0 > 1$, which will be seen soon as the relevant range, the density ratio is limited to $\frac{\rho_1^\infty}{\rho_0} = \frac{\gamma+1}{\gamma-1}$ (for $\mathrm{Ma}_0 \to \infty$), while pressure and temperature grow with Ma_0, see Fig. 8.10.

8.3.4 Entropy Generation

In our figures we have implicitly assumed that the incoming flow is supersonic, $\mathrm{Ma}_0 > 1$. The reason for this is that the entropy generation in a shock must be non-negative. Since we ignore heat transfer outside the shock, the shock itself is adiabatic, hence the entropy generation rate is

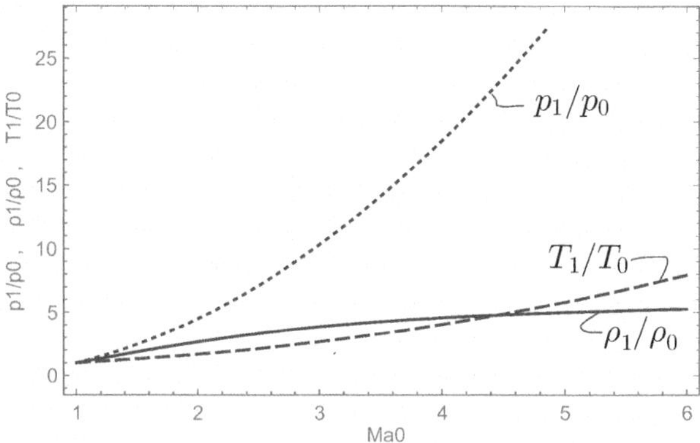

Fig. 8.10 Density, pressure, and temperature ratios across a shock over incoming Mach number Ma_0 for $\mathrm{Ma}_0 \geq 1$.

$$\Sigma_S = \rho_1 v_1 s_1 - \rho_0 v_0 s_0 = \rho_0 v_0 \left(s_1 - s_0\right) \ . \tag{8.42}$$

Entropy generation in the shock must be non-negative, hence the entropy change must be positive, $s_1 - s_0 = \frac{\Sigma_S}{\rho_0 v_0} \geq 0$. For the ideal gas with constant specific heats, the entropy change across the shock is

$$s_1 - s_0 = c_p \ln \frac{T_1}{T_0} - R \ln \frac{p_1}{p_0} = \frac{R}{\gamma - 1} \ln \left[\frac{p_1}{p_0} \left(\frac{\rho_0}{\rho_1} \right)^{\gamma} \right] \tag{8.43}$$

As Fig. 8.11 shows, the entropy change is positive only for Mach numbers $\mathrm{Ma}_0 > 1$. Of course, this can also be seen from the explicit equations. That is, for a shock to occur the inflow must be supersonic, $\mathrm{Ma}_0 > 1$.

For the Mach number ratio we find

$$\frac{\mathrm{Ma}_1}{\mathrm{Ma}_0} = \frac{v_1}{v_0} \sqrt{\frac{T_0}{T_1}} = \frac{\rho_0}{\rho_1} \sqrt{\frac{p_0}{p_1} \frac{\rho_1}{\rho_0}} = \sqrt{\frac{p_0}{p_1} \frac{\rho_0}{\rho_1}} \tag{8.44}$$

hence the Mach number behind the shock is below unity (for $\mathrm{Ma}_0 > 1$),

$$\mathrm{Ma}_1 = \sqrt{1 - \frac{(\gamma + 1)\left(\mathrm{Ma}_0^2 - 1\right)}{1 + \gamma\left(2\mathrm{Ma}_0^2 - 1\right)}} = \sqrt{\frac{1 + \frac{\gamma - 1}{2} \mathrm{Ma}_0^2}{\gamma \mathrm{Ma}_0^2 - \frac{\gamma - 1}{2}}} \leq 1 \ , \tag{8.45}$$

that isin a normal shock we observe a sudden change from supersonic to subsonic flow.

For $\mathrm{Ma}_0 \gtrsim 1$ the shock is weak, and a Taylor series in $(\mathrm{Ma}_0 - 1)$ gives the dimensionless entropy change as

$$\frac{s_1 - s_0}{R} = \frac{16}{3} \frac{\gamma}{(\gamma + 1)^2} \left(\mathrm{Ma}_0 - 1\right)^3 + \mathcal{O}\left(\left(\mathrm{Ma}_0 - 1\right)^4\right) \tag{8.46}$$

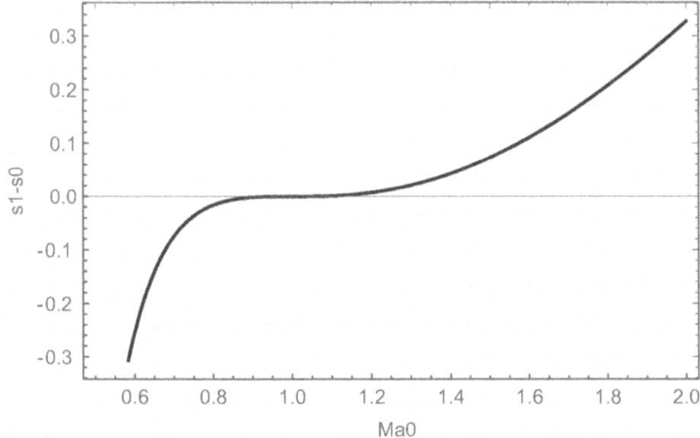

Fig. 8.11 Entropy generation across the shock $(s_1 - s_0)/R$ over Mach number Ma_0. Entropy is generated only for $M_0 > 1$.

hence weak shocks are almost reversible since their entropy generation is of third order.

Entropy generation implies work loss. As well, the flow behavior changes, as can be studied when considering shocks in Laval nozzles or supersonic diffusers. In the converging part of a Laval nozzle, i.e., past the throat, as long as the flow is supersonic, the increase of cross section leads to acceleration of the flow. A shock within the nozzle reduces flow speed to subsonic, and the flow after the shock will slow down further as the cross section increases towards the exit (the diverging part acts as diffuser). Accordingly, loss of thrust is not only due to the shock, but as well due to subsequent deceleration instead of acceleration, and increase of pressure, see Fig. 8.5. Similarly, a shock in the converging part of a supersonic diffuser will lead to subsonic flow after the shock, from when on the diffuser will act as a Laval nozzle, accelerating the flow instead of decelerating to increase pressure.

8.4 Oblique Shocks

In oblique shocks, flow velocity and shock are not perpendicular, and velocity changes not only its absolute value, but also its direction. This happens for instance when the gas flows against a wedge, as shown in Fig. 8.12, where θ is the wall angle, and β is the shock angle, which are both measured against the direction of the incoming flow. Note that we ignore wall friction effects, which would create a thin boundary layer at the wedge.

The flow is best described relative to the shock, and from the figure we identify the components of incoming and outgoing velocity normal ($\perp$) and tangential ($||$) to the shock as

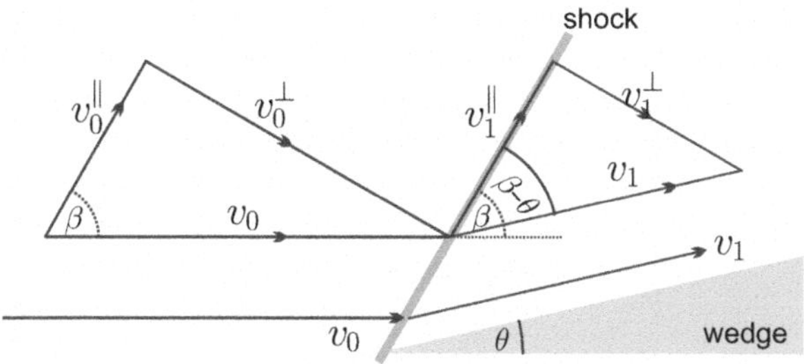

Fig. 8.12 Geometry of an oblique shock and decomposition of velocities parallel and perpendicular to the shock.

$$v_0^{\perp} = v_0 \sin\beta \quad , \quad v_0^{\parallel} = v_0 \cos\beta \tag{8.47}$$

and

$$v_1^{\perp} = v_1 \sin(\beta - \theta) \quad , \quad v_1^{\parallel} = v_1 \cos(\beta - \theta) \tag{8.48}$$

Again we consider the jump conditions for the conservation laws, where now we have to account for the tangential and normal velocities.

Mass conservation gives continuity of the mass flux perpendicular to the shock,

$$\rho_0 v_0^{\perp} = \rho_1 v_1^{\perp} \ . \tag{8.49}$$

Momentum conservation gives two relations, for normal and tangential momentum fluxes,

$$\rho_0 \left(v_0^{\perp}\right)^2 + p_0 = \rho_1 \left(v_1^{\perp}\right)^2 + p_1 \ , \tag{8.50}$$

$$\rho_0 v_0^{\parallel} v_0^{\perp} = \rho_1 v_1^{\parallel} v_1^{\perp} \ . \tag{8.51}$$

Combining the tangential momentum with the mass balance, it follows that tangential velocity remains unchanged,

$$v_0^{\parallel} = v_1^{\parallel} \ . \tag{8.52}$$

Since $v^2 = \left(v_0^{\perp}\right)^2 + \left(v_0^{\parallel}\right)^2$, and tangential velocity remains, conservation of energy gives

$$h_0 + \frac{1}{2}\left(v_0^{\perp}\right)^2 = h_1 + \frac{1}{2}\left(v_1^{\perp}\right)^2 \ . \tag{8.53}$$

For ρ, p, and $v^{\perp}$ we therefore have just the same equations as for the normal shock. However, since the tangential component is non-zero, the relation between pre-shock Mach number and incoming velocity is

$$\frac{v_0^{\perp}}{a_0} = \frac{v_0 \sin\beta}{a_0} = \mathrm{Ma}_0 \sin\beta \tag{8.54}$$

Hence, all previous results remain valid if we replace Ma_0 by $\mathrm{Ma}_0 \sin\beta$,

$$\frac{\rho_1}{\rho_0} = \frac{v_0^\perp}{v_1^\perp} = \frac{\frac{\gamma+1}{2}\mathrm{Ma}_0^2 \sin^2\beta}{1 + \frac{\gamma-1}{2}\mathrm{Ma}_0^2 \sin^2\beta} , \tag{8.55}$$

$$\frac{p_1}{p_0} = \frac{2\gamma}{\gamma+1}\mathrm{Ma}_0^2 \sin^2\beta - \frac{\gamma-1}{\gamma+1} , \tag{8.56}$$

and, as before,

$$\frac{T_1}{T_0} = \frac{p_1}{p_0}\frac{\rho_0}{\rho_1} , \tag{8.57}$$

$$s_1 - s_0 = \frac{R}{\gamma-1}\ln\left[\frac{p_1}{p_0}\left(\frac{\rho_0}{\rho_1}\right)^\gamma\right] \geq 0 \tag{8.58}$$

The necessary condition for entropy increase across the shock now reads

$$\mathrm{Ma}_0 \sin\beta \geq 1 , \tag{8.59}$$

which defines the smallest possible shock angle, known as the Mach angle, for given Mach number.

The maximum shock angle is $\beta = \frac{\pi}{2}$ for the normal shock, so that possible shock angles are in the range

$$\arcsin\frac{1}{\mathrm{Ma}_0} \leq \beta \leq \frac{\pi}{2} . \tag{8.60}$$

Next we consider the relation between deflection angle θ, shock angle β, and Mach number. From the geometry we have

$$\tan\beta = \frac{v_0^\perp}{v_0^\parallel} \quad , \quad \tan(\beta-\theta) = \frac{v_1^\perp}{v_1^\parallel} \tag{8.61}$$

hence

$$\frac{\tan(\beta-\theta)}{\tan\beta} = \frac{v_1^\perp}{v_0^\perp} = \frac{\rho_0}{\rho_1} = \frac{1 + \frac{\gamma-1}{2}\mathrm{Ma}_0^2 \sin^2\beta}{\frac{\gamma+1}{2}\mathrm{Ma}_0^2 \sin^2\beta} \tag{8.62}$$

To simplify we use some trigonometric relations[2], to find (after some lines of algebra)

$$\tan\theta = \frac{2}{\tan\beta}\frac{\mathrm{Ma}_0^2 \sin^2\beta - 1}{\mathrm{Ma}_0^2(\gamma + \cos 2\beta) + 2} . \tag{8.63}$$

For the Mach number behind the shock we once more can use the previous result for the normal shock, by accounting for the angles, which gives

[2] Specifically, $\tan(\beta-\theta) = \frac{\tan\beta - \tan\theta}{1+\tan\beta\tan\theta}$, $\sin\beta = \tan\beta\cos\beta$, $\sin^2\beta + \cos^2\beta = 1$, and $2\cos^2\beta = 1 + \cos 2\beta$. Separate $\tan\theta$ and then simplify step by step.

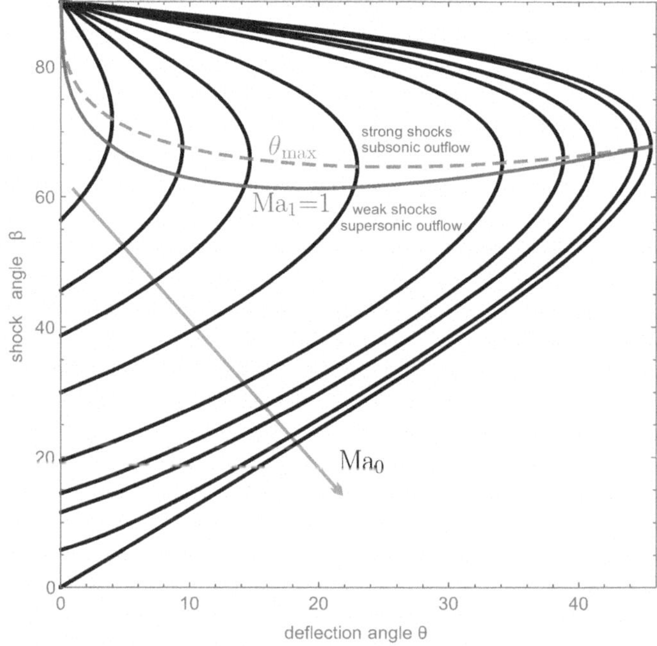

Fig. 8.13 Relation between shock angle β and deflection angle θ for $\mathrm{Ma}_0 = 1.2, 1.4, 1.6, 2, 3, 4, 5, 10, \infty$. The dashed curve connects the points of maximum deflection angle $\theta_{\max}$, which distinguishes between weak shocks with small β and (mostly) supersonic outflow, and strong shocks with large subsonic outflow. Normal shocks are strong shocks with $\beta = \frac{\pi}{2}$, $\theta = 0$ (upper left corner). The curve $\mathrm{Ma}_1 = 1$ distinguishes between sub- and supersonic outflow.

$$\mathrm{Ma}_1^2 \sin^2(\beta - \theta) = \frac{1 + \frac{\gamma-1}{2}\mathrm{Ma}_0^2 \sin^2\beta}{\gamma \mathrm{Ma}_0^2 \sin^2\beta - \frac{\gamma-1}{2}} \,. \tag{8.64}$$

Figure 8.13 shows curves of shock angle β over deflection angle θ for a variety of Mach numbers between $\mathrm{Ma}_0 = 1.2$ and $\mathrm{Ma}_0 = \infty$. Sonic outflow states, where $\mathrm{Ma}_1 = 1$, lie on the continuous horizontal curve, which separates solutions with supersonic outflow (below the curve) and subsonic outflow (above the curve).

For each inflow Mach number Ma_0 there is a maximum reflection angle $\theta_{\max}$. In the figure, the maxima are connected through the dashed horizontal curve. For $\theta < \theta_{\max}$, there are two possible solutions for the shock angle β. One distinguishes between *strong shocks*, for which the shock angle is large, and the outflow is subsonic, and *weak shocks*, where outflow is (mostly) supersonic, the shock angle is smaller, and entropy generation is smaller as well. Generally, weak shocks are more likely to occur, unless the downstream pressure is high independent of the shock.

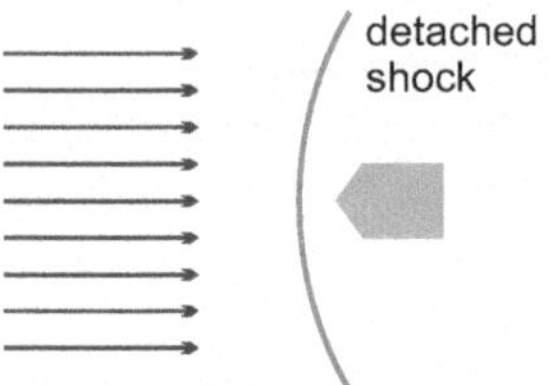

Fig. 8.14 Detached shock in front of a blunt wedge with angle $\theta > \theta_{\max}$.

Fig. 8.15 NASA photo of shocks and expansion waves of a supersonic aircraft in flight.

When the supersonic flow hits a body with a larger angle θ, then no solution of this kind exists. Instead, one will observe a detached shock as depicted in Fig. 8.14, which must be determined numerically. At the center of the aircraft's nose, blunt shocks are normal shocks, hence strong shocks with large entropy generation. Supersonic aircraft have long noses with small angles, so that shocks do not detach, and weak shocks with smaller dissipation occur.

While shocks result from compression due to upward edges, downward edges induce change of flow direction by isentropic expansion. The NASA photo in Fig. 8.15 shows the numerous shocks and expansion waves around a supersonic aircraft in flight. Shocks are irreversible, hence they induce work loss which must be accounted for in aircraft propulsion.

The interested reader is referred to the vast literature on compressible gas dynamics for further insight into this fascinating topic.[3]

[3] E.g., H.W. Liepmann, A. Roshko: Elements of Gasdynamics, Wiley, New York 1957

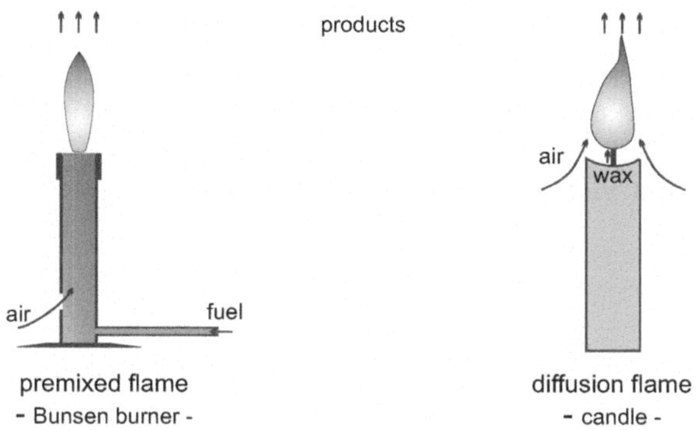

Fig. 8.16 Bunsen burner and candle as examples for premixed and diffusion flames.

8.5 Flames

8.5.1 Premixed and Diffusion Flames

Flows involving chemical reactions are quite complex, due to the interplay of chemical reactions, diffusion, heat transfer, and flow processes.

One distinguishes between diffusion flames and premixed flames. In a premixed flame, fuel and oxidizer—the reactants—are well mixed, and the reaction zone travels through the mixture, leaving reaction products in its wake. Figure 8.16 shows a Bunsen burner as a familiar example from laboratory use (see discussion in Sec. 8.5.5).

Combustion processes in the cylinder of Otto engines are premixed flames: the air-fuel mixture is drawn into the cylinder, and the spark plug triggers the exothermic reaction traveling through air-fuel mixture in the cylinder. Note that a certain temperature must be reached for the combustion to start—cold air fuel mixture is in a metastable state—hence the spark plug is needed to initiate the process, and then the heat provided by the exothermic reaction keeps the reaction going.

In a diffusion flame, the oxidizer must diffuse towards the fuel, hence fuel consumption is limited by the amount of oxidizer that can be provided. A simple everyday example is the flame of a candle, see Fig. 8.16, where the fuel is the evaporating wax at the wick, and the oxidizer is oxygen from the surrounding air. Due to the flow triggered by the buoyancy of the hot product, the candle flame is affected by convection of the surrounding air—not at all a trivial problem! Heat from the combustion process serves to melt more wax.

In Diesel engines, the fuel is sprayed into the hot compressed air in the cylinder, so that some of the fuel evaporates, and some remains as small droplets. Due to the high temperature in the compressed air the reaction starts without a spark. The ensuing combustion process is multi-facetted, with the combustion of droplets dominated by diffusion, and the combustion of evaporated fuel being similar to premixed flames.

The simple flame model describes basic flame propagation modes but does not suffice for a realistic picture of combustion processes. Most importantly, the model does not yield flame velocity, which depends on reaction rates and the local state.

8.5.2 A Simple Model for Premixed Flames

In the following, in order to obtain basic understanding, we study a rather simplified model for premixed flames based on the following assumptions:

a) We do not distinguish individual reactants (such as fuel and oxidizer) and products (such as carbon dioxide and water), but consider only an overall reaction $A \rightarrow B$. One might think of A as a homogeneous mixture of reactants, and of B as a homogeneous mixture of products.
b) Reactant A and product B have the same molar mass, and the same constant specific heat, so that

$$R_A = R_B = R \quad , \quad c_{P,A} = c_{p,B} = c_p \; . \tag{8.65}$$

c) Reactant and product are ideal gases, so that

$$p = \rho RT \quad , \quad h_A = c_p T + h_A^0 \quad , \quad h_B = c_p T + h_B^0 \tag{8.66}$$

With this, the heat of reaction is independent of temperature and pressure, and determined only through the enthalpy constants,

$$\Delta h_R = h_B - h_A = h_B^0 - h_A^0 \tag{8.67}$$

The heat of reaction is defined as the amount of heat that must be exchanged to keep temperature and pressure constant over the reaction. Hence, for an exothermic reaction, the heat of reaction is negative, $\Delta h_R < 0$, and for an endothermic reaction it is positive, $\Delta h_R > 0$.

In a premixed flame, the reaction occurs in a reaction zone through which the concentration of reactants decreases, and the concentration of products increases, see Fig. 8.17. The width of the zone depends on reaction rate. Since only some of the collisions between particles lead to a reaction, the reaction mean free path, which determines the width of the flame zone, is larger than the collision mean free path. Compared to the macroscopic length scale the zone is small, and can be considered as a singular surface, just as for shocks.

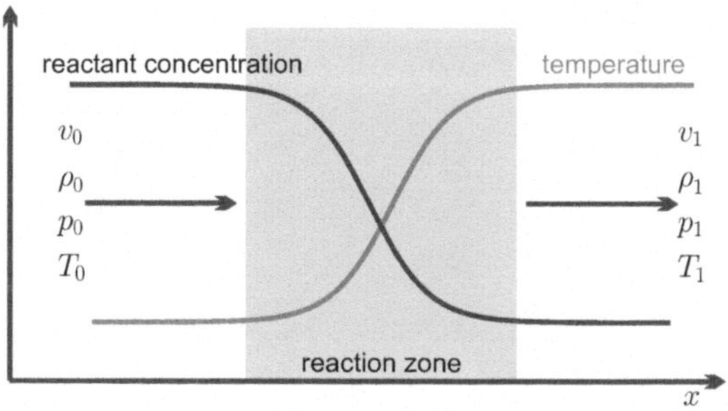

Fig. 8.17 Reaction zone in a premixed flame.

8.5.3 Flame Conditions

We consider only flames where velocity is normal to the reaction zone, and we do not resolve the reaction zone, but treat it as a singular surface. Similar to normal shocks, we find the jump conditions from the conservation laws

$$\begin{aligned} \rho_0 v_0 &= \rho_1 v_1 \\ \rho_0 v_0^2 + p_0 &= \rho_1 v_1^2 + p_1 \\ \frac{\gamma}{\gamma-1}\frac{p_0}{\rho_0} + h_A^0 + \frac{1}{2}v_0^2 &= \frac{\gamma}{\gamma-1}\frac{p_1}{\rho_1} + h_B^0 + \frac{1}{2}v_1^2 \end{aligned} \tag{8.68}$$

The only difference to normal shocks is in the energy balance, due to the difference of enthalpy constants for reactants and products. Accordingly, mass and momentum balance again combine to the Rayleigh line,

$$\left.\frac{p_1}{p_0}\right|_R = 1 + \gamma \mathrm{Ma}_0^2 \left(1 - \frac{\rho_0}{\rho_1}\right) \tag{8.69}$$

while the equation for the Hugoniot curve has an additional term related to the heat of reaction,

$$\left.\frac{p_1}{p_0}\right|_H = \frac{\frac{\gamma+1}{\gamma-1} - \frac{\rho_0}{\rho_1} + \frac{2\gamma}{\gamma-1}\mathsf{q}}{\frac{\gamma+1}{\gamma-1}\frac{\rho_0}{\rho_1} - 1} \tag{8.70}$$

where the dimensionless heat of reaction, which is defined to be positive as

$$\mathsf{q} = -\frac{\Delta h_R}{\frac{\gamma}{\gamma-1}\frac{p_0}{\rho_0}} = -\frac{\Delta h_R}{c_p T_0} > 0\,, \tag{8.71}$$

serves as parameter. For $\mathsf{q} = 0$, the heat of reaction vanishes, and we find the expression for normal shocks.

We have found before that momentum and mass balances give

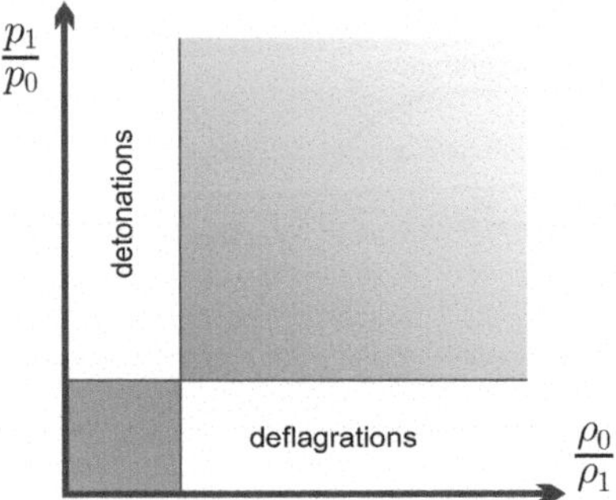

Fig. 8.18 Allowed (white) and excluded (grey) areas for flame solutions in the $\left[\frac{p_1}{p_0}, \frac{\rho_0}{\rho_1}\right]$-diagram. Accessible areas give either detonations or deflagrations as indicated.

$$v_1^2 = \frac{p_0 - p_1}{\rho_1 - \frac{\rho_1^2}{\rho_0}} = \frac{p_0}{\rho_1} \frac{1 - \frac{p_1}{p_0}}{\frac{p_1}{p_0}\frac{\rho_0}{\rho_1} - 1} . \tag{8.72}$$

Since the left hand side is positive, the right hand side must be positive as well, that is only solutions are allowed in which either

$$\frac{p_1}{p_0} > 1 \text{ and } \frac{\rho_0}{\rho_1} < 1 \qquad \text{or} \qquad \frac{p_1}{p_0} < 1 \text{ and } \frac{\rho_0}{\rho_1} > 1 . \tag{8.73}$$

The accessible regions in the $\left[\frac{p_1}{p_0}, \frac{\rho_0}{\rho_1}\right]$-diagram are shown in Fig. 8.18, where the solutions in the two accessible regions describe either deflagrations or detonations as discussed below.

8.5.4 Flame Propagation

For the above evaluation we have considered a frame of reference resting with the flame front, such that v_0 and v_1 are velocities of reactant and product relative to the front. It is instructive, however, to look at a flame from a frame of reference in which the reactant is at rest, and the flame front moves into the resting reactants, leaving products behind.

With the reactant on the left, and at rest, the front moves with the velocity $(-v_0)$ to the left. We denote the flow velocities in this frame as $\hat{v} = v - v_0$, so that for reactant and flame $\hat{v}_0 = v_0 - v_0 = 0 \;\; , \;\; \hat{v}_f = 0 - v_0 \; = -v_0$. For the velocity of the products we find, with the mass balance,

$$\hat{v}_1 = v_1 - v_0 = v_0 \left(\frac{v_1}{v_0} - 1\right) = v_0 \left(\frac{\rho_0}{\rho_1} - 1\right) \lessgtr 0 . \tag{8.74}$$

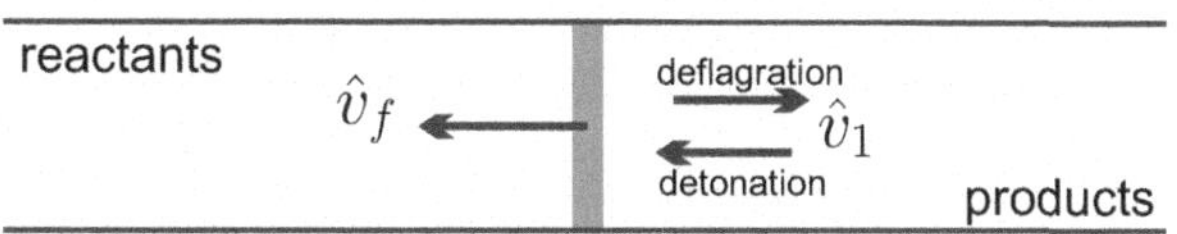

Fig. 8.19 Flame front travelling into resting reactant. The products follow the flame front for a detonation, and move away from the front for a deflagration.

For a *deflagration*, the product moves away from the flame, i.e., to the right in Fig. 8.19, and pressure and density decrease,

$$\text{deflagration:} \quad \frac{p_1}{p_0} < 1, \ \frac{\rho_0}{\rho_1} > 1 \quad \Rightarrow \quad \hat{v}_1 > 0 \ . \tag{8.75}$$

For a *detonation* the product follows the flame, i.e., moves to the left, and pressure and density increase,

$$\text{detonation:} \quad \frac{p_1}{p_0} > 1, \ \frac{\rho_0}{\rho_1} < 1 \quad \Rightarrow \quad \hat{v}_1 < 0 \ . \tag{8.76}$$

Shocks are induced through flow geometry, as we have seen in the discussion of oblique shocks at wedges, or in examples concerning normal shocks in Laval nozzles. Then, the inflow Mach number Ma_0 is known, since the shock is stationary at the wedge or in the nozzle.

For a flame propagating into the reactant, as discussed here, the flame velocity $\hat{v}_f = -v_0$, and hence the Mach number Ma_0 develops as a result of the details of the processes in the reaction zone, in particular reaction rates, as well as the overall flow conditions (e.g., is a pipe in which the process occurs open or closed? How does geometry affect the flow of products?). Indeed, one would expect that low reaction rates lead to rather slow flame propagation.

For the solutions discussed here, the Mach number Ma_0 is, instead, considered as an input. With that, the discussion must focus on possible modes for given Ma_0.

8.5.5 Deflagrations ($\mathrm{Ma}_0 < 1$)

Figure 8.20 shows the Rayleigh line for $\mathrm{Ma}_0 = 0.3$ and two Hugoniot curves. The Hugoniot curve for $\mathsf{q} = 0$ passes through the point $\mathsf{0}$ at $(1, 1)$ and its intersection with the Rayleigh line (which is outside the range shown in the figure) is not related to accessible solutions, since the second law allows normal shocks only for $\mathrm{Ma}_0 > 1$. The Hugoniot curve for $\mathsf{q} = 1$ intersects the Rayleigh line at points A and B, which are the possible flame solutions for this Mach number.

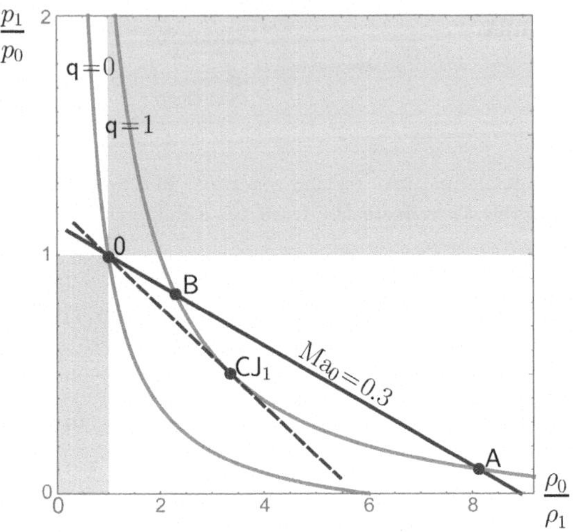

Fig. 8.20 Deflagration: Rayleigh line for $\mathrm{Ma}_0 = 0.3$ (through AB), Hugoniot curves for $q = 0$ and $q = 1$, and tangent through the Chapman-Jouguet point (CJ_1).

Solution A describes a *strong deflagration*, for which the products are supersonic, $\mathrm{Ma}_{1,\mathrm{A}} > 1$. Such solutions, where a subsonic flow becomes supersonic through (chemical) heat addition, are not observed.

Solution B describes a *weak deflagration*, for which the products are subsonic, $\mathrm{Ma}_{1,\mathrm{B}} < 1$. Such solutions are observed, but might require a flame holder for stabilization.

For instance, in a Bunsen burner (Fig. 8.16) fuel and oxidizer are premixed, and flow is laminar. The rim of the burner provides a heat sink to stabilize the flame. Reaction rates and flame velocity are smaller at lower temperatures. If the location of the flame moves away from the burner, the temperature grows, such increasing the flame velocity, so that the flame moves back to the burner. When the flame enters the burner, the temperature drops, the flame velocity is lower, and the flame is pushed back to the rim. This results in stationary flames at the burner exit.

Stabilization fails if the flow velocity is not well adjusted. The flame velocity must be considered relative to the inflow of reactant. Mismatched flow velocity will lead to blowout, where the flame is blown away from the burner (flow velocity > flame velocity), or flashback, where the flame enters the burner (flow velocity < flame velocity).

Flame velocity depends on reaction rate, hence on the fuel used, and thus burners must be adjusted to the fuel. Therefore, when considering conversion of a combustion system to sustainable fuels, one cannot simply change the operation of, e.g., a gas turbine from natural gas to hydrogen, which has about five times faster flame velocity, nor to ammonia, which has about one-fifth of the flame velocity, but must exchange the combustion system.

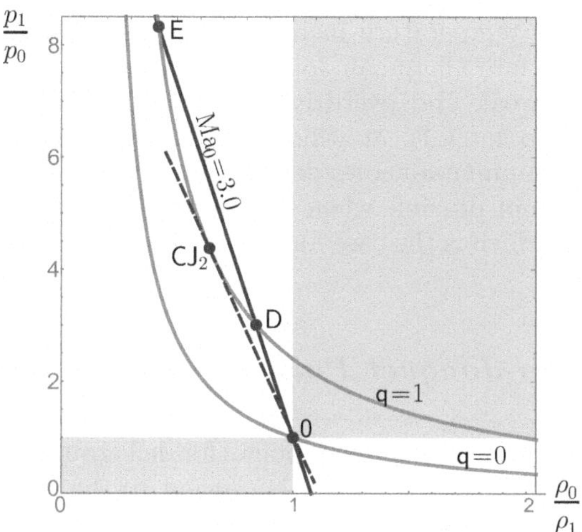

Fig. 8.21 Detonation: Rayleigh line for $\mathrm{Ma}_0 = 3.0$ (through DE) and Hugoniot curves for $\mathsf{q} = 0$ and $\mathsf{q} = 1$, as well as the tangent through the Chapman-Jouguet point (CJ_2)

The slope of the Rayleigh line depends on the Mach number. At the Chapman-Jouguet point CJ_1 the Rayleigh line is a tangent to the Hugoniot curve. These corresponding conditions will be discussed further below, where it will be seen that the Chapman-Jouguet flame has sonic outflow, $\mathrm{Ma}_{1,\mathsf{CJ}} = 1$. Hence the point CJ_1 separates weak and strong deflagrations.

Deflagrations are not possible when the Rayleigh line is too steep to interject the Hugoniot line, which is the case for inflow Mach numbers $\mathrm{Ma}_0 > \mathrm{Ma}_{0,\mathsf{CJ}_1}$.

8.5.6 Detonations ($\mathbf{Ma_0 > 1}$)

Detonations are shock waves followed by combustion waves. Figure 8.21 shows Hugoniot curves and Rayleigh line for detonations, where inflow is supersonic, $\mathrm{Ma}_0 = 3.0$. The Hugoniot curve for $\mathsf{q} = 0$ passes through the point $\mathsf{0}$ at $(1, 1)$ and its single intersection with the Rayleigh line (outside the figure) gives the normal shock. The Hugoniot curve for $\mathsf{q} = 1$ (or any other $\mathsf{q} > 0$) intersects the Rayleigh line at points D and E, which are the possible flame solutions at this Mach number.

State D describes a *weak detonation*, where the product outflow is supersonic, $\mathrm{Ma}_{1,\mathsf{D}} > 1$. Such solutions seems not to occur.

State E describes an *overdriven detonation*, with subsonic outflow, $\mathrm{Ma}_{1,\mathsf{E}} < 1$. This mode of propagation can be observed in special cases, for instance as a result of strong shocks in the reactant that lead to auto-ignition of the mixture due to the temperature increase. Generally, these detonations are unstable and decay into Chapman-Jouguet detonations.

The regions of weak and overdriven detonations are separated by the Chapman-Jouguet point CJ_2, at which the Rayleigh line is tangent to the Hugoniot curve. Chapman-Jouguet detonations are stable.

Detonations are not possible when the Rayleigh line is too flat to interject the Hugoniot line, which is the case for inflow Mach numbers $\mathrm{Ma}_0 < \mathrm{Ma}_{0,\mathsf{CJ}_1}$.

8.5.7 Chapman-Jouguet Points

We have identified Chapman-Jouguet points for deflagrations and detonations as the points where the Rayleigh line is tangent to the Hugoniot curve. For their evaluation, we use compact notation with the abbreviations $\varpi = \frac{\rho_0}{\rho_1}$, $\pi = \frac{p_1}{p_0}$, so that the two pressure functions for the Rayleigh line and Hugoniot curve are

$$\pi_{|R} = 1 + \gamma \mathrm{Ma}_0^2 (1 - \varpi) \quad , \quad \pi_{|H} = \frac{\frac{\gamma+1}{\gamma-1} - \varpi + \frac{2\gamma}{\gamma-1}\mathsf{q}}{\frac{\gamma+1}{\gamma-1}\varpi - 1} \tag{8.77}$$

with the slopes

$$\frac{d\pi_{|R}}{d\varpi} = -\gamma \mathrm{Ma}_0^2 \quad , \quad \frac{d\pi_{|H}}{d\varpi} = -\frac{\frac{\gamma+1}{\gamma-1}\pi_{|H} + 1}{\frac{\gamma+1}{\gamma-1}\varpi - 1} \tag{8.78}$$

Equating the slopes, $\frac{d\pi_{|R}}{d\varpi}_{|\mathsf{CJ}} = \frac{d\pi_{|H}}{d\varpi}_{|\mathsf{CJ}}$ gives after a short calculation

$$\frac{\gamma+1}{\gamma-1}\left[\gamma \mathrm{Ma}_{0,\mathsf{CJ}}^2 \varpi_{|\mathsf{CJ}} - \pi_{\mathsf{CJ}}\right] = 1 + \gamma \mathrm{Ma}_{0,\mathsf{CJ}}^2 = \pi_{\mathsf{CJ}} + \gamma \mathrm{Ma}_{0,\mathsf{CJ}}^2 \varpi_{|\mathsf{CJ}}$$

where the second equality follows from the Rayleigh line. This simplifies to the Mach number of the flame for the Chapman-Jouguet point

$$\mathrm{Ma}_{0,\mathsf{CJ}}^2 = \left[\frac{\pi}{\varpi}\right]_{|\mathsf{CJ}} = \left[\frac{p_1}{p_0}\frac{\rho_1}{\rho_0}\right]_{|\mathsf{CJ}} . \tag{8.79}$$

From the definition of the Mach number, the ideal gas law, and the mass balance results a relation between the pre- and post-flame Mach numbers,

$$\mathrm{Ma}_1 = \frac{v_1}{\sqrt{\gamma \frac{p_1}{\rho_1}}} = \frac{\rho_0 v_0}{\sqrt{\gamma \frac{p_1}{\rho_1}}\rho_1} = \frac{v_0}{\sqrt{\gamma \frac{p_0}{\rho_0}}} \frac{\frac{\rho_0}{\rho_1}}{\sqrt{\frac{\rho_0}{\rho_1}\frac{p_1}{p_0}}} = \mathrm{Ma}_0 \sqrt{\frac{\varpi}{\pi}} \tag{8.80}$$

hence, at a Chapman-Jouguet point the product flow behind a deflagration or detonation is sonic,

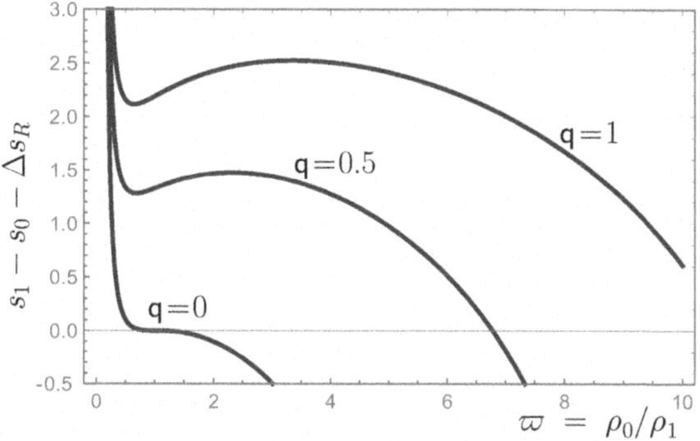

Fig. 8.22 Entropy change across a flame over density ratio $\varpi = \rho_0/\rho_1$, $\mathsf{q} = 0, 0.5, 1$. Minima and maxima of the curves are the Chapman-Jouguet points for detonations and deflagrations, respectively.

$$\mathrm{Ma}_{1,\mathsf{CJ}} = \mathrm{Ma}_{0,\mathsf{CJ}} \sqrt{\frac{\varpi}{\pi}}_{|\mathsf{CJ}} = 1\ . \tag{8.81}$$

The incoming velocity v_0, and hence the flame velocity $\hat{v}_f = -v_0$, follows from (8.68) by solving for $\{\rho_1, p_1, v_0\}$ with $v_1 = \sqrt{\gamma \frac{p_1}{\rho_1}}$ and given inflow conditions $\{\rho_0, p_0\}$.

8.5.8 Chapman-Jouguet Points and the Second Law

The entropy change across the flame is

$$s_1 - s_0 = \frac{R}{\gamma - 1} \ln \left[\pi \varpi^{\gamma}\right] + \Delta s_R \tag{8.82}$$

where Δs_R is the entropy of reaction. All flame solutions lie on the Hugoniot curve, hence we can study the entropy change over the density ratio as

$$\mathsf{s}(\varpi) = \left[s_1 - s_0 - \Delta s_R\right]_{|H} = \frac{R}{\gamma - 1} \ln \left[\pi_{|H} \varpi^{\gamma}\right]\ . \tag{8.83}$$

Figure 8.22 shows three curves, for $\mathsf{q} = 0, 0.5, 1$. For $\mathsf{q} > 0$ we observe two local extrema, a minimum and a maximum, which for $\mathsf{q} = 0$ deteriorate into an inflection point. We are interested in the extrema. The derivative of the curve is

$$\frac{d\mathsf{s}(\varpi)}{d\varpi} = \frac{R\gamma}{\gamma - 1} \frac{1}{\varpi} + \frac{R}{\gamma - 1} \frac{1}{\pi_{|H}} \frac{d\pi_{|H}}{\partial \varpi}\ . \tag{8.84}$$

Instead of evaluating for the extrema by setting the derivative to zero, we evaluate at the Chapman-Jouguet points, where $\frac{d\pi_{|H}}{\partial\varpi}_{|\mathsf{CJ}} = \frac{d\pi_{|R}}{d\varpi}_{|\mathsf{CJ}} = -\gamma \mathrm{Ma}_{0,\mathsf{CJ}}^2$, $\pi_{|H} = \pi_{|\mathsf{CJ}}$ to find

$$\frac{ds(\varpi)}{d\varpi} = \frac{R\gamma}{\gamma-1}\frac{1}{\pi_{|\mathsf{CJ}}}\left[\left[\frac{\pi}{\varpi}\right]_{|\mathsf{CJ}} - \mathrm{Ma}_{0,\mathsf{CJ}}^2\right] = 0\,. \tag{8.85}$$

That is the Chapman-Jouguet points are the extrema of entropy generation. Comparing with the figure, we identify the Chapman-Jouguet detonation (where from Fig. 8.21 $\varpi = \frac{\rho_0}{\rho_1} < 1$) as the flame with the smallest entropy increase, while the Chapman-Jouguet deflagration (for which Fig. 8.20 shows $\varpi = \frac{\rho_0}{\rho_1} > 1$) is the flame with the largest entropy increase.

Problems

8.1. Rockets

1. The balance of momentum in integral form reads, for an observer in an inertial system,

$$\frac{d}{dt}\int_{V(t)} \rho v_i dV = -\oint_{\partial V(t)} \rho v_i (v_k - u_k) n_k dA + \oint_{\partial V(t)} t_{ik} n_k dA\,.$$

   Reduce and simplify this equation for the case of a rocket, to find

$$\frac{dM}{dt} = \dot{m}(v_E - v_R) + (p_E - p_0) A_E$$

   where $M = m_R v_R$ is momentum of the rocket of total mass m_R travelling at velocity v_R, $\dot{m}$ is exiting mass flow of propellant, v_E and p_E are velocity and pressure of the exiting flow, p_0 is the pressure of the surrounding environment, and A_E is nozzle exit area. State all steps, and all assumptions and simplifications required to find this relation.
   Note: the velocity v_E is the exit velocity as seen from the (accelerating) rocket, whereas $(v_E - v_R)$ is the exit velocity as seen from an observer in an inertial frame—carefully explain how these relate to v_i, u_i in the balance.
2. Due to expulsion of propellant at rate $\dot{m}$, the mass of the rocket changes. Show that with help of the mass balance the above reduces to the equation of motion for the rocket in the form

$$m_R \frac{dv_R}{dt} = F = \dot{m} v_E + (p_E - p_0) A_E$$

with the thrust F.

3. Assume constant values for mass flux $\dot{m}$ and the exit velocity v_E, ignore the pressure difference, and determine the velocity of the rocket as function of time. Assume that 85% of the initial rocket mass is propellant, and determine the ratio between final rocket velocity and exit velocity v_E. Plot rocket mass and velocity over time from start to exhaustion of fuel. Discuss multistage rockets.
4. Multiply the balance of momentum with rocket velocity, and write as balance of kinetic energy, in the equivalent forms

$$m_R \frac{d\frac{1}{2}v_R^2}{dt} = \cdots \quad \text{or} \quad \frac{d\frac{m_R}{2}v_R^2}{dt} = \cdots$$

Determine the right hand sides for both equations, and identify propulsive power.

8.2. Nozzle Flow

Consider a reversible Laval nozzle for rocket propulsion. Pressure and temperature in the combustion chamber are $p_s = 10\,\text{bar}$ and $T_s = 2500\,\text{K}$, respectively. The mass flow through the nozzle is $30\frac{\text{kg}}{\text{s}}$ of combustion product (ideal gas, constant specific heats, $R = 0.287\frac{\text{kJ}}{\text{kg K}}$, $\gamma = 1.4$), the cross section at the end of the nozzle is $A_e = 700\,\text{cm}^2$ and the flow is isentropic throughout the nozzle. The environmental pressure is 0.9 bar.

1. Do you expect supersonic or subsonic flow at the outlet? Why?
2. Determine the area of the throat of the nozzle.
3. Find pressures and gas velocities at throat and end.
4. Discuss the flow behind the nozzle

8.3. Rocket Engine

A converging-diverging nozzle is fed from a combustion chamber at temperature $T_s = 2200\,\text{K}$. The flow through the nozzle is isentropic, and the outflow is supersonic with the velocity $v = 1400\frac{\text{m}}{\text{s}}$. The pressure in the throat is measured as $p^* = 4\,\text{bar}$. The gas flowing through the nozzle can be considered as an ideal gas with constant specific heats, $c_p = 0.98\frac{\text{kJ}}{\text{kg K}}$, $\gamma = \frac{c_p}{c_v} = 1.4$.

1. Determine the pressure in the combustion chamber.
2. Determine temperature and pressure at the nozzle exit.
3. Determine the speed of sound at the nozzle exit.

8.4. Ramjet Diffuser

A ramjet diffuser is to be designed with the following characteristics: The ramjet is to fly at high altitude, where the temperature is 210 K and the pressure is 0.35 bar. The flight velocity is $\text{Ma}_I = 3.5$, and the inlet diameter is $0.4\,\text{m}^2$. Consider air as ideal gas with constant specific heats, $R = 0.287\frac{\text{kJ}}{\text{kg K}}$, $c_p = 1.004\frac{\text{kJ}}{\text{kg K}}$.

1. Determine the mass flow into the engine.
2. Determine stagnation temperature, stagnation pressure, and stagnation density.
3. Determine the cross section and radius of the throat, and pressure, temperature, mass density and velocity in the throat.
4. When the velocity at the entry of the combustion chamber is $200\frac{\text{m}}{\text{s}}$, determine the corresponding temperature, pressure, and cross section.
5. Determine the velocity at the combustion chamber inlet, for the case that the cross section is the same as the inlet cross section.

8.5. Ramjet

In a simple thermodynamic model a ramjet operates as follows (seen from an observer resting with the aircraft): Air enters the diffuser where it is decelerated and the pressure increases. Next, the air is heated by combustion of fuel, and then the compressed hot gas is expanded through a nozzle.

Consider this process for air as ideal gas with variable specific heats, and the following sub-processes:
1-2: Outside air at $T_1 = 240\,\text{K}$, $p_1 = 0.2\,\text{bar}$ enters the diffuser with velocity $v_1 = 3040\frac{\text{km}}{\text{h}}$; this is the velocity of the aircraft relative to the air
2-3: The compressed air is isobarically heated to $2300\,\text{K}$
3-4: The heated air is expanded through an isentropic nozzle to $p_4 = p_1$

1. Make a sketch of this series of processes in p-v- and T-s-diagrams.
2. Determine temperature T_2 and pressure p_2 at the diffuser exit. The diffuser can be considered to operate on an adiabatic reversible process, and at the diffuser exit the flow velocity is negligibly small.
3. Determine the heat added per unit mass of air flowing through, q_{23}.
4. Determine the exit velocity v_4.
5. The mass flow of air is $\dot{m} = 1\frac{\text{kg}}{\text{s}}$. Determine the cross sections at diffuser inlet and nozzle exit, A_1 and A_4.
6. Determine the thrust of the engine, $F = \dot{m}\,(v_4 - v_1)$ and the propulsive power $\dot{W} = F v_1$.

8.6. Normal Shock

A normal shock occurs in the exit cross section ($A_e = 0.75\,\text{m}^2$) of a rocket nozzle, due to high back pressure. The incoming flow properties are $T_0 = 400\,^\circ\text{C}$, $p_0 = 0.1\,\text{kPa}$, $\text{Ma}_0 = 3.5$.

1. Determine the state after the shock, ρ_1, p_1, T_1, Ma_1.
2. Determine the entropy generation rate $\dot{S}_{gen}$.

8.7. Shock in Diverging Channel

Air at $1.5\,\text{bar}$, $300\,\text{K}$ enters a diverging channel with a velocity of $800\frac{\text{m}}{\text{s}}$. The channel exit area is twice the inlet area, and a normal shock occurs directly at the inlet.

Consider air as ideal gas with constant specific heats, $R = 0.287\frac{\text{kJ}}{\text{kg K}}$, $\gamma = 1.4$. For the solution, use the flow function $\psi\left(\frac{p}{p_s}\right)$ or its plot. Show all units, and present all calculations clearly, with comments.

1. Make a sketch with numbered states.
2. Determine the state behind the shock.
3. Determine the back pressure.
4. What happens if the channel is inverted? Is a reversible flow possible? If not, is a shock still possible? Where?

8.8. A Diffuser

Air at $p_1 = 0.4\,\text{bar}$, $T_1 = 260\,\text{K}$ enters a diverging duct at Mach number $\text{Ma}_{in} = 3$, where the cross section is $0.5\,\text{m}^2$. A normal shock occurs directly at the inlet (to state 2), while the remaining flow through the duct is reversible, and reaches local speed of sound at the exit.

Consider air as ideal gas with constant specific heats, $R = 0.287\frac{\text{kJ}}{\text{kg K}}, \gamma = 1.4$. Show all units, and present all calculations clearly, with comments. Feel free to use the plot of the flow function $\psi\left(\frac{p}{p_s}\right)$.

1. Make a sketch with numbered states.
2. Determine state 2 behind the shock: pressure, density, temperature, velocity, Mach number.
3. Determine the new stagnation state for state 2 after the shock.
4. Determine the exit state: pressure, density, temperature, velocity, cross section.
5. Determine the entropy generation rate.

8.9. Laval Nozzle with Shock

In a Laval nozzle, a normal shock is observed at a location where the cross section is $0.5\,\text{m}^2$. The flow before the shock has a velocity $\mathcal{V}_1 = 1500\frac{\text{m}}{\text{s}}$, while pressure and temperature are $p_1 = 0.2\,\text{bar}$, $T_1 = 1200\,\text{K}$.

For the following, consider air as an ideal gas with constant specific heat $c_p = \frac{7}{2}R$. All flows apart from the shock are isentropic.

1. Determine the Mach number Ma_1 and density ρ_1 before the shock, and the mass flow.
2. Determine the stagnation state p_{s1}, T_{s1} for the flow before the shock.
3. Determine the cross section of the throat.
4. Determine the conditions just behind the shock, $\mathcal{V}_2$, p_2, T_2, ρ_2, Ma_2.
5. Determine the entropy generation rate across the shock.
6. Determine the stagnation state p_{s2}, T_{s2} for the flow after the shock. Compare to p_{s1}, T_{s1} and discuss.
7. For the above flow conditions, the back pressure of the nozzle is $p_b = 1.25\,\text{bar}$. Determine the exit cross section of the nozzle, and the corresponding exit velocity, temperature and Mach number

8. At what back pressure is the flow in the nozzle isentropic (no shock)? Determine the corresponding exit velocity, temperature and Mach number.

8.10. Reversible and Irreversible Laval Nozzles

The stagnation state for the combustion gas in a rocket engine is $T_s = 3000\,\mathrm{K}$, $p_s = 400\,\mathrm{kPa}$. The hot gas is expanded in Laval nozzles A and B which both have throat cross section $A_{th} = 0.05\,\mathrm{m}^2$, but different exit cross sections A_e^{A}, A_e^{B}. The back pressure (i.e., the pressure of the external environment) is $p_b = 40\,\mathrm{kPa}$. The gas can be described as air with constant specific heats ($R = 0.287\frac{\mathrm{kJ}}{\mathrm{kg\,K}}$, $\gamma = 1.4$), all flows except the shock are reversible.

1. Nozzle A is designed to allow for reversible expansion and acceleration so that the exit pressure is equal to the back pressure, $p_e^{\mathsf{A}} = p_b$. Determine for the exiting flow: temperature T_e^{A}, mass density ρ_e^{A}, velocity v_e^{A}, Mach number $\mathrm{Ma}_e^{\mathsf{A}}$, exit cross section A_e^{A}, mass flow $\dot{m}_{\mathsf{A}}$ and thrust $F^{\mathsf{A}} = \dot{m}_{\mathsf{A}} v_e^{\mathsf{A}}$.
2. Nozzle B is a rather failed design, where the flow is isentropic in the nozzle up to the exit cross section, where a normal shock in the exit cross section occurs such that the exiting flow is at $p_e^{\mathsf{B}} = p_b$.

   a. By combining equations for reversible nozzle flow and shocks find the equation to determine the pressure p_1^{B}, i.e., the pressure before the shock. Determine p_1^{B} from numerical solution.
   b. Determine for the state before the shock: temperature T_1^{B}, mass density ρ_1^{B}, velocity v_1^{B}, Mach number $\mathrm{Ma}_1^{\mathsf{B}}$, exit cross section A_e^{B}, mass flow $\dot{m}_{\mathsf{B}}$.
   c. Determine for the state behind the shock: temperature T_2^{B}, mass density ρ_2^{B}, velocity v_2^{B}, Mach number $\mathrm{Ma}_2^{\mathsf{B}}$, mass flow $\dot{m}_{\mathsf{B}}$ and thrust $F^{\mathsf{B}} = \dot{m}_B v_b^{\mathsf{B}}$.
   d. Determine the stagnation properties for state 2, and discuss the difference compared to the reversible case.
   e. Determine the entropy generation rate in the shock, $\dot{S}_{gen} = \dot{m}_{\mathsf{B}}\left(s_2^{\mathsf{B}} - s_1^{\mathsf{B}}\right)$.

3. Compare thrust and exit velocities of both nozzles.

8.11. Irreversible Laval Nozzles

In continuation of the previous problem, consider nozzle C which has the same stagnation state, throat cross section and back pressure, but its exit corss section lies above that of nozzle B, $A_e^{\mathsf{C}} > A_e^{\mathsf{B}}$. In this case, the shock appears not in the exit cross section, but somewhere within the diverging part of the Laval nozzle.

For a given A_e^{C} determine the cross section at the shock, as well as velocity, Mach number, and thrust at the exit. Plot these for a range of values. Use a suitable computer software.

8.12. Irreversible Laval Nozzle

The stagnation state for the combustion gas in a rocket engine is $T_s = 2500\,\mathrm{K}$, $p_s = 200\,\mathrm{kPa}$. The hot gas is expanded in a Laval nozzle with throat cross section $A_{th} = 0.1\,\mathrm{m}^2$, and exit cross section $A_e = 1.25\,\mathrm{m}^2$. The gas can be described as air with constant specific heats ($R = 0.287\frac{\text{kJ}}{\text{kg K}}$, $\gamma = 1.4$), all flows except the shock are reversible.

1. For which range of back pressure p_b will there be a normal shock within the nozzle?
2. For back pressures in this range, determine cross section, temperatures, pressures, velocities and Mach numbers at the location of the shock, and at the nozzle exit. Plot the results over back pressure. Use a suitable computer software.

8.13. Oblique Shock

An incoming flow of $\text{Ma}_0 = 3.0$, $p_0 = 0.7\,\text{bar}$, $T_0 = 280\,\text{K}$, with $\gamma = 1.4$, impinges in the direction of the symmetry axis on a wedge with opening angle of $40°$.

Determine the two possible shock angles, and the flow conditions v_1, p_1, T_1, ρ_1, Ma_1 behind the shock for weak and strong shock. For both cases, determine parallel and perpendicular flow velocities before and behind the shock. Compare the results, and discuss.

8.14. Chapman-Jouguet Points

Show that the Mach number behind a Chapman-Jouguet deflagration or detonation is equal to 1.

8.15. Chapman-Jouguet Points

Consider a reacting gas mixture, which fulfills the assumptions made when discussing Chapman-Jouguet theory (constant specific heats, same values of R and γ for products and reactants, etc.).

The reactant state is $p_0 = 1.2\,\text{bar}$, $T_0 = 300\,\text{K}$, the gas constant of the mixture is $R = 0.60\frac{\text{kJ}}{\text{kg K}}$, the ratio of specific heats is $\gamma = 1.25$, and the dimensionless heat of reaction is $\mathsf{q} = -\frac{\Delta h_R}{c_p T_0} = 8$.

1. Assume that the flame front is at a Chapman-Jouguet point, hence $\text{Ma}_1 = 1$, and determine pressure p_1, density ρ_1, temperature T_1, flame velocity v_0, product velocity v_1, and and the corresponding Mach number Ma_0 for detonation and deflagration. Use the computer to obtain the solution, but show and explain the procedure.
2. Consider the velocities in the rest frame of the reactant, and show that for the deflagration the product moves away from the front, but for a detonation it follows the front.
3. Plot the Hugoniot curve and the two Rayleigh lines for Chapman-Jouguet detonation and deflagration.

8.16. Deflagrations

Consider deflagrations for a reactant at $p_0 = 2\,\text{bar}$, $T_0 = 500\,\text{K}$, with $R = 0.4\frac{\text{kJ}}{\text{kg K}}$, $\gamma = 1.3$, $\mathsf{q} = -\frac{\Delta h_R}{c_p T_0} = 4$.

1. Determine velocity and Mach number of the reactant for a Chapman-Jouguet deflagration.

2. For a reactant Mach number $\mathrm{Ma}_0 = 0.15$, determine the product state (p_1, ρ_1, T_1, v_1, Ma_1) for weak and strong deflagration.
3. Determine the change of entropy across the flame for both.

8.17. Detonations

Consider detonations for a reactant at $p_0 = 3\,\mathrm{bar}$, $T_0 = 800\,\mathrm{K}$, with $R = 0.5\frac{\mathrm{kJ}}{\mathrm{kg\,K}}$, $\gamma = 1.1$, $\mathsf{q} = -\frac{\Delta h_R}{c_p T_0} = 4$.

1. Determine velocity and Mach number of the reactant for a Chapman-Jouguet detonation.
2. For a reactant Mach number $\mathrm{Ma}_0 = 8$, determine the product state (p_1, ρ_1, T_1, v_1, Ma_1) for weak and overdriven detonation.
3. Determine the change of entropy across the flame for both.

Chapter 9
Mixtures in Nonequilibrium

Due to cross-coupling and, possibly, a large number of components, transport processes in mixtures are rather rich. This chapter aims at giving some insight by studying binary mixtures in simple settings. Driving forces for diffusion are discussed, including separation of gas mixtures in centrifuges. Combined convective and diffusive transport across semi-permeable membranes for desalination and pressure retarded osmosis is used as an instructive example. The chapter closes with a motivation of the Arrhenius law for reaction rates by extension of the principles of Linear Irreversible Thermodynamics towards nonlinear expressions.

9.1 Binary Mixtures

9.1.1 Basic Equations

Above, we have, through a number of simplifying assumptions, reduced the transport equations for mixtures to the simple diffusion equation. To gain further insight into the rich array of transport processes occurring, we study the transport equations for an ideal non-reacting binary mixture. An ideal mixture has no mixing volume, and no enthalpy of mixing.

Collecting from earlier sections, the transport equations are:
mass balance, with overall mass density ρ, center of mass velocity v_k,

$$\frac{D\rho}{Dt} + \rho\frac{\partial v_k}{\partial x_k} = 0 \tag{9.1}$$

partial mass balance, with mass fraction $\mathsf{c} = \mathsf{c}_1 = 1 - \mathsf{c}_2$, and diffusion flux $J_k = J_k^1 = -J_k^2$,

$$\rho\frac{D\mathsf{c}}{Dt} + \frac{\partial J_k}{\partial x_k} = 0 \tag{9.2}$$

H. Struchtrup, *A Thermodynamic Introduction to Transport Phenomena*,
https://doi.org/10.1007/978-3-031-61868-0_9

momentum balance, with stress tensor t_{ik},

$$\rho\frac{Dv_i}{Dt}-\frac{\partial t_{ik}}{\partial x_k}=\rho f_i \tag{9.3}$$

energy balance, with heat flux q_k,

$$\rho\frac{Du}{Dt}+\frac{\partial q_k}{\partial x_k}=t_{ij}\frac{\partial v_i}{\partial x_j}\,. \tag{9.4}$$

Here, $u\left(\rho,T,\mathsf{c}\right)$ is the caloric equation of state for the mixture, and the fluxes are given by the phenomenological equations

$$t_{ij}=-p\delta_{ij}+\lambda\frac{\partial v_k}{\partial x_k}\delta_{ij}+2\eta\frac{\partial v_{\langle i}}{\partial x_{j\rangle}} \tag{9.5}$$

$$q_k=-\kappa\frac{\partial T}{\partial x_k}-B_1\frac{\partial}{\partial x_k}\left(\frac{\mu_1-\mu_2}{T}\right) \tag{9.6}$$

$$J_k=-\frac{B_1}{T^2}\frac{\partial T}{\partial x_k}-B_{11}\frac{\partial}{\partial x_k}\left(\frac{\mu_1-\mu_2}{T}\right) \tag{9.7}$$

with thermal equation of state $p\left(\rho,T,\mathsf{c}\right)$, chemical potentials $\mu_\alpha\left(\rho,T,\mathsf{c}\right)$, and phenomenological coefficients $\lambda,\eta,\kappa,B_1,B_{11}$ that all depend on (ρ,T,c). Positive definiteness of the coefficient matrix implies $\kappa T^2>0$, $B_{11}>0$ and $\left[\kappa T^2B_{11}-B_1^2\right]>0$; the coefficient B_1 might be positive or negative.

9.1.2 Ideal Mixture

The chemical potential for an ideal mixture is

$$\mu_\alpha=g_\alpha\left(T,p\right)+R_\alpha T\ln X_\alpha \tag{9.8}$$

where $g_\alpha\left(T,p\right)$ is the Gibbs free energy of component α alone at mixture temperature T and pressure p, and X_α is the mole fraction, with $X=X_1=1-X_2$. The mole fraction is related to mass fractions through partial mole and mass densities, as

$$X=\frac{n_1}{n_1+n_2}=\frac{\frac{\rho_1}{M_1}}{\frac{\rho_1}{M_1}+\frac{\rho_2}{M_2}}=\frac{\frac{\mathsf{c}_1}{M_1}}{\frac{\mathsf{c}_1}{M_1}+\frac{\mathsf{c}_2}{M_2}}=\frac{\frac{\mathsf{c}}{M_1}}{\frac{\mathsf{c}}{M_1}+\frac{1-\mathsf{c}}{M_2}} \tag{9.9}$$

To proceed, we need some derivatives of the chemical potentials. The Gibbs equation for a component alone at (T,p) reads $dg_\alpha=-s_\alpha dT+\mathsf{v}_\alpha dp$, where $\mathsf{v}_\alpha\left(T,p\right)=\left(\frac{\partial g_\alpha}{\partial p}\right)_T$ and $s_\alpha\left(T,p\right)=-\left(\frac{\partial g_\alpha}{\partial T}\right)_p$ are specific volume and entropy of α alone at (T,p). Considering (T,p,c) as variables, we find the following derivatives:

$$\frac{\partial}{\partial T}\left(\frac{\mu_\alpha}{T}\right)=\frac{1}{T}\left(\frac{\partial g_\alpha}{\partial T}\right)_p-\frac{g_\alpha}{T^2}=-\frac{s_\alpha}{T}-\frac{h_\alpha-Ts_\alpha}{T^2}=-\frac{h_\alpha(T,p)}{T^2} \tag{9.10}$$

$$\frac{\partial}{\partial p}\left(\frac{\mu_\alpha}{T}\right)=\frac{1}{T}\left[\left(\frac{\partial g_\alpha}{\partial p}\right)_T\right]=\frac{\mathsf{v}_\alpha(T,p)}{T} \tag{9.11}$$

$$\frac{\partial}{\partial \mathsf{c}}\left(\frac{\mu_\alpha}{T}\right)=R_\alpha\frac{d\ln X_\alpha}{d\mathsf{c}}=R_\alpha\frac{d}{d\mathsf{c}}\ln\left(\frac{\frac{\mathsf{c}_\alpha}{M_\alpha}}{\frac{\mathsf{c}}{M_1}+\frac{1-\mathsf{c}}{M_2}}\right)=\frac{\bar{R}\frac{1}{\mathsf{c}_\alpha}\frac{d\mathsf{c}_\alpha}{d\mathsf{c}}}{(1-c)M_1+cM_2} \tag{9.12}$$

These are now used to study the fluxes further.

For the heat flux we obtain

$$q_k=-\left[\kappa+B_1\left(\frac{h_1-h_2}{T^2}\right)\right]\frac{\partial T}{\partial x_i}-B_1\left[\frac{\mathsf{v}_1-\mathsf{v}_2}{T}\right]\frac{\partial p}{\partial x_i}-B_1\left[\frac{\bar{R}}{\mathsf{c}(1-\mathsf{c})\left[(1-\mathsf{c})M_1+\mathsf{c}M_2\right]}\right]\frac{\partial \mathsf{c}}{\partial x_i}, \tag{9.13}$$

where h_α, v_α denote properties of the constituents alone at (T,p). Here, the first term describes heat transfer due to temperature gradients, the usual thermal conduction. The second term describes heat transfer through pressure gradients, and the third term describes the Dufour, or diffusion-thermo, effect, that is heat transfer due to gradients in composition.

For the diffusion flux we obtain

$$J_k=-\left[\frac{B_1}{T^2}+B_{11}\left(\frac{h_1-h_2}{T^2}\right)\right]\frac{\partial T}{\partial x_i}-B_{11}\left[\frac{\mathsf{v}_1-\mathsf{v}_2}{T}\right]\frac{\partial p}{\partial x_i}-B_{11}\left[\frac{\bar{R}}{\mathsf{c}(1-\mathsf{c})\left[(1-\mathsf{c})M_1+\mathsf{c}M_2\right]}\right]\frac{\partial \mathsf{c}}{\partial x_i} \tag{9.14}$$

where the first term describes the Soret, or thermo-diffusion, effect, i.e., diffusion driven by temperature gradients. The second term describes mass transfer through pressure gradients. The third term describes conventional mass diffusion, which is driven by gradients of composition.

The pressure effects are interesting. Since $B_{11}>0$, component 1 is pushed towards the region of lower pressures (against the pressure gradient), if it has larger specific volume (alone), $\mathsf{v}_1(T,p)<\mathsf{v}_2(T,p)$. For a mixture of ideal gases, the pressure contribution to mass flux is, with the ideal gas law $\mathsf{v}_\alpha=\frac{\bar{R}T}{pM_\alpha}$,

$$J_k^p=-B_{11}\left[\frac{\mathsf{v}_1-\mathsf{v}_2}{T}\right]\frac{\partial p}{\partial x_i}=-B_{11}\bar{R}\left[\frac{1}{M_1}-\frac{1}{M_2}\right]\frac{\partial\ln p}{\partial x_i}\,. \tag{9.15}$$

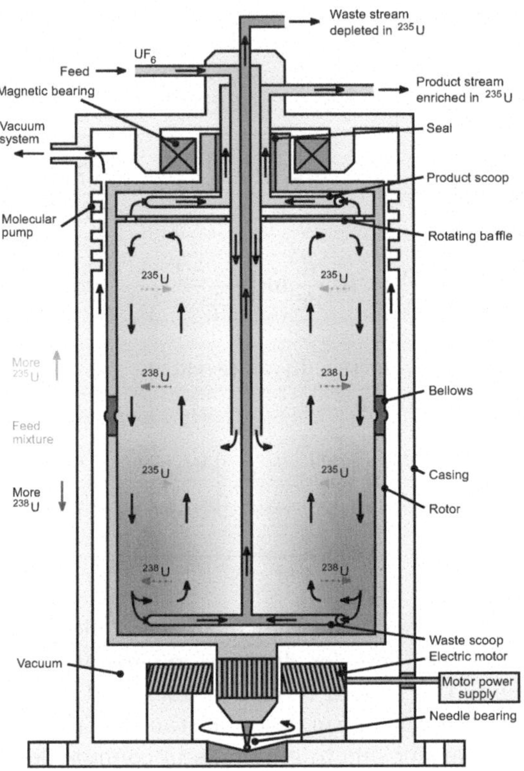

Fig. 9.1 Gas centrifuge for separation of uranium isotopes ^{235}U and ^{238}U via gaseous UF_6 (https://en.wikipedia.org/wiki/Gas_centrifuge#).

Accordingly, the gas with lower molar mass is pushed towards the region of lower pressure, the heavier gas is pushed towards higher pressures. This effect is used in centrifugal gas separation, where the pressure gradients are established by fast rotation.[1] With many centrifuges in series, it is possible to separate gases with small differences in molar mass, for instance separation of uranium isotope ^{235}U from the dominant ^{238}U using gaseous uranium hexafluoride, UF_6. The temperature contribution to mass flux can be used to enhance centrifuge performance, with pressure gradient in radial direction, and temperature gradient in vertical direction, see Fig. 9.1.

In heat transfer and diffusion problems pressure typically varies little, and heat and diffusion fluxes are driven dominantly by gradients in temperature and composition. Nevertheless, pressure gradients might drive the flow, as expressed in the momentum balance.

[1] The gas in the centrifuge is subject to the centrifugal body force $f_{\text{cent}} = \omega^2 r$ (radius $r = \sqrt{x^2 + y^2}$, angular velocity ω), see Prob. A.20.

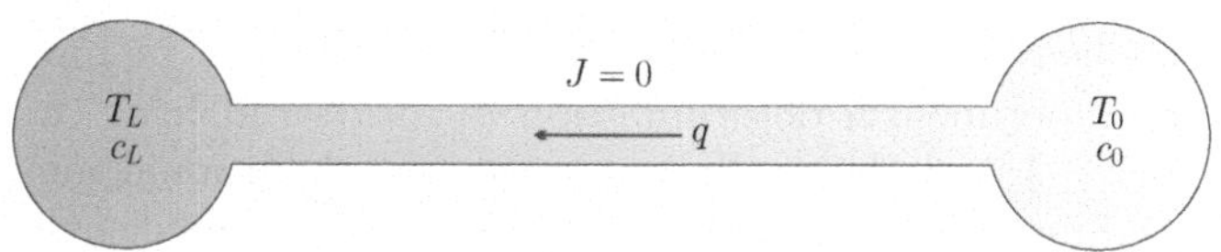

Fig. 9.2 A binary mixture in temperature gradient at steady state exhibits a concentration gradient.

Since B_1 has no prescribed sign, Soret and Dufour effects might be in either direction, depending on the transport properties of the mixture.

9.1.3 Steady State Heat Transfer

As a simple example, we consider a binary mixture enclosed between two boundaries at different temperatures. There will be no macroscopic flow, $v_i = 0$, hence, from the balance of momentum, $p = const.$

Heat and diffusion fluxes can be written in compact form as

$$q_i = -\hat{\kappa}\frac{\partial T}{\partial x_i} - \kappa_D \frac{\partial \mathsf{c}}{\partial x_i} \quad , \quad J_i = -D_T \frac{\partial T}{\partial x_i} - D\frac{\partial \mathsf{c}}{\partial x_i} \ . \tag{9.16}$$

where we introduced the abbreviations for heat transfer, diffusion-thermo, thermo-diffusion, and diffusion coefficients as

$$\begin{aligned} \hat{\kappa} &= \kappa + B_1 \frac{h_1 - h_2}{T^2} \quad , \quad \hat{\kappa}_D = B_1 \frac{\mathsf{v}_1 - \mathsf{v}_2}{T} \\ D_T &= \frac{B_1}{T^2} + B_{11}\frac{h_1 - h_2}{T^2} \quad , \quad D = \frac{\bar{R}\, B_{11}}{\mathsf{c}\,(1-\mathsf{c})\left[(1-\mathsf{c})\, M_1 + \mathsf{c} M_2\right]} \end{aligned} \tag{9.17}$$

Note that with this compact notation Onsager symmetry becomes invisible in the equations, one must rely on the above list (9.17) to recognize that these four coefficients are determined by three transport coefficients κ, B_{11}, B_1 and thermal and caloric equations of state.

We consider a one-dimensional steady state setting with impermeable boundaries for all components, as sketched in Fig. 9.2. The diffusion flux must vanish, $J_x = 0$, hence the composition gradient is linked to the temperature gradient as

$$\frac{d\mathsf{c}}{dx} = -\frac{D_T}{D}\frac{dT}{dx} \ . \tag{9.18}$$

With this, the heat flux becomes

$$q_x = -\left(\hat{\kappa} - \kappa_D \frac{D_T}{D}\right)\frac{dT}{dx} \ . \tag{9.19}$$

Assuming that heat flux and the gradients of T and c can be measured, this setting allows experimental determination of an overall heat transfer coefficient $\left(\hat{\kappa} - \kappa_D \frac{D_T}{D}\right)$ and the coefficient ratio $\frac{D_T}{D}$. Additional measurements must be devised in order to measure all four transport coefficients $\hat{\kappa}$, κ_D, D_T, D of which three are independent.

This simple example shows that cross-coupling makes it quite difficult to measure transport coefficients in mixtures. Additional difficulty will arise from buoyancy effects due to variation of mixture density ρ, which induce convection currents affecting the system. Note that all coefficients depend on the local state $\{\rho, T, \mathsf{c}\}$.

9.2 Osmotic Processes

9.2.1 Semi-permeable Membranes

We study local osmotic transport processes. Specifically, we consider semi-permeable membranes that divide salt solutions of different salinity. The membranes are designed to let water pass, but not salt, although in practice some salt passes through as well. The main area of application for such semi-permeable membranes is in desalination of saltwater—mostly seawater—to produce drinking water. They are also discussed in the context of osmotic power production, where power is drawn from mixing flows at different salinity.

Proper understanding and design of such systems requires models for water and salt transport across the membrane, which we discuss under simplifying assumptions.

We assume that the salt solutions are incompressible ideal mixtures. Then, the chemical potential of water in the mixture is

$$\mu_{\mathrm{w}}\left(T, p, X_{\mathrm{w}}\right) = g_{\mathrm{w}}\left(T, p\right) + R_{\mathrm{w}} T \ln X_{\mathrm{w}} \;, \tag{9.20}$$

where X_{w} is the mole fraction of water in the solutions. We note that in water NaCl dissociates into Na^+ and Cl^- ions, hence the mole fraction is

$$X_{\mathrm{w}} = \frac{n_{\mathrm{w}}}{n_{\mathrm{w}} + n_{\mathrm{Na}^+} + n_{\mathrm{Cl}^-}} = \frac{n_{\mathrm{w}}}{n_{\mathrm{w}} + 2n_{\mathrm{NaCl}}} \;, \tag{9.21}$$

where n_α are mole numbers (or mole densities) of the components. The mole fraction of salt is $X_s = 1 - X_{\mathrm{w}}$.

We consider two solutions A and B of different concentration and pressure but at the same temperature separated by a membrane, as shown in Fig. 9.3. For now, we consider only water transport across the membrane, with the mass flux J_{w}.

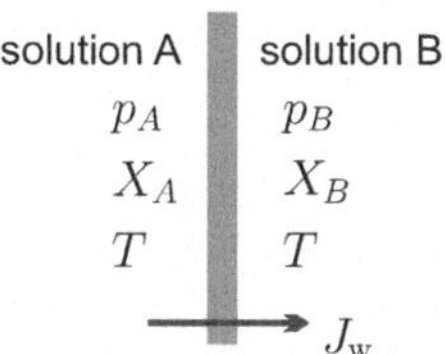

Fig. 9.3 Two water-salt solutions in contact through a semi-permeable membrane that allows water to pass, but not the salt. Water passes the membrane until the chemcial potential is equilibrated.

The membrane must balance the hydraulic pressure difference $\Delta p = p_B - p_A$, hence has internal stresses which appear as supply terms in the balance—this mechanical problem is ignored. Note that the membrane material itself cannot carry these stresses, hence the actual membrane material is attached to a porous support layer, which affects transport behavior.

9.2.2 Water Transport across Membrane

Considering the membrane as a singular surface, mass and energy balance for transport of water across the membrane read

$$J_\mathrm{w} = \rho_\mathrm{w}^A v_\mathrm{w}^A = \rho_\mathrm{w}^B v_\mathrm{w}^B \quad , \quad h_\mathrm{w}^A J_\mathrm{w} + q^A = h_\mathrm{w}^B J_\mathrm{w} + q^B \; . \tag{9.22}$$

The second law accounts for transfer of entropy and entropy generation as

$$J_\mathrm{w}\left(s_\mathrm{w}^B - s_\mathrm{w}^A\right) + \frac{q^B - q^A}{T} = \Sigma_s \geq 0 \; , \tag{9.23}$$

where we have assumed homogeneous temperature. Note that dissipative processes in the membrane might lead to heating, the heat fluxes $q^{A,B}$ describe the heat removed from the membrane into the solutions A and B.

As we did often before, we combine the conservation laws with the second law; eliminating $q^B - q^A$ gives

$$J_\mathrm{w}\left[\left(h_\mathrm{w}^A - T s_\mathrm{w}^A\right) - \left(h_\mathrm{w}^B - T s_\mathrm{w}^B\right)\right] = T\Sigma_s \geq 0 \; , \tag{9.24}$$

or, with the chemical potential $\mu_\mathrm{w} = h_\mathrm{w} - T s_\mathrm{w}$,

$$J_\mathrm{w}\left(\mu_\mathrm{w}^A - \mu_\mathrm{w}^B\right) = T\Sigma_s \geq 0 \; . \tag{9.25}$$

In the context of Linear Irreversible Thermodynamics, this relation suggests a phenomenological equation for water transport of the form

$$J_\mathrm{w} = \hat{\mathrm{A}}\left(\mu_\mathrm{w}^A - \mu_\mathrm{w}^B\right) \; , \tag{9.26}$$

that is the water flux across the membrane is proportional to the difference of the chemical potential of water with positive transport coefficient $\hat{A}$.

For an ideal mixture, and the water as incompressible liquid, introducing the hydraulic pressure difference $\Delta p = p_B - p_A$ we find[2]

$$\mu_w^A = g_w(T, p_A) + R_w T \ln X_{w,A} , \tag{9.27}$$

$$\mu_w^B = g_w(T, p_B) + R_w T \ln X_{w,B} = g_w(T, p_A) + \frac{1}{\rho_w}\Delta p + R_w T \ln X_{w,B} . \tag{9.28}$$

With this, the mass flux of water across the membrane can be written as

$$J_w = \frac{\hat{A}}{\rho_w}\left(\rho_w R_w T \ln \frac{X_{w,A}}{X_{w,B}} - \Delta p\right) , \tag{9.29}$$

where ρ_w denotes the mass density of water alone at (T, p).

9.2.3 Osmotic Pressure

We study equilibrium across the membrane for the special case that solution A is pure water, $X_A = 1$. In equilibrium flux and force must vanish, so that $J_{w|E} = 0$ and $\mu_w^A = \mu_w^B$. Then, to maintain equilibrium between the two systems, the hydraulic pressure difference must equal the osmotic pressure of solution B, which is thus defined as

$$\pi_B = \Delta p_{|E} = -\rho_w R_w T \ln X_{w,B} . \tag{9.30}$$

For dilute solutions, where water mole fraction X differs only little from unity, the expression for osmotic pressure can be somewhat simplified. Specifically, we can write, with mole density (or molar concentration[3]) $\bar{C}_\alpha$,

$$-\ln X_w = -\ln \frac{\bar{C}_w}{\bar{C}_w + 2\bar{C}_{NaCl}} = \ln\left(1 + \frac{2\bar{C}_{NaCl}}{\bar{C}_w}\right) \simeq \frac{2\bar{C}_{NaCl}}{\bar{C}_w} , \tag{9.31}$$

In a dilute solution the molar concentration of water $\bar{C}_w$ in the mixture differs only little from the molar density of water alone, hence $M_w \bar{C}_w \simeq \rho_w$. We also use $R_w = \bar{R}/M_w$ to express the osmotic pressure through either the partial mole density (molar concentration) $\bar{C}_{NaCl}$ or the partial mass density (mass

[2] Once more we use $\left(\frac{\partial g}{\partial p}\right)_T = \frac{1}{\rho}$. The expanision is exact for incompressible substances ($\rho = const.$), since all higher contributions vanish, $\left(\frac{\partial^n g}{\partial p^n}\right)_T = \left(\frac{\partial^{n-1} \frac{1}{\rho}}{\partial p^{n-1}}\right)_T = 0$.

[3] In this chapter we use notation typical for this field. The term *concentration* refers to an amount per volume, just as *density*. Often authors simply refer to "concentration", and leave it to the reader to find out whether it is mass concentration, or mole concentration, or number concentration, or ...

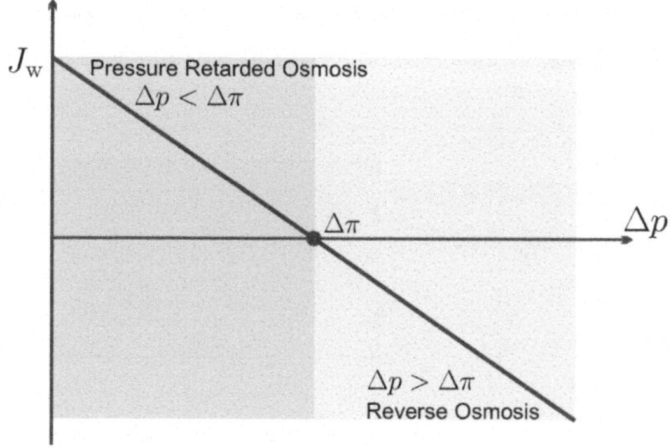

Fig. 9.4 Osmotic and hydrostatic pressure differences and flow directions.

concentration) C_{NaCl} of salt

$$\pi = \rho_{\mathrm{w}} R_{\mathrm{w}} T \frac{2\bar{C}_{\mathrm{NaCl}}}{\bar{C}_{\mathrm{w}}} = 2\bar{C}_{\mathrm{NaCl}} \bar{R} T = 2 C_{\mathrm{NaCl}} R_{\mathrm{NaCl}} T \,. \tag{9.32}$$

For other solutions one writes

$$\pi = \mathsf{i} \bar{C}_{\mathrm{s}} \bar{R} T = \mathsf{i} C_{\mathrm{s}} R_{\mathrm{s}} T \tag{9.33}$$

where $\bar{C}_{\mathrm{s}}$, C_{s} are mole and mass density (concentrations) of the dissolved substance and i is the van't Hoff factor that indicates the number of ions into which the dissolved substance dissociates ($\mathsf{i} = 2$ for NaCl, $\mathsf{i} = 3$ for $CaCl_2$).

9.2.4 Osmotic Flow Processes

With the definition of osmotic pressure, we can rewrite (9.29) for the volume flux $\hat{J}_{\mathrm{w}}$ across the membrane, i.e., the throughput velocity ($\frac{\mathrm{m}}{\mathrm{s}} = \frac{\mathrm{m}^3}{\mathrm{m}^2\,\mathrm{s}}$), as

$$\hat{J}_{\mathrm{w}} = \frac{J_{\mathrm{w}}}{\rho_{\mathrm{w}}} = \frac{\hat{\mathrm{A}}}{\rho_{\mathrm{w}}^2} \left(\pi_B - \pi_A - \Delta p\right) = \mathrm{A}\left(\Delta\pi - \Delta p\right) \,. \tag{9.34}$$

The coefficient $\mathrm{A} = \hat{\mathrm{A}}/\rho_{\mathrm{w}}^2$ is known as the water permeation coefficient with SI unit $[\mathrm{A}] = \frac{\mathrm{m}}{\mathrm{Pa\,s}}$.

Our signs are such that flow from system A to system B is positive ($J_{\mathrm{w}} > 0$), and $\Delta p - p_B - p_A$, $\Delta\pi = \pi_B - \pi_A$. We thus expect the following, see Fig. 9.4:

a) Hydraulic flow: If $\Delta\pi_{\mathrm{osm}} = 0$ (no difference in concentration) and $\Delta p > 0$, then $\hat{J}_{\mathrm{w}} < 0$: water will flow from system B to A, that is from high to

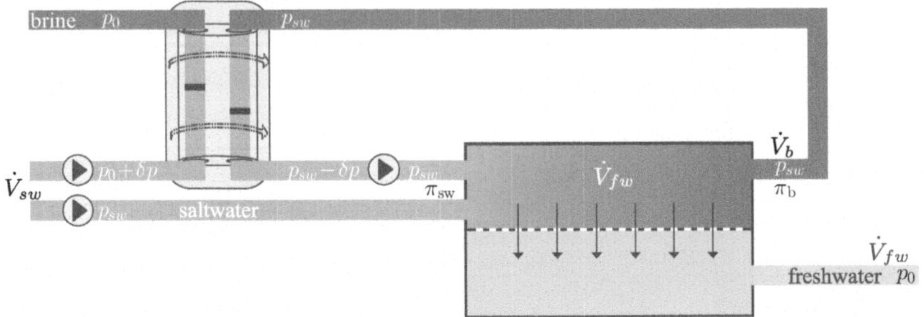

Fig. 9.5 Schematic of a reverse osmosis (RO) desalination plant with pressure exchanger, pump, and booster pumps.

low hydraulic pressure. When no salt is present, the resulting relation $\hat{J}_\mathrm{w} = -\mathrm{A}\Delta p$ is Darcy's law for flow through a porous medium.

b) Direct Osmosis: If $\Delta p = 0$ (no difference in hydraulic pressure) and $\Delta\pi > 0$, then $\hat{J}_\mathrm{w} > 0$: water will flow from system A to B, that is water is drawn into the saltier solution.
c) Pressure Retarded Osmosis (PRO): If $\Delta\pi > 0$, and $\Delta p > 0$, but $\Delta p < \Delta\pi$, then $\hat{J}_\mathrm{w} > 0$: water will flow from A to B, that is from the less salty solution to the more salty solution, but against the hydraulic pressure gradient.
d) Reverse Osmosis (RO): If $\Delta\pi > 0$, and $\Delta p > 0$, but $\Delta p > \Delta\pi$, then $\hat{J}_\mathrm{w} < 0$: water will flow from B to A, i.e., from the system with higher hydraulic pressure and higher salt concentration to the system with lower pressure and salt concentration.

9.2.5 Applications: RO and PRO

Reverse osmosis is now the prevalent method to produce freshwater. Figure 9.5 shows a simplified schematic of a reverse osmosis (RO) seawater desalination plant, where incoming seawater is pressurized from environmental pressure p_0 to a pressure $p_{sw} > \pi_\mathrm{sw}$ and then runs through a membrane unit where freshwater passes through the membrane. The pressurized brine (with $\pi_\mathrm{b} > \pi_\mathrm{sw}$) leaving the membrane is used to pressurize part of the incoming seawater in a pressure exchanger, while the other part of the seawater is pressurized by a pump.

The pump work required will be larger in actual processes with irreversible losses. For a membrane-based desalination system there are losses to friction in the pipes, the pump, in filters, and as water is pressed through the membrane. Below we will discuss losses due to non-uniform salt concentration: salt might accumulate in front of the membrane and this *concentration polarization* increases the local osmotic pressure and thus the work requirement.

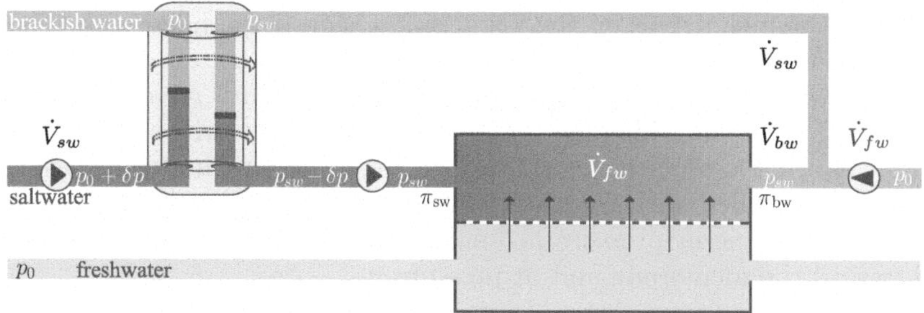

Fig. 9.6 Pressure retarded osmosis power plant with pressure exchanger, booster pumps and turbine.

Typical values for separation work in commercial membrane desalination systems are about 10 kJ per litre of freshwater.

In the inverse process, pressure retarded osmosis (PRO) can produce power as shown in Fig. 9.6. Saltwater is pressurized to a pressure p_{sw} between the environmental and osmotic pressures, $(p_{sw} - p_0) < \pi_{\rm sw}$, so that freshwater at the environmental pressure p_0 is drawn through the semipermeable membrane. This increases the flow of pressurized brackish ($\pi_{\rm bw} < \pi_{\rm sw}$) water, which is split into two streams. The first stream runs through a turbine to produce power, while the other stream is used to pressurize the incoming seawater in a pressure exchanger. Irreversible pressure losses in pipes and pressure exchanger reduce the amount of power produced. Moreover, careful pre-treatment (filtering etc.) of the feed solutions is required to avoid membrane fouling, and demands a considerable amount of work. Due to the low energy density of the process, the significant irreversible losses in membranes, and costly pre-treatment, osmotic power generation appears to be not viable for commercial deployment.[4]

9.2.6 Pressure Exchangers

Since seawater, brine, and brackish water are incompressible, direct pressure exchangers (PX) are commonly employed, which aim at direct pressure transfer between a high-pressure stream and a low-pressure stream of incompressible liquids. Pressure exchangers use flow channels in a rotating drum, so that the channels are sometimes in contact with the low pressure flows, and sometimes with the high pressure flows. We discuss this for the flows of brine and saltwater in Fig. 9.5, where pressure is to be exchanged between the exiting high pressure brine flow $\dot{V}_b$ at p_{sw} and the incoming low pressure saltwater flow $\widehat{\dot{V}}_{sw}$ (the portion of $\dot{V}_{sw}$ flowing through the PX) at $p_0 + \delta p$.

[4] Z. Wang, L. Wang, M. Elimelech, Joule **8**, 334-343 (2024); doi: 10.1016/j.joule.2023.12.015

When the channel arrives on the low pressure side (left), filled with brine b, it is brought into contact with incoming slightly pressurized saltwater at pressure $p_0 + \delta p = p_{sw}$ on one side, and brine at pressure p_0 at the other side. Due to the pressure difference δp, brine is pushed into the environment at p_0, that is brine is replaced by saltwater. Next, the rotating drum brings the channel, now filled with saltwater, into contact with the high pressure streams (right). Brine at pressure p_{sw} enters the channel while saltwater is pushed toward the membrane unit at pressure $p_{sw} - \delta p$.

The irreversible pressure losses δp ensure flow in charge and discharge, they result from hydrodynamic losses; the typical value for these losses is $\delta p = 0.5\,\text{bar}$.

Since the liquids replace each other in charge/discharge, the volume flow rates for the two entering streams must be equal, that is $\widehat{\dot{V}}_{sw} = \dot{V}_b$. Some pressure exchangers use a small amount of leakage to ensure lubrication, but generally leakage is small, and is ignored in the discussion below.

Because of the pressure loss δp the exchange of pressures between the two streams is not perfect. To correct for the losses, feed and booster pumps are required, as shown in Fig. 9.5, which together require the total work input

$$\delta \dot{W}_{PX} = 2\,\delta P\,\dot{V}_b\,. \tag{9.35}$$

The loss factor of the pressure exchanger is defined as the ratio of extra work required relative to the target work exchange,[5]

$$\gamma_{PX} = \frac{\delta \dot{W}_{PX}}{\dot{V}_b\,(p_{sw} - p_0)} = \frac{2\,\delta P}{p_{sw} - p_0}\,. \tag{9.36}$$

With the pressure loss being approximately constant, the loss factor decreases significantly with increasing pressure difference, hence pressure exchangers are particularly useful at large pressure differences. In RO plants, which operate with pressure differences of $\sim 60\,\text{bar}$, the loss factor is only $\gamma_{PX} = 1.7\%$.

The loss factor for turbine-pump pairs operating between pressures p_H and p_L follows from their isentropic efficiencies $\eta_{T,P}$ as[6]

$$\gamma_{TP} = \frac{\delta \dot{W}_{TP}}{\dot{V}\,(p_H - p_L)} = \frac{\left|\dot{W}_P\right| - \dot{W}_T}{\dot{V}\,(p_H - p_L)} = \frac{\left|\frac{\dot{W}_P^{rev}}{\eta_P} + \eta_T \dot{W}_T^{rev}\right|}{\dot{V}\,(p_H - p_L)} = \frac{1}{\eta_P} - \eta_T \tag{9.37}$$

so that for $\eta_T = \eta_P = 0.9$ the loss factor is $\gamma_{TP} = 21\%$, independently of the pressure difference.

---

[5] D. Bharadwaj, T.M. Fyles, H. Struchtrup, J. Non-Eq. Thermodyn. **41**, 327-347 (2016); doi: 10.1515/jnet-2016-0017.

[6] Reversible power is $\dot{W} = -\dot{V}\Delta p$ with $\Delta p = (p_{out} - p_{in}) = (p_L - p_H) < 0$ for a turbine and $\Delta p = (p_{out} - p_{in}) = (p_H - p_L) > 0$ for a pump [TEC, Chap. 9].

9.2.7 Combined Water and Salt Transfer[7]

With ideal semi-permeable membranes, equilibrium at $\Delta p = \Delta\pi \neq 0$ is possible. However, real semi-permeable membranes do not only allow water to pass, but also small amounts of salt, hence transport processes will be ongoing. This transport of salt is typically described as Fickian diffusion through the membrane, with diffusion mass flux of salt as

$$J_{\rm s}^{\rm m} = {\rm B}\left(C_{{\rm s},A} - C_{{\rm s},B}\right) = -{\rm B}\Delta C_{\rm s} \ . \tag{9.38}$$

The coefficient B is the salt permeability coefficient with SI unit $[{\rm B}] = \frac{\rm m}{\rm s}$. The sign in the above equation for $J_s^{\rm m}$ is such that positive direction is from $A \to B$.

In the adjacent bulk phases, the transport of salt is due to convective transfer and diffusion. From the definition of the diffusion flux (3.36) we have

$$C_{\rm s}v_{\rm s} = C_{\rm s}v + J_{\rm s} \quad , \quad \rho_{\rm w}v_{\rm w} = \rho_{\rm w}v - J_{\rm s} \tag{9.39}$$

Here, $C_{\rm s}v_{\rm s}$ is the local mass flux of salt, $C_{\rm s}v$ is the convective mass flux of salt, travelling with the velocity v of the mixture, and $J_{\rm s} = -D\frac{dC_{\rm s}}{dx}$ is the diffusion flux according to Fick's law, written for mass concentration.

The equation for water transport is simplified by using the mass density of water alone, instead of the partial mass density in the mixture. In steady state the mass flux in the bulk phases must equal the salt flux through the membrane, $J_{\rm s}^{\rm m}$. Eliminating mixture velocity v between (9.39), and with the water volume flux through the membrane $\hat{J}_{\rm w} = v_{\rm w}$, the membrane salt flux is

$$J_{\rm s}^{\rm m} = C_{\rm s}v_{\rm s} = C_{\rm s}\hat{J}_{\rm w} - D_{\rm s}\frac{dC_{\rm s}}{dx} = const. \ , \tag{9.40}$$

where $D_{\rm s} = \frac{\rho_{\rm s}+\rho_{\rm w}}{\rho_{\rm w}}D$ is the diffusion coefficient of salt relative to water (and not the mixture);[8] we shall consider this coefficient to be constant.

Integration between concentrations $C_{{\rm s},a}$ and $C_{{\rm s},b}$ over a distance δ yields

$$C_{{\rm s},b} = C_{{\rm s},a}e^{\frac{\hat{J}_{\rm w}}{D/\delta}} + \frac{J_{\rm s}^{\rm m}}{\hat{J}_{\rm w}}\left(1 - e^{\frac{\hat{J}_{\rm w}}{D/\delta}}\right) \tag{9.41}$$

Written with concentrations, the volume flow of water through the membrane (9.34) is

$$\hat{J}_{\rm w} = {\rm A}\left(2R_{\rm s}T\Delta C_{\rm s} - \Delta p\right) = {\rm A}\left(-2R_{\rm s}T\frac{J_{\rm s}^{\rm m}}{\rm B} - \Delta p\right) \tag{9.42}$$

[7] The arguments in the following sections follow K.L. Lee, R.W. Baker, H.K. Lonsdale, J. Membrane Science **8**(2), 141-171 (1981).

[8] This is the viewpoint of the Maxwell-Stefan equations, see Sec. 4.6.

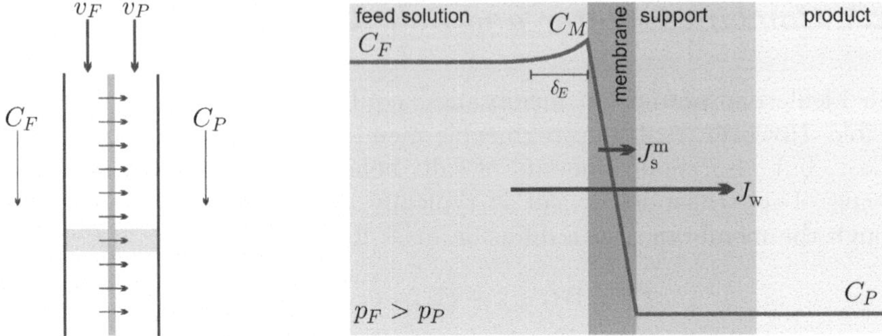

Fig. 9.7 Reverse Osmosis: Water and salt are transferred from a highly pressurized feed solution across a semipermeable membrane. The product flow is mainly water with a bit of salt. Salt concentration C_M at the membrane is elevated due to polarization effects. The active membrane is supported by a porous material. The left insert shows the co-flow arrangement of feed and product flows along the membrane surface.

We proceed to use these relations to determine transport of water and salt through membranes for reverse osmosis and pressure retarded osmosis.

9.2.8 Reverse Osmosis

In reverse osmosis, a salty feed solution (F) is pressurized above the osmotic pressure, to produce freshwater (P), see Fig. 9.7. Feed and product flow might pass the membrane in a co-flow arrangement to reduce the salinity difference along the membrane, as shown in the insert, but other geometries, such as cross-flow are used as well.

We study only the local exchange in the highlighted cell. Salt concentrations C_F, C_P are measured in the bulk of the flows, in some distance of the membrane. As the water flux J_{w} passes the membrane, left behind salt accumulates in front of the membrane, so that the local concentration C_M is above the bulk concentration C_F of the feed. This external polarization increases the concentration difference across the membrane, and thus reduces water flux. The concentration C_M is balanced by diffusion towards the bulk feed flow, which is observed in the layer of thickness δ_E. One might use active mixing, e.g., by spacer elements that divert the flow—which is laminar, hence mixing is difficult—to reduce it further. Such spacers contribute to pressure loss of the feed and draw flows (velocities v_F, v_D in Fig. 9.7). Some salt enters the product with the flux J_s^{m} driven by the concentration difference across the membrane.

Using the equations of the previous section, the diffusion in the external polarization layer relates fluxes and salt concentrations as

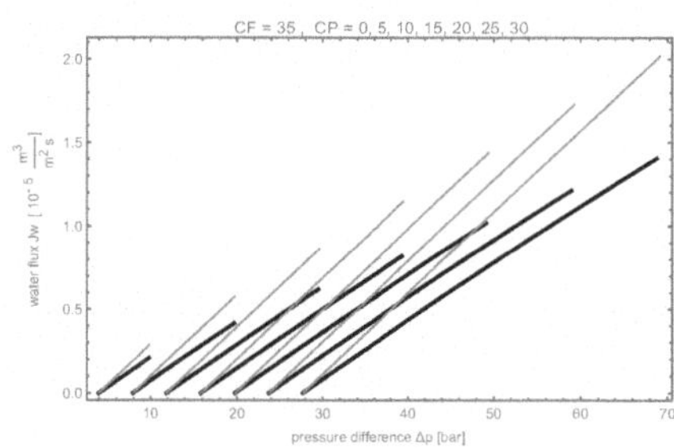

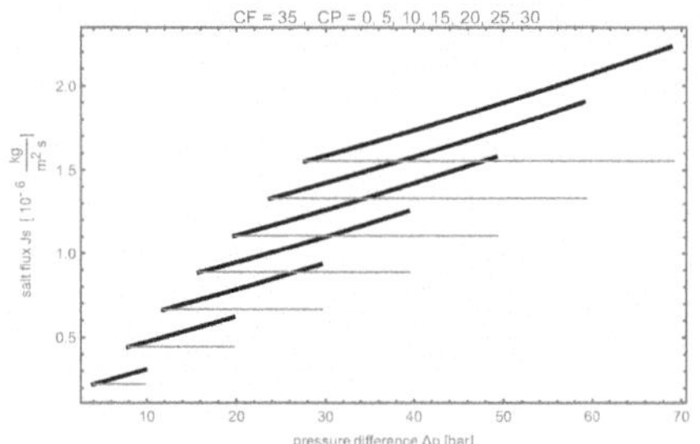

Fig. 9.8 Volume flux of water $\hat{J}_\mathrm{w}$ and mass flux of salt J_s^m for reverse osmosis, with (black) and without (grey) polarization over pressure difference $\Delta p = p_F - p_P$. Feed solution is seawater with $C_F = 35\frac{\mathrm{kg}}{\mathrm{m}^3}$, $\pi_F = 27.6\,\mathrm{bar}$, product solution with various concentrations C_P. Pressure range is up to $2.5\Delta\pi$. Coefficients: $\mathrm{A} = 4.86\times10^{-12}\frac{\mathrm{m}}{\mathrm{s\,Pa}}$, $\mathrm{B} = 4.44\times10^{-8}\frac{\mathrm{m}}{\mathrm{s}}$, $k_E = 3.85\times10^{-5}\frac{\mathrm{m}}{\mathrm{s}}$.

$$C_M = C_F e^{\frac{\hat{J}_\mathrm{w}}{k_E}} + \frac{J_\mathrm{s}^\mathrm{m}}{\hat{J}_\mathrm{w}}\left(1 - e^{\frac{\hat{J}_\mathrm{w}}{k_E}}\right) , \tag{9.43}$$

where $k_E = D/\delta_E$ is the external mass transfer coefficient, and water and salt fluxes across the membrane are given by

$$\hat{J}_\mathrm{w} = \mathrm{A}\left(-2R_sT\frac{J_\mathrm{s}^\mathrm{m}}{\mathrm{B}} - (p_P - p_F)\right) \quad , \quad J_\mathrm{s}^\mathrm{m} = \mathrm{B}\left(C_M - C_P\right) . \tag{9.44}$$

Combining these relations to eliminate C_M gives the salt flux in terms of the water flux,

$$J_\mathrm{s}^\mathrm{m} = \hat{J}_\mathrm{w} C_F \frac{e^{\frac{\hat{J}_\mathrm{w}}{k_E}} - \frac{C_P}{C_F}}{\frac{\hat{J}_\mathrm{w}}{\mathrm{B}} + e^{\frac{\hat{J}_\mathrm{w}}{k_E}} - 1} , \tag{9.45}$$

and an implicit relation for the water flux (with $2C_FR_sT = \pi_F$)

$$\hat{J}_\mathrm{w} = \mathrm{A}\left[(p_F - p_P) - \pi_F\frac{\hat{J}_\mathrm{w}}{\mathrm{B}}\frac{e^{\frac{\hat{J}_\mathrm{w}}{k_E}} - \frac{C_P}{CF}}{\frac{\hat{J}_\mathrm{w}}{\mathrm{B}} + e^{\frac{\hat{J}_\mathrm{w}}{k_E}} - 1}\right] , \tag{9.46}$$

which must be solved numerically (Prob. 9.18).

Polarization reduces the water flux, and thus increases the power requirement for reverse osmosis. In the limit $\delta \to 0$, i.e., $k_E \to \infty$, polarization vanishes, and the flux is limited only through the water permeation coefficient of the membrane,

$$\hat{J}_\mathrm{w} = \mathrm{A}\left[(p_F - p_P) - (\pi_F - \pi_P)\right] . \tag{9.47}$$

Figure 9.8 shows volume flux of water and mass flux J_s^m of salt for transfer between seawater ($C_F = 35\frac{\mathrm{kg}}{\mathrm{m}^3}$) and a product side with different concentrations C_P, comparing results for fluxes with and without polarization, where the latter are considerable smaller. With polarization salt flux increases with pressure difference, since with the larger mass flux of water concentration C_M at the membrane is increasing.

For RO the power required in the membrane process is $\dot{W}_{\rm m} = \dot{V}\Delta p$, where $\dot{V} = A\hat{J}_{\rm w}$ is the volume flow through the membrane of area A. Hence, the work per volume of freshwater produced (unit kJ/ m^3 = kPa) is

$$w_{\rm fw} = \frac{\dot{W}_{\rm m}}{\dot{V}} = \Delta p = \frac{\hat{J}_{\rm w}}{\rm A}\left[1 + \pi_F \frac{\rm A}{\rm B} \frac{e^{\frac{\hat{J}_{\rm w}}{k_E}} - \frac{C_P}{CF}}{\frac{\hat{J}_{\rm w}}{\rm B} + e^{\frac{\hat{J}_{\rm w}}{k_E}} - 1}\right] \tag{9.48}$$

For given volume flux $\hat{J}_{\rm w}$ polarization implies higher work demand. As the figure shows, for given $\hat{J}_{\rm w}$ the pressure difference Δp for the case with polarization is significantly larger than for the case without polarization.

The power considered here refers only to the local membrane process. The overall performance of a membrane unit results from a detailed model for the unit, and integration along the membrane, where the local fluxes across the membrane are as above. The actual power available will be larger, due to irreversibilities in pumps, turbines, and pressure exchangers, as well as pressure losses in the flows.

9.2.9 Pressure Retarded Osmosis

In pressure retarded osmosis, water is drawn from a low salinity feed solution (F) into a high salinity draw solution (D) at elevated pressure, $p_D > p_F$, so that the pressurized volume flow increases, see Fig. 9.9. Feed and draw flow pass the membrane in counter-flow arrangement to reduce the salinity difference along the membrane, as shown in the insert. Salt diffuses in the opposite direction, from draw to feed solution, and convective transfer leads to an accumulation of salt on the feed side, and an decrease of concentration on the draw side, with external polarization effects in the solution in front of the membrane assembly, and internal polarization in the porous support layer. External polarization can be reduced by active mixing through flow diversion as discussed for RO. Internal polarization in the support layer is difficult to avoid, and depends on the geometry of the material.

With mass transfer coefficients k_{E_1}, k_{E_2}, and k_S, for external and internal polarization, the following equations relate concentrations C_F, C_1, C_2, C_3,

$$C_1 = C_F e^{\frac{\hat{J}_{\rm w}}{k_{E1}}} + \frac{J_{\rm s}^{\rm m}}{\hat{J}_{\rm w}}\left(1 - e^{\frac{\hat{J}_{\rm w}}{k_{E1}}}\right) \tag{9.49}$$

$$C_2 = C_1 e^{\frac{\hat{J}_{\rm w}}{k_S}} + \frac{J_{\rm s}^{\rm m}}{\hat{J}_{\rm w}}\left(1 - e^{\frac{\hat{J}_{\rm w}}{k_S}}\right) \tag{9.50}$$

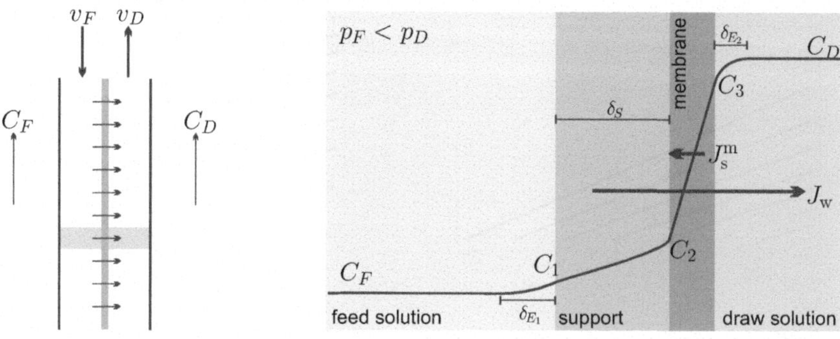

Fig. 9.9 Pressure Retarded Osmosis: Water is drawn from the low-salinity feed against the pressure gradient into the salty draw solution, while salt diffuses in the opposite direction. External (in draw and feed) and internal (in the support layer) polarization effects reduce the concentration difference across the membrane. The left insert shows the counter-flow arrangement of feed and draw flows along the membrane surface.

$$J_{\mathrm{s}}^{\mathrm{m}} = \mathrm{B}\left(C_2 - C_3\right) \tag{9.51}$$

$$C_D = C_3 e^{\frac{\hat{J}_{\mathrm{w}}}{k_{E_2}}} + \frac{J_{\mathrm{s}}^{\mathrm{m}}}{\hat{J}_{\mathrm{w}}}\left(1 - e^{\frac{\hat{J}_{\mathrm{w}}}{k_{E_2}}}\right) \tag{9.52}$$

Elimination of C_1, C_2, C_3 yields the salt flux as

$$J_{\mathrm{s}}^{\mathrm{m}} = -\hat{J}_{\mathrm{w}} C_D \frac{1 - \frac{C_F}{C_D} e^{\frac{\hat{J}_{\mathrm{w}}}{k_{E1}} + \frac{\hat{J}_{\mathrm{w}}}{k_S} + \frac{\hat{J}_{\mathrm{w}}}{k_{E_2}}}}{\frac{\hat{J}_{\mathrm{w}}}{\mathrm{B}} e^{\frac{\hat{J}_{\mathrm{w}}}{k_{E_2}}} + e^{\frac{\hat{J}_{\mathrm{w}}}{k_{E_1}} + \frac{\hat{J}_{\mathrm{w}}}{k_S} + \frac{\hat{J}_{\mathrm{w}}}{k_{E_2}}} - 1}\,, \tag{9.53}$$

and the water flux again obeys an implicit equation,

$$\hat{J}_{\mathrm{w}} = \mathrm{A}\left(\pi_D \frac{\hat{J}_{\mathrm{w}}}{\mathrm{B}} \frac{1 - \frac{C_F}{C_D} e^{\frac{\hat{J}_{\mathrm{w}}}{k_{E1}} + \frac{\hat{J}_{\mathrm{w}}}{k_S} + \frac{\hat{J}_{\mathrm{w}}}{k_{E_2}}}}{\frac{\hat{J}_{\mathrm{w}}}{\mathrm{B}} e^{\frac{\hat{J}_{\mathrm{w}}}{k_{E_2}}} + e^{\frac{\hat{J}_{\mathrm{w}}}{k_{E_1}} + \frac{\hat{J}_{\mathrm{w}}}{k_S} + \frac{\hat{J}_{\mathrm{w}}}{k_{E_2}}} - 1} - \left(p_D - p_F\right)\right) \tag{9.54}$$

Also for PRO we observe reduction of both fluxes due to polarization effects. Figure 9.10 shows volume flux of water and mass flux $J_{\mathrm{s}}^{\mathrm{m}}$ of salt for seawater as draw solution ($C_F = 35\frac{\mathrm{kg}}{\mathrm{m}^3}$) and a feed side with different concentrations C_P, comparing results for fluxes with and without polarization, where the latter are, again, considerably smaller.

The power gain from the PRO membrane process is $\dot{W} = \dot{V}\Delta p = A\hat{J}_{\mathrm{w}}\Delta p$, the power per membrane area is $\frac{\dot{W}}{A} = \hat{J}_{\mathrm{w}}\Delta p$. The figure shows that for a given pressure difference Δp polarization reduces the mass flux, hence the power production is significantly reduced. The power considered here refers only to the membrane process. The actual power available will be smaller, due to irreversibilities in pumps, turbines, and pressure exchangers, etc.

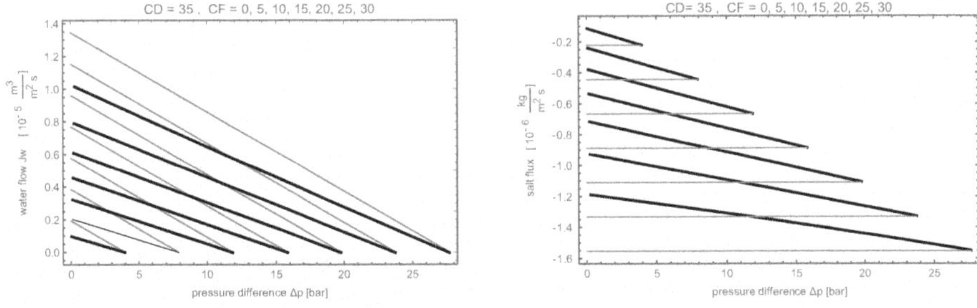

Fig. 9.10 Volume flux of water J_{w} and mass flux of salt J_s^{m} for PRO, with (black) and without (grey) polarization over pressure difference $\Delta p = p_F - p_P$. Feed solution is seawater with $C_F = 35 \frac{\mathrm{kg}}{\mathrm{m}^3}$, $\pi_F = 27.6\,\mathrm{bar}$, product solution with various concentrations C_P. Coefficients: A $= 4.86 \times 10^{-12} \frac{\mathrm{m}}{\mathrm{s\,Pa}}$, B $= 4.44 \times 10^{-8} \frac{\mathrm{m}}{\mathrm{s}}$, $k_{E_1} = k_S = k_{E_2} = 3.85 \times 10^{-5} \frac{\mathrm{m}}{\mathrm{s}}$.

9.3 Chemical Reactions and Nonlinear Reaction Rates

9.3.1 Entropy Generation and Affinity

For reaction rates, the usual linear force-flux relations of Linear Irreversible Thermodynamics fail. Indeed, linear force-flux relations are expected for relatively small deviation from equilibrium, but not for strong nonequilibrium. Chemical reactions, in which a mixture of reactants turns into products, initially are far from equilibrium states, hence linear relations for reaction rates are not applicable. Here, we use some simple arguments that lead to non-linear flux-force relations for reaction rates.

We recall the contribution of n reaction rates λ^a to entropy generation (4.17),

$$\Sigma_R = -\frac{1}{T} \sum_{a=1}^{n} \lambda^a \left(\sum_{\alpha=1}^{\nu} \gamma_\alpha^a \bar{\mu}_\alpha \right) \tag{9.55}$$

where γ_α^a are the stoichiometric coefficients, and $\bar{\mu}_\alpha = \mu_\alpha M_\alpha$ are the molar chemical potentials. Cross-coupling to the divergence of velocity is ignored.

For simplicity, we proceed by studying a single reaction mechanism with reaction rate λ, so that reactive entropy generation is given by

$$\Sigma_R = -\frac{1}{T} \lambda \sum_{\alpha=1}^{\nu} \gamma_\alpha \bar{\mu}_\alpha = -\frac{1}{T} \lambda \bar{\mathcal{A}} \,, \tag{9.56}$$

where $\bar{\mathcal{A}} = \sum_{\alpha=1}^{\nu} \gamma_\alpha \bar{\mu}_\alpha$ is the affinity. Non-zero values of the affinity are the thermodynamic force that describes the deviation from reactive equilibrium. The reaction rate is the associated thermodynamic flux—reactions take place in order to bring the affinity to zero, and the system into reactive equilibrium.

A simple linear phenomenological force-flux relation as in Chap. 4 yields $\lambda = -L_R \bar{\mathcal{A}}$ with a phenomenological coefficient $L_R > 0$, but, as stated above, such a relation fails when the reacting mixture is far from equilibrium, as is often the case.

9.3.2 Ideal Mixtures

For ideal mixtures, the molar chemical potentials are

$$\bar{\mu}_\alpha = \bar{g}_\alpha (T, p) + \bar{R} T \ln X_\alpha \tag{9.57}$$

where $\bar{g}_\alpha (T, p)$ are the molar Gibbs free energies of the components alone at temperature T and pressure p of the reacting mixture, $\bar{R}$ is the universal gas constant, and X_α are the mole fractions of the components. The corresponding affinity for a single reaction is

$$\bar{\mathcal{A}} = \Delta \bar{g}_R + \bar{R} T \ln \prod_{\alpha=1}^{\nu} X_\alpha^{\gamma_\alpha} , \tag{9.58}$$

with the Gibbs free energy of reaction

$$\Delta \bar{g}_R (T, p) = \sum_{\alpha=1}^{\nu} \gamma_\alpha \bar{g}_\alpha (T, p) . \tag{9.59}$$

Before we develop a model for reaction rates, we have a look at chemical equilibrium states, for which the affinity vanishes,

$$\bar{\mathcal{A}}_{|E} = \sum_{\alpha=1}^{\nu} \gamma_\alpha \bar{\mu}_{\alpha|E} = 0 . \tag{9.60}$$

This is the law of mass action, which for an ideal mixture gives

$$\prod_{\alpha=1}^{\nu} X_{\alpha|E}^{\gamma_\alpha} = \exp \left[-\frac{\Delta \bar{g}_R}{\bar{R} T} \right] = K_X (T, p) . \tag{9.61}$$

The chemical constant $K_X (T, p)$ depends on temperature and pressure through the Gibbs free energy of reaction. See [TEC, Chap. 23] for discussion and applications.

9.3.3 Reaction Rates

Now we turn our attention to reaction rates. We recall that the stoichiometric coefficients are negative ($\gamma_\alpha < 0$) for reactants, and positive ($\gamma_\alpha > 0$) for products. The following discussion is meaningful for elementary reactions, i.e., there are no hidden intermediate steps.

For a reaction to take place, all reactants must be present. The probability to find a molecule of type α in a volume element is proportional to the local value of the partial mole density $\bar{\rho}_\alpha$. Since probabilities multiply, the probability to find $|\gamma_\alpha|$ particles of type α at one location is $\bar{\rho}_\alpha^{|\gamma_\alpha|}$ and the probability to have all reactant particles required at one location is $\prod_{\alpha^-} \bar{\rho}_\alpha^{|\gamma_\alpha|}$, where the product is taken over all reactants, i.e., negative γ_α. Introducing k_f as a measure for the probability that a forward reaction takes place when all particles meet, the forward reaction rate assumes the form

$$\lambda_f = k_f \prod_{\alpha^-} \bar{\rho}_\alpha^{|\gamma_\alpha|} = k_f \bar{\rho}^{\gamma_f} \prod_{\alpha^-} X_\alpha^{|\gamma_\alpha|} \,, \tag{9.62}$$

where $\bar{\rho}$ is the mole density of the mixture, and $\gamma_f = \sum_{\alpha^-} |\gamma_\alpha|$.

For the backward reaction, we expect correspondingly

$$\lambda_b = k_b \prod_{\alpha^+} \bar{\rho}_\alpha^{\gamma_\alpha} = k_b \bar{\rho}^{\gamma_b} \prod_{\alpha^+} X_\alpha^{\gamma_\alpha} \,, \tag{9.63}$$

where the product is taken over all products, i.e., positive stoichiometric coefficients, k_b is the corresponding reaction probability and $\gamma_b = \sum_{\alpha^+} \gamma_\alpha$.

The net reaction rate is the difference

$$\lambda = \lambda_f - \lambda_b = k_f \bar{\rho}^{\gamma_f} \prod_{\alpha^-} X_\alpha^{|\gamma_\alpha|} - k_b \bar{\rho}^{\gamma_b} \prod_{\alpha^+} X_\alpha^{\gamma_\alpha} \,. \tag{9.64}$$

This simple model for reaction rates is compatible with the law of mass action for ideal mixtures. Indeed, setting $\lambda = 0$, we find the equilibrium relation

$$\prod_{\alpha=1}^{\nu} X_{\alpha|E}^{\gamma_\alpha} = \frac{k_f}{k_b} \frac{\bar{\rho}^{\gamma_f}}{\bar{\rho}^{\gamma_b}} = K_X = \exp\left[-\frac{\Delta \bar{g}_R}{\bar{R}T}\right] \,, \tag{9.65}$$

that is the chemical constant $K_X(T,p)$ is related to the ratio of the probability coefficients.

To proceed, we split the affinity into contributions for reactants and products, as

$$\bar{\mathcal{A}}_f = -\bar{g}_f + \bar{R}T \ln \prod_{\alpha^-} X_\alpha^{|\gamma_\alpha|} \,, \tag{9.66}$$

$$\bar{\mathcal{A}}_b = -\bar{g}_b + \bar{R}T \ln \prod_{\alpha^+} X_\alpha^{\gamma_\alpha} \,, \tag{9.67}$$

so that

$$\bar{\mathcal{A}} = \bar{\mathcal{A}}_b - \bar{\mathcal{A}}_f \quad \text{and} \quad \Delta \bar{g}_R = \bar{g}_f - \bar{g}_b \ . \tag{9.68}$$

Here, $\bar{g}_f$, $\bar{g}_b$ are not simply given by summing over the Gibbs free energies of the components involved in the forward and backward reactions, but include an energy shift E_s

$$-\bar{g}_f = \sum_{\alpha^-} |\gamma_\alpha| \bar{g}_\alpha - E_s \quad , \quad -\bar{g}_b = \sum_{\alpha^+} \gamma_\alpha \bar{g}_\alpha - E_s \ . \tag{9.69}$$

From the definitions of $\bar{\mathcal{A}}_{f,b}$ follows

$$\exp\left[\frac{\bar{\mathcal{A}}_f + \bar{g}_f}{\bar{R}T}\right] = \prod_{\alpha^-} X_\alpha^{-\gamma_\alpha} \quad , \quad \exp\left[\frac{\bar{\mathcal{A}}_b + \bar{g}_b}{\bar{R}T}\right] = \prod_{\alpha^+} X_\alpha^{\gamma_\alpha} \ , \tag{9.70}$$

which gives the reaction rate in terms of affinities,

$$\lambda = k_f \bar{\rho}^{\gamma_f} \exp\left[\frac{\bar{\mathcal{A}}_f + \bar{g}_f}{\bar{R}T}\right] - k_b \bar{\rho}^{\gamma_b} \exp\left[\frac{\bar{\mathcal{A}}_b + \bar{g}_b}{\bar{R}T}\right] \ . \tag{9.71}$$

Finally, with the law of mass action (9.65) this assumes the form

$$\lambda = \hat{k}\left(\exp\frac{\bar{\mathcal{A}}_f}{\bar{R}T} - \exp\frac{\bar{\mathcal{A}}_b}{\bar{R}T}\right) \tag{9.72}$$

with the abbreviation

$$\hat{k} = k_f \bar{\rho}^{\gamma_f} \exp\left[\frac{\bar{g}_f}{\bar{R}T}\right] = k_b \bar{\rho}^{\gamma_b} \exp\left[\frac{\bar{g}_b}{\bar{R}T}\right] \ . \tag{9.73}$$

The corresponding entropy generation is, with $\bar{\mathcal{A}} = \bar{\mathcal{A}}_b - \bar{\mathcal{A}}_f$,

$$\Sigma_R = -\frac{1}{T}\lambda\bar{\mathcal{A}} = \frac{\hat{k}}{T}\left(\exp\frac{\bar{\mathcal{A}}_f}{\bar{R}T} - \exp\frac{\bar{\mathcal{A}}_b}{\bar{R}T}\right)\left(\bar{\mathcal{A}}_f - \bar{\mathcal{A}}_b\right) \geq 0 \tag{9.74}$$

Evidently, the entropy generation rate is strictly non-negative, and vanishes only for $\bar{\mathcal{A}}_f = \bar{\mathcal{A}}_b$, that is for the equilibrium condition $\bar{\mathcal{A}} = 0$. Thus, the nonlinear reaction rate is constructed such that the second law of thermodynamics is always fulfilled. Note that elsewhere we constructed phenomenological laws based on the entropy generation rate as linear relations between forces and fluxes. Instead, here we used heuristic arguments to find an expression for the reaction rate that stands in agreement with the second law. In case that $\frac{|\bar{\mathcal{A}}_{f,b}|}{\bar{R}T} \ll 1$, Eq. (9.72) reduces to a linear law, $\lambda = \frac{\hat{k}}{\bar{R}T}\left(\bar{\mathcal{A}}_f - \bar{\mathcal{A}}_b\right)$

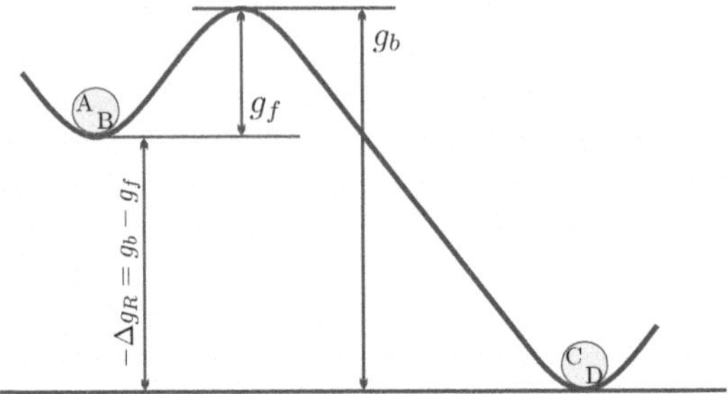

Fig. 9.11 Gibbs free energies of activation.

9.3.4 Arrhenius Model

With (9.73) the directed reaction rates can be written as

$$k_f = \frac{\hat{k}}{\bar{\rho}^{\gamma_f}} \exp\left[-\frac{\bar{g}_f}{\bar{R}T}\right] = \hat{k}_f \exp\left[-\frac{\bar{g}_f}{\bar{R}T}\right] , \tag{9.75}$$

$$k_b = \frac{\hat{k}}{\bar{\rho}^{\gamma_b}} \exp\left[-\frac{\bar{g}_b}{\bar{R}T}\right] = \hat{k}_b \exp\left[-\frac{\bar{g}_b}{\bar{R}T}\right] . \tag{9.76}$$

so that the overall reaction rate (9.64) becomes

$$\lambda = \hat{k}_f \exp\left[-\frac{\bar{g}_f}{\bar{R}T}\right] \prod_{\alpha^-} \bar{\rho}_\alpha^{|\gamma_\alpha|} - \hat{k}_b \exp\left[-\frac{\bar{g}_b}{\bar{R}T}\right] \prod_{\alpha^+} \bar{\rho}_\alpha^{\gamma_\alpha} . \tag{9.77}$$

Both coefficients are of the form $\exp\left[-\frac{\bar{e}}{\bar{R}T}\right]$, where $\bar{e}$ is an energy; here, the Gibbs free energy of activation. The barometric formula is of the same form, only with the potential energy Mgz. The factor $\exp\left[-\frac{\bar{e}}{\bar{R}T}\right]$ is known as the *Boltzmann factor*, it is the typical form of a probability in a thermally activated system..

Figure 9.11 shows the Gibbs free energies of activation in an energy landscape for a reaction $\mathrm{A} + \mathrm{B} \rightleftharpoons \mathrm{C} + \mathrm{D}$, with $\gamma_b = \gamma_f = 2$. The Gibbs free energy of reaction, $-\Delta\bar{g}_R$, is a measure for the energy difference between the reactants (A, B) and the products (C, D). As depicted, the equilibrium of the reaction lies towards the products. For the forward reaction $(\mathrm{A}, \mathrm{B}) \to (\mathrm{C}, \mathrm{D})$ to occur the activation energy $\bar{g}_f$ must be overcome, while for the backward reaction $(\mathrm{C}, \mathrm{D}) \to (\mathrm{A}, \mathrm{B})$ the larger activation energy $\bar{g}_b$ must be overcome.

In equilibrium, according to the law of mass action (9.65) the same number of reactions takes place in both directions, $\lambda_f = \lambda_b$, or $k_f \bar{\rho}_\mathrm{A} \bar{\rho}_\mathrm{B} = k_b \bar{\rho}_\mathrm{C} \bar{\rho}_\mathrm{D}$. Due to the larger activation energy $\bar{g}_b$, the likelihood for a backward reaction is less $(k_b = \frac{\hat{k}}{\bar{\rho}^2} e^{-\frac{\bar{g}_b}{\bar{R}T}})$ than the likelihood for the forward reaction $(k_f = \frac{\hat{k}}{\bar{\rho}^2} e^{-\frac{\bar{g}_f}{\bar{R}T}})$. With $k_b < k_f$ and $k_f \bar{\rho}_\mathrm{A} \bar{\rho}_\mathrm{B} = k_b \bar{\rho}_\mathrm{C} \bar{\rho}_\mathrm{D}$ follows $\bar{\rho}_\mathrm{A} \bar{\rho}_\mathrm{B} < \bar{\rho}_\mathrm{C} \bar{\rho}_\mathrm{D}$: there must be more particles on the product side.

The individual activation energies $\bar{g}_f, \bar{g}_b$ do not affect the equilibrium, which is described by their difference, the Gibbs free energy of reaction $\Delta\bar{g}_R = \bar{g}_f - \bar{g}_b$. However, the individual activation energies determine how fast the equilibrium will be reached. For large activation energies, reactions are unlikely, reactants will persist in a metastable state, and make no progress towards the final equilibrium state.

The smaller the activation energies divided by temperature, $\frac{\bar{g}_{f,b}}{RT}$, are, the faster the equilibrium will be reached. Thus, equilibrium is reached faster at higher temperatures. Fuel in contact with oxygen must be ignited—that is heated to higher T—to overcome the energy barrier at the point of ignition. The released heat provides the energy to overcome the barrier in the neighborhood, a fast chain reaction occurs.

Catalysts do not take part in the reaction, but their presence lowers the activation energies. Ammonia synthesis provides an example for a reaction that is stuck in a metastable state, and will come to reaction only with the help of catalysts. Even then, to overcome the energy barrier, the temperature must be increased.

The above form of reaction rates is well-known in reaction kinetics, where one typically does not write Gibbs free energies but activation energies $\bar{E}_{f,b}$. For elementary reactions as discussed here, the exponents of concentrations are the stoichiometric coefficients. For more complex reactions, which involve the production and consumption of intermediates, the stoichiometric coefficients are replaced by general exponents ξ_α, which give the reaction order for component α in the respective (forward or backward) reaction,

$$\lambda = \hat{k}_f e^{-\frac{\bar{E}_f}{RT}} \prod_{\alpha^-} \bar{\rho}_\alpha^{|\xi_\alpha|} - \hat{k}_b e^{-\frac{\bar{E}_b}{RT}} \prod_{\alpha^+} \bar{\rho}_\alpha^{\xi_\alpha} \ . \tag{9.78}$$

In any case, the rate constants $\hat{k}_f$ and $\hat{k}_b$ depend on temperature and pressure of the reaction, and are linked through their ratio in equilibrium, which is determined by the law of mass action. Further details can be found in the specialist literature.

9.3.5 Simplified Equations for a Flame

To close, we write the simplified transport equations for a reactive binary mixture as discussed in the consideration of flames in Chap. 8, with the reaction equation $A \rightleftharpoons B$.

The conservation laws for a binary mixture are

$$\frac{\partial \rho}{\partial t} + \frac{\partial \rho v_k}{\partial x_k} = 0 \ , \tag{9.79}$$

$$\frac{\partial \rho \mathsf{c}}{\partial t} + \frac{\partial}{\partial x_k} \left[\rho v_k \mathsf{c} + J_k \right] = \tau \ , \tag{9.80}$$

$$\frac{\partial \rho v_i}{\partial t} + \frac{\partial}{\partial x_k} \left[\rho v_i v_k - t_{ik} \right] = 0 \ , \tag{9.81}$$

$$\frac{\partial}{\partial t}\left[\rho\left(u+\frac{1}{2}v^2\right)\right] + \frac{\partial}{\partial x_k}\left[\rho\left(u+\frac{1}{2}v^2\right)v_k + q_k - v_i t_{ik}\right] = 0 \ . \tag{9.82}$$

These must now be specified through appropriate constitutive equations.

We base the mass fraction on component A, so that

$$\mathsf{c} = \mathsf{c}_\mathrm{A} = \frac{\rho_\mathrm{A}}{\rho} \quad , \quad \mathsf{c}_\mathrm{B} = 1 - \mathsf{c} = \frac{\rho_\mathrm{B}}{\rho} \ . \tag{9.83}$$

The ideal gases A and B have the same molar mass M, gas constant R and constant specific heat c_p, and only differ in their enthalpy constants, so that

$$h_A = c_p T + h_A^0 \quad , \quad h_B = c_p T + h_B^0 \ , \tag{9.84}$$

hence the heat of reaction is independent of temperature and pressure, and determined through the enthalpy constants,

$$\Delta h_R = h_B - h_A = h_B^0 - h_A^0 \ . \tag{9.85}$$

The enthalpy density ρh of the mixture is

$$\rho h = \rho_\mathrm{A} h_\mathrm{A} + \rho_\mathrm{B} h_\mathrm{B} = \rho \left(\mathsf{c} h_\mathrm{A} + (1-\mathsf{c})\, h_\mathrm{B} \right) = \rho \left(c_p T + h_B^0 - \mathsf{c} \Delta h_R \right)$$

For transport we ignore cross-coupling of fluxes and ignore viscosities, so that diffusion flux, heat flux and stress tensor are

$$J_k = J_k^\mathrm{A} = -J_k^\mathrm{B} = -D \frac{\partial \mathsf{c}}{\partial x_k} \ , \tag{9.86}$$

$$q_k = -\kappa \frac{\partial T}{\partial x_k} \ , \tag{9.87}$$

$$t_{ik} = -p \delta_{ik} \ . \tag{9.88}$$

A flame reaction has the equilibrium state on the product side, hence it suffices to consider only the forward reaction rate, where we allow arbitrary order of the reaction through the exponents α, β,

$$\tau = -k_b \rho_\mathrm{A}^\alpha \rho_\mathrm{B}^\beta e^{-\frac{E}{RT}} = -k_b \rho^{\alpha+\beta} \mathsf{c}^\alpha \left(1-\mathsf{c}\right)^\beta e^{-\frac{E}{RT}} \tag{9.89}$$

Combining all of the above for the special case of a one-dimensional steady state reaction process yields

$$\frac{d\rho v}{dx} = 0 \ , \tag{9.90}$$

$$\frac{d}{dx}\left[\rho v \mathsf{c} - D\frac{d\mathsf{c}}{dx}\right] = -k_b \rho^{\alpha+\beta} \mathsf{c}^{\alpha} \left(1-\mathsf{c}\right)^{\beta} e^{-\frac{E}{\bar{R}T}} \ , \tag{9.91}$$

$$\frac{d}{dx}\left[\rho v^2 + \rho R T\right] = 0 \ , \tag{9.92}$$

$$\frac{d}{dx}\left[\rho v \left(c_p T - \mathsf{c}\Delta h_R + \frac{1}{2}v^2\right) - \kappa\frac{dT}{dx}\right] = 0 \ . \tag{9.93}$$

These equations determine the resolved processes in the flame zone, that we ignored earlier. Indeed, integrating the above for states far before ($\mathsf{c} = 1$) and after ($\mathsf{c} = 0$) the reaction zone, where $\frac{d\mathsf{c}}{dx} = \frac{dT}{dx} = 0$, we recover the equations (8.68) used in the discussion of flames. It is left to the interested reader to proceed with the integration of these equations based on some reasonable choices of parameters and study the details of the flame structure in this simple setting.

Problems

9.1. Gas Mixture in Gravitational Field

Consider an isothermal mixture of ideal gases at rest in the gravitational field. Show that the transport equations reduce to

$$\frac{\partial p}{\partial x_i} = \rho f_i \quad , \quad J_i^{\alpha} = -B_{\alpha\beta}\frac{\partial \frac{\mu_\alpha - \mu_\nu}{T}}{\partial x_i} = 0 \ ,$$

where $f_i = \{0, 0, -\mathsf{g}\}_i$. Solve the above equations and show that for each component the density obeys the barometric formula

$$\rho_\alpha = \rho_\alpha^0 \exp\left[-\frac{M_\alpha \mathsf{g} z}{\bar{R}T}\right] \ .$$

Consider (a) air as binary mixture of oxygen and nitrogen with partial mass densities at ground level ($z = 0$) given as $\rho^0_{O_2} = 0.27\frac{\mathrm{kg}}{\mathrm{m}^3}$ and $\rho^0_{N_2} = 0.89\frac{\mathrm{kg}}{\mathrm{m}^3}$, (b) a mixture of oxygen and helium which is equimolar at $z = 0$. For both, plot densities, mole fraction of the lighter gas, and total pressure as function of height. Chose $T = 298\,\mathrm{K}, \mathsf{g} = 9.81\frac{\mathrm{m}}{\mathrm{s}^2}$.

9.2. Binary Gas Mixture

Consider a mixture of two ideal gases. Specify the equations from the lecture for a binary mixture. You should get the balance laws

$$\frac{D\rho}{Dt} + \rho\frac{\partial v_k}{\partial x_k} = 0 \quad , \quad \rho\frac{D\mathsf{c}}{Dt} + \frac{\partial J_k}{\partial x_k} = 0$$

$$\rho\frac{Dv_i}{DT} - \frac{\partial t_{ik}}{\partial x_k} = \rho G_i \quad , \quad \rho\frac{Du}{Dt} + \frac{\partial q_k}{\partial x_k} = t_{ij}\frac{\partial v_i}{\partial x_j}$$

and the constitutive equations

$$q_i = -\kappa \frac{\partial T}{\partial x_i} + B \frac{\partial \frac{\mu_1 - \mu_2}{T}}{\partial x_i} \quad , \quad J_i = -\frac{B}{T^2} \frac{\partial T}{\partial x_i} - \hat{D} \frac{\partial \frac{\mu_1 - \mu_2}{T}}{\partial x_i}$$

$$t_{ij} = -p\delta_{ij} + \nu \frac{\partial v_k}{\partial x_k} \delta_{ij} + 2\eta \frac{\partial v_{\langle i}}{\partial x_{j\rangle}}$$

$$\rho u = \rho_1 u_1 + \rho_2 u_2 = \rho \left(c u_1 + (1 - c)\, u_2 \right)$$

1. Note the occurrence of the parameter B in the equations for q_i and J_i due to Onsager symmetry. Show that the given form is proper, by checking units.
2. According to Fick's law, the diffusion flux is $J_i = -D \frac{\partial c}{\partial x_i}$. Find and discuss the assumptions that are required to reduce the general theory to Fick's law. What is the relation between the coefficients D and $\hat{D}$?
3. Consider a mixture of two ideal gases at rest in a closed vessel. The vessel of length L is subjected to a temperature difference ΔT. Use the balance equations for a mixture together with the constitutive equations from LIT to compute temperature profile and composition profile of the mixture. Treat the coefficients as constants to solve the problem. Discuss under which circumstances temperature gradient and concentration gradient have the same or the opposite sign.

9.3. Physiological Solution for Use in Hospitals
The osmotic pressure in bodily fluids of mammals is 7.7 atm at 36 °C. Determine the amount of salt (NaCl) that must be added to 1 litre water to give a solution with the same osmotic pressure. $M_{\mathrm{NaCl}} = 58.5 \frac{\mathrm{kg}}{\mathrm{kmol}}$. Note: NaCl dissociates in solution.

9.4. Curing Meat
A 400 g piece of meat contains 75% of salty water at osmotic pressure of 8 bar (at 20 °C), plus fat, protein, etc. The meat is rubbed with salt (NaCl), which draws water out of the cells, so that the salt can dissolve and dissociate. Theoretically, how many grams of salt are required to draw half the water from the meat? Consider the final equilibrium state only. Why would you use more salt?

9.5. Osmotic Equilibrium
A semipermeable membrane which allows only water to pass, divides two containers. There is 1.2 litre of water in total, and 20 g of (dissociated!) NaCl in each container. One container is kept at a pressure of 10 bar and the other at 20 bar; the temperature of both is 300 K.

1. Show that in equilibrium the pressure difference between the two containers equals the difference in osmotic pressures.
2. Set up the equation needed to determine how water is distributed between the two containers, and determine the water masses in both containers.

3. Will the equilibrium change when the temperature is lowered to 20 °C? If so, in which direction does water move, low to high pressure, or high to low?

9.6. Osmotic Equilibrium with Temperature Difference

Two piston-cylinder systems are connected by a semipermeable membrane that allows only water to pass. Both cylinders contain water and NaCl. The left cylinder is pressurized to 15 bar and its temperature is maintained at 320 K. The right cylinder is maintained at a temperature of 325 K.

Determine the pressure that must be exerted on the right cylinder so that in chemical equilibrium the mole fraction of dissolved salt in both containers is $X_{s,L} = X_{s,R} = 0.025$.

9.7. Saltwater in a Piston Cylinder Device

Saltwater at 20 °C, 1 bar is pressurized isothermally and pushed against a semi-permeable membrane, which allows only fresh water to pass. The freshwater at the other side of the membrane is at 1.5 bar. When the pressure has reached 42 bar, the pressure is released and the remaining brine is pushed out at a pressure of 1 bar.

One litre of the incoming saltwater contains 45 g of sodium chloride ($M_{\mathrm{NaCl}} = 58.5 \frac{\mathrm{kg}}{\mathrm{kmol}}$), and the mass densities of saltwater, freshwater and brine are $\rho_{sw} = 1033 \frac{\mathrm{kg}}{\mathrm{m}^3}$, $\rho_{fw} = 1000 \frac{\mathrm{kg}}{\mathrm{m}^3}, \rho_b = 1050 \frac{\mathrm{kg}}{\mathrm{m}^3}$.

Assume that the temperature remains constant throughout the device, that the salt-water mixture is always well mixed, and that all irreversibilities can be ignored.

1. Determine the pressure p_1 at which freshwater starts to pass the membrane.
2. Determine the mole fraction of salt in the brine.
3. For 1 litre of incoming saltwater, determine the corresponding volumes of freshwater and brine produced.
4. Determine the work required for the process, per litre of salt water.

9.8. Salt Solution in Gravitational Field

Consider a dilute salt solution in the gravitational field and determine the salt concentration and the mass density of the salt water as a functions of height when concentration at $z = 0$ is C_{s}^0.

Note: While the salt dissociates, the ions can locally move independently due to screening by the polar water molecules, but the mixture remains locally electroneutral since the Coulomb forces are strong. The result should be a barometric formula, similar to ideal gases.

9.9. Desalination by Gravity

A pipe is closed at one end with a semipermeable membrane which only allows water to pass. The pipe is pressed vertically into an ocean, as sketched. The temperature of the ocean is 4 °C, and the salt water contains 35 g sodium chloride (NaCl) per litre, moreover $\rho_{sw} = 1030 \frac{\mathrm{kg}}{\mathrm{m}^3}$, $\rho_{fw} = 1000 \frac{kg}{m^3}$

1. To what height H_1 must the pipe be immersed, before fresh water passes the membrane?
2. At what height H_2 have both, sea water and fresh water, the same height, $h = H$? Explain why this is possible.
3. When $h > H_2$ one can run a water wheel with the fresh water leaving the pipe. Of course, this setup is not a perpetual motion engine. Why? (for some discussion see: Scientific American, June 1971, 124-125, and April 1972, 110-111). The above calculation assumed constant density of the seawater, and a homogeneous salt content. Can that be expected?

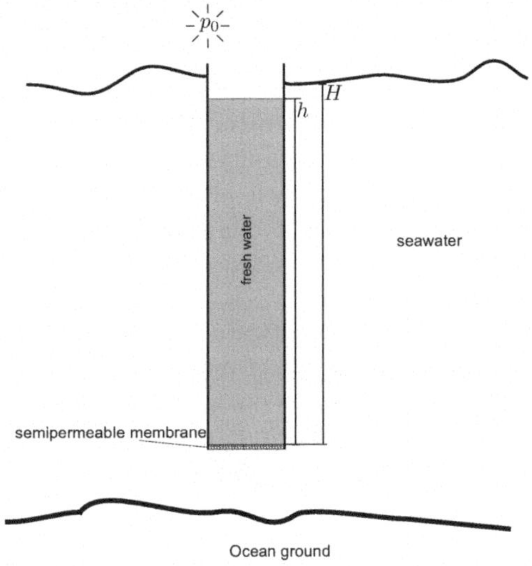

9.10. Desalination by Gravity with Gradients

The solution of the previous problem assumed constant density of the seawater. Consider the same problem for the case that the salt concentration follows the barometric law, so that the mass density of the saltwater is

$$\rho_{sw} = \rho_{fw} + \rho_{s,0} \exp\left[-\frac{gz}{2R_sT}\right] ,$$

where R_s is the gas constant for the salt (the factor 2 accounts for dissociation) and $\rho_{s,0}$ is the mass density at the surface (at $z = 0$). Note that z points upwards, so that the density is larger at greater depth.

1. Discuss the salt concentration profile in the oceans. Under what circumstances would you expect the exponential and the constant profile, respectively?
2. Determine the difference $h - H$ for the exponential profile.
3. Discuss your findings.

9.11. Reverse Osmosis

Consider the continuous desalination device depicted below. Fresh seawater at 15 °C, 1 bar is pumped isothermally to a pressure of 35 bar, and then flows past a semipermeable membrane which allows only freshwater to pass. The exiting brine drives a turbine, and leaves the system at 1 bar. Assume that the temperature remains constant throughout the device, and that all processes are reversible.

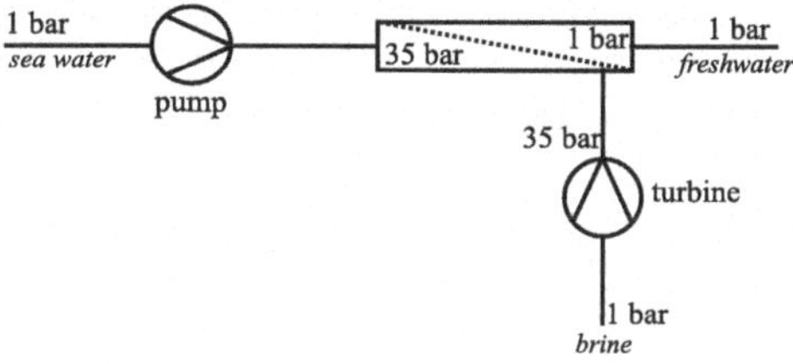

The mass densities of sea water, freshwater and brine are given as: $\rho_{sw} = 1030\frac{\text{kg}}{\text{m}^3}$, $\rho_{fw} = 1000\frac{\text{kg}}{\text{m}^3}$, $\rho_b = 1040\frac{\text{kg}}{\text{m}^3}$, and the seawater contains $35\frac{\text{g}}{\text{litre}}$ of sodium chloride ($M_{\text{NaCl}} = 58.5\frac{\text{kg}}{\text{kmol}}$).

1. Determine the volume flow of freshwater produced per volume flow of sea water.
2. Determine the work required to drive the pump, and the work that can be recovered by the turbine. Compute the net work required per litre of freshwater.
3. Assume now that pump and turbine are irreversible, with efficiencies of 85%, and determine the net work.

9.12. Osmotic Power Plant with Irreversibilities

Saltwater at 10°C, 1bar is pumped adiabatically to a pressure p_M and then passes a semi-permeable membrane through which freshwater enters and dilutes the saltwater. The exiting solution (diluted saltwater) drives an adiabatic turbine, and leaves the system at 1 bar. Assume that the temperature remains constant throughout the device.

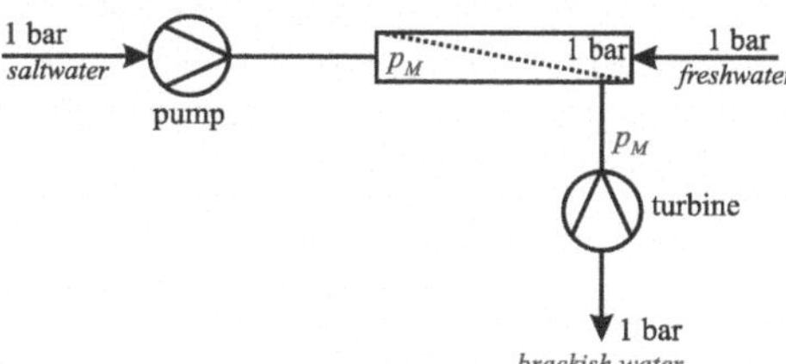

The incoming saltwater contains 70 g of NaCl ($M_{\text{NaCl}} = 58.5\frac{\text{kg}}{\text{kmol}}$) per litre. To simplify calculations, assume that there is no mixing volume, and the mass densities of NaCl and water alone are $\rho_{\text{NaCl}} = 2165\frac{\text{kg}}{\text{m}^3}$, $\rho_{fw} = 1000\frac{\text{kg}}{\text{m}^3}$. Treat all flows as incompressible.

1. Determine the mass density ρ_{sw} and the osmotic pressure π_{sw} of the incoming saltwater and chose $p_M - p_0 = 0.5\ \pi_{sw}$.
2. The volume flow of saltwater is 1000 litres/minute. Determine the volume flows of the freshwater drawn in, and of the brackish (i.e., diluted) saltwater that enters the turbine. Assume that the exiting mixture has an osmotic pressure of $0.55\ \pi_{sw}$.
3. Determine the entropy generation in the membrane module.
4. Determine the power required to drive the pump, and the power produced by the turbine. Compute the net work produced per litre of freshwater drawn in for a) reversible operation, b) isentropic efficiencies of 0.9 for both devices.
5. Compare to reversible work and discuss the source and size of irreversibilities.

9.13. Salt Transport

The entropy generation associated with salt flux from solution A to solution B across a membrane that does not fully prevent salt from passing is $\Sigma_s = (\mu_B - \mu_A) J_s$. For diluted solutions, the chemical potential difference reduces to $\mu_B - \mu_A = -R_s T \ln \frac{C_B}{C_A}$, where C_A and C_B are the salt concentrations (partial mass densities) on both sides.

Which of the following suggestions for the salt flux are acceptable choices? State your reasons, and give the range of allowed values for the constants β_α!

1. $J_1 = \beta_1 (C_B - C_A)$
2. $J_2 = \beta_2 (C_B - C_A)^2$
3. $J_3 = \beta_3 (\ln C_A - \ln C_B)$
4. $J_4 = \frac{\beta_4}{\ln C_A - \ln C_B}$
5. $J_5 = \beta_5 e^{C_B - C_A}$
6. $J_6 = \beta_6 \left(e^{C_B - C_A} - 1\right)$

9.14. Permeation Coefficient

A semi-permeable membrane which allows only water to pass, divides two containers I and II; the membrane area is $10\,\mathrm{cm}^2$, both containers are kept at 0.95 bar and 15 °C. Initially each container is filled with 750 g of water. Then, 40 g of calcium chloride are dissolved in container I, and water passes the membrane. Water transfer is measured by means of a scale, and the process runs until 2 g of water have passed.

1. $CaCl_2$ dissociates into Ca^{2+} and Cl^-, its molar mass is $M_{CaCl_2} = 111 \frac{\mathrm{kg}}{\mathrm{kmol}}$, and its mass density as a solid is $2.15 \frac{\mathrm{g}}{\mathrm{cm}^3}$. Determine the osmotic pressure in container I after the salt is dissolved.
2. How long does the process take, when the permeation coefficient is $A = 10^{-12} \frac{\mathrm{m}}{\mathrm{s\,Pa}}$? Ignore the small change of osmotic pressure during the process, and polarization.

9.15. Measurement of Permeation Coefficient

A semi-permeable membrane which allows only water to pass, divides two containers; the membrane area is $1\,\text{cm}^2$. In thermodynamic equilibrium at 300 K, container I holds 750 g of water and 45 g of NaCl (dissociates into Na^+ and Cl^-, $M_{\text{NaCl}} = 58.5\frac{\text{kg}}{\text{kmol}}$) at a pressure of 25 bar, and container II holds 600 g of water and 21 g of $MgCl_2$ (dissociates into Mg^{2+} and Cl^-, $M_{\text{MgCl}_2} = 95.2\frac{\text{kg}}{\text{kmol}}$).

To measure the water permeation coefficient of the membrane, container I is pressurized to 50 bar, while container B remains at 25 bar. By means of a scale, it is found that after 15 min a mass of 1.35 g of water has passed from I to II. Show all units, and present all calculations clearly, with comments.

1. Determine the initial osmotic pressures of the two solutions (careful with the dissociation!)
2. Determine the equilibrium pressure in container II.
3. Use the above data to determine the permeation coefficient A in m/(s Pa).
4. Discuss/criticize the measurement method: where would you see possible problems?

9.16. RO and PRO in a Piston-Cylinder System

Consider a piston-cylinder system, where the end of the cylinder is closed by a semipermeable membrane of area A_{m}, which is also the cross section of the cylinder. The cylinder of volume $V(t)$ is filled with saltwater of mass concentration $C_{\text{s}}(t) = \frac{m_s}{V}$. Initial volume and concentration are V_0 and C_s^0. The outside of the cylinder is exposed to freshwater at p_0.

For the following evaluation ignore polarization effects, and assume that the water flux across the membrane is given by (permeability A)

$$\hat{J}_{\text{w}} = \text{A}\left[\pi_{\text{s}} - (p_{\text{s}} - p_0)\right]$$

such that $\hat{J}_{\text{w}} > 0$ for pressure retarded osmosis, and $\hat{J}_{\text{w}} < 0$ for reverse osmosis.

1. Make a sketch of the system indicating all relevant properties. Express osmotic pressure π_s through cylinder volume and initial concentration.
2. Set up the equation that relates the volume change $\frac{dV}{dt}$ to the pressure difference $\Delta p = p_{\text{s}} - p_0$. Use the initial volume V_0 and outside pressure p_0 to non-dimensionalize the equation. Introduce dimensionless time τ in a suitable manner, so that no coefficients remain in the final equation.
3. Assume constant $\Delta p = \gamma \pi_0$, where π_0 is the initial osmotic pressure in the cylinder. Show that the final volume in the cylinder is $V_E = \frac{1}{\gamma} V_0$. Determine the work for the process with $\gamma = 0.5$ for RO and $\gamma = 2$ for PRO.
4. Determine and plot volume over time for $\gamma = \frac{1}{2}, 2$.

9.17. RO and PRO in a Piston-Cylinder System

Consider the same set-up as in the previous problem, but now assume that the piston moves with constant speed so that volume flow $\dot{V}$ is constant.

1. Determine and plot pressure over time for RO and PRO with $v_p = \mp 0.01, \mp 0.02$ or other values.
2. Determine work for the process in dependence of volume flow by integrating power over duration of the process. Assume that the final volume is $V_E = \gamma V_0$ with $\gamma = 0.5$ for RO and $\gamma = 2$ for PRO.
3. Determine average power by dividing total work for the process over duration. Plot as function of $\dot{V}$ and determine the optimal rates and physical piston speeds.

9.18. RO Membrane

For the volume flow of water through a semi-permeable RO membrane, we found the simple relation

$$\hat{J}_{\mathrm{w}} = \mathrm{A}\left(p_F - p_P - (\pi_F - \pi_P)\right)$$

and the corrected expression

$$\hat{J}_{\mathrm{w}} = \mathrm{A}\left[(p_F - p_P) - \pi_F \frac{\hat{J}_{\mathrm{w}}}{\mathrm{B}} \frac{e^{\frac{\hat{J}_{\mathrm{w}}}{k_E}} - \frac{C_P}{CF}}{\frac{\hat{J}_{\mathrm{w}}}{\mathrm{B}} + e^{\frac{\hat{J}_{\mathrm{w}}}{k_E}} - 1}\right] .$$

Here, $\hat{J}_{\mathrm{w}}$ denotes volumes flux of water with SI unit $\left[\frac{\mathrm{m}^3}{\mathrm{m}^2\,\mathrm{s}}\right]$, C_{s} is salt concentration in $\left[\frac{\mathrm{kg}}{\mathrm{m}^3}\right]$, $\pi = 2C_{\mathrm{s}}R_{\mathrm{s}}T$ is osmotic pressure, Δp is the difference of hydraulic pressure between both sides of the membrane, and A, B, k_E are parameters that describe transport behavior.

Parameters are: $\mathrm{A} = 4.86 \times 10^{-12} \frac{\mathrm{m}}{\mathrm{s\,Pa}}$, $\mathrm{B} = 4.44 \times 10^{-8} \frac{\mathrm{m}}{\mathrm{s}}$, $k = 3.85 \times 10^{-5} \frac{\mathrm{m}}{\mathrm{s}}$, $C_F = 35 \frac{\mathrm{kg}}{\mathrm{m}^3}$. The salt is NaCl and the temperature is 288 K.

For $C_P = 0$ and $C_P = 17.5 \frac{\mathrm{kg}}{\mathrm{m}^3}$, determine the water flux for $\Delta p = 2(\pi_D - \pi_F)$ for both cases. Explain how you solved the implicit equation, show code, figures, etc.

9.19. PRO Membrane

The power gain from the PRO membrane process is $\dot{W} = \Delta p \dot{V}$, where $\Delta p = p_D - p_F$ and $\dot{V} = A\hat{J}_{\mathrm{w}}$ is volume flow of water through membrane area A, so that the power that can be obtained per membrane area is $\frac{\dot{W}}{A} = \Delta p \hat{J}_{\mathrm{w}}$. Parameters are: $\mathrm{A} = 4.86 \times 10^{-12} \frac{\mathrm{m}}{\mathrm{s\,Pa}}$, $\mathrm{B} = 4.44 \times 10^{-8} \frac{\mathrm{m}}{\mathrm{s}}$, $k_E = 3.85 \times 10^{-5} \frac{\mathrm{m}}{\mathrm{s}}$, $C_D = 35 \frac{\mathrm{kg}}{\mathrm{m}^3}$, $C_F = 0 \frac{\mathrm{kg}}{\mathrm{m}^3}$. The salt is NaCl and the temperature is 288 K.

1. For the case without polarization, the volume flux of water through the membrane is $\hat{J}_{\mathrm{w}} = \mathrm{A}(\Delta\pi - \Delta p)$, where $\Delta\pi = \pi_D - \pi_F$. Determine the optimal pressure difference $\Delta p = p_D - p_F$ by maximizing power. Determine the corresponding volume flux and power per area.

2. Determine for this case membrane area and volume flow of water across the membrane for a power output of $\dot{W} = 1\,\mathrm{MW}$. Plot power per area $\dot{W}/A$ over pressure difference Δp.
3. With polarization effects included the mass flux obeys the implicit equation (simplified here for just one polarization effect)

$$\hat{J}_{\mathrm{w}} = \mathrm{A}\left(\pi_D \frac{\hat{J}_{\mathrm{w}}}{\mathrm{B}} \frac{1 - \frac{C_F}{C_D} e^{\frac{\hat{J}_{\mathrm{w}}}{k_E}}}{\frac{\hat{J}_{\mathrm{w}}}{\mathrm{B}} e^{\frac{\hat{J}_{\mathrm{w}}}{k}} + e^{\frac{\hat{J}_{\mathrm{w}}}{k}} - 1} - \Delta p\right)$$

To determine optimum pressure difference and mass flux for this case, it is easiest to use a graphical approach: Plot power and pressure difference as functions of volume flux, and read the required data from the graphs.
4. Also for this case determine membrane area and volume flow of water across the membrane for a power output of $\dot{W} = 1\,\mathrm{MW}$. Add the curve for power per area W/A over pressure difference Δp to the earlier plot without polarization.
5. Discuss the influence of polarization on this process.

9.20. PRO Membrane

For the volume flow of water through a semi-permeable membrane, we found the simple relation

$$\hat{J}_{\mathrm{w}} = \mathrm{A}\left(\pi_D - \pi_F - (p_D - p_F)\right)$$

and the corrected expression (simplified here for just one polarization effect)

$$\hat{J}_{\mathrm{w}} = \mathrm{A}\left(\pi_D \frac{\hat{J}_{\mathrm{w}}}{\mathrm{B}} \frac{1 - \frac{C_F}{C_D} e^{\frac{\hat{J}_{\mathrm{w}}}{k_E}}}{\frac{\hat{J}_{\mathrm{w}}}{\mathrm{B}} e^{\frac{\hat{J}_{\mathrm{w}}}{k}} + e^{\frac{\hat{J}_{\mathrm{w}}}{k}} - 1} - (p_D - p_F)\right)$$

Here, $\hat{J}_{\mathrm{w}}$ denotes volumes flux of water with SI unit $\left[\frac{\mathrm{m}^3}{\mathrm{m}^2\,\mathrm{s}}\right]$, C_{s} is salt concentration in $\left[\frac{\mathrm{kg}}{\mathrm{m}^3}\right]$, $\pi = 2C_{\mathrm{s}}R_{\mathrm{s}}T$ is osmotic pressure, Δp is the difference of hydraulic pressure between both sides of the membrane, and A, B, k_E are parameters that describe transport behavior.

Parameters are: $\mathrm{A} = 4.86 \times 10^{-12} \frac{\mathrm{m}}{\mathrm{s\,Pa}}$, $\mathrm{B} = 4.44 \times 10^{-8} \frac{\mathrm{m}}{\mathrm{s}}$, $k = 3.85 \times 10^{-5} \frac{\mathrm{m}}{\mathrm{s}}$, $C_D = 35 \frac{\mathrm{kg}}{\mathrm{m}^3}$.

1. For $C_F = 0$ and $C_F = 17.5 \frac{\mathrm{kg}}{\mathrm{m}^3}$, determine the water flux for $\Delta p = \frac{1}{2}(\pi_D - \pi_F)$ for both cases. Explain how you solved the implicit equation, show code, figures, etc.
2. Plot $\hat{J}_{\mathrm{w}}$ over Δp for $C_F = \{0, 5, 10, 15, 20, 25, 30, 35\} \frac{\mathrm{kg}}{\mathrm{m}^3}$ to reproduce the figure shown in the text. Explain how you solved the problem, show code.

9.21. Law of Mass Action

The chemical potential of an ideal gas mixture is $\mu_\alpha = g_\alpha(T,p) - R_\alpha T \ln X_\alpha$ where $g_\alpha(T,p) = h_\alpha(T) - T\left(s_\alpha^0(T) - R_\alpha \ln \frac{p}{p_0}\right)$. In chemical equilibrium, the law of mass action holds as $\sum_\alpha \mu_\alpha M_\alpha \gamma_\alpha = 0$, where γ_α is the stoichiometric coefficient for component α.

1. Introduce the partial pressure $p_\alpha = pX_\alpha$ and show that for an ideal gas the law of mass action can be written as
$$\prod_\alpha \left(\frac{p_\alpha}{p_0}\right)^{\gamma_\alpha} = K_p(T) \ ,$$
where $K_p(T)$ is the chemical constant. Identify the chemical constant.
2. Consider the dissociation of oxygen, with the reaction equation $\frac{1}{2}O_2 \rightleftharpoons O$. Use data from the property tables in TEC (heat and entropy of formation, enthalpy $h(T)$ etc.) to determine the chemical constant for this reaction at 3200 K. Note that atomic oxygen (O) can be approximated as an ideal gas with constant specific heat $c_p = \frac{5}{2}R$.
3. Compare your value with that from a suitable data table, and discuss why there is a difference.
4. Determine the pressure at which both gases have equal mole fractions (3200 K).
5. Now the pressure is quadrupled: determine the mole fractions of both gases.

9.22. A Simple Reaction

Consider the reaction A $\rightleftharpoons$ B which obeys the rate law

$$\frac{d\bar{\rho}_A}{dt} = -\frac{d\bar{\rho}_B}{dt} = k_f \bar{\rho}_A^{\alpha_f} C_B^{\beta_f} e^{-\frac{\bar{E}_f}{\bar{R}T}} - k_b \bar{\rho}_A^{\alpha_b} \bar{\rho}_B^{\beta_b} e^{-\frac{\bar{E}_b}{\bar{R}T}}$$

where $\bar{\rho}_A$, $\bar{\rho}_B$ denote mole densities of components, $k_{f,b}$ are rate constants for forward and backward reactions, the exponents $\alpha_{f,b}, \beta_{f,b}$ determine the order of the reaction, and $\bar{E}_{f,b}$ are activation energies.

A and B have the same specific heat and gas constant, but different energy constants, so that $\bar{u}_\alpha = \bar{c}_v(T - T_0) + \bar{u}_\alpha^0$, where $\bar{u}_A^0 - \bar{u}_B^0 = \bar{E}_f - \bar{E}_b = \Delta E$.

The reaction proceeds from the initial state $\bar{\rho}_A(t=0) = \bar{\rho}_A^0$, $\bar{\rho}_B(t=0) = \bar{\rho}_B^0$, $T(t=0) = T_0$ towards the final equilibrium state.

1. Show that $\bar{\rho}_A(t) + \bar{\rho}_B(t) = \bar{\rho}_A^0 + \bar{\rho}_B^0$.
2. The system is isolated at constant volume. Express the temperature $T(t)$ through $\bar{\rho}_A(t), \bar{\rho}_B(t)$ and the initial conditions. You should find
$$T = T_0\left(1 + \frac{\bar{\rho}_A^0 - \bar{\rho}_A}{\bar{\rho}_A^0 + \bar{\rho}_B^0}\frac{\Delta E}{\bar{c}_v T_0}\right)$$
The reaction is exothermic for $\Delta E = \bar{u}_A^0 - \bar{u}_B^0 > 0$.

3. Introduce dimensionless properties

$$\hat{\rho}_A = \frac{\bar{\rho}_A}{\bar{\rho}_A^0}\ ,\quad \hat{\rho}_B = \frac{\bar{\rho}_B}{\bar{\rho}_A^0}\ ,\quad \hat{T} = \frac{T}{T_0}\ ,\quad \hat{t} = t\ k_f\left(\bar{\rho}_A^0\right)^{\alpha_f+\beta_f-1}$$

as well as the dimensionless rate ratio $\hat{k} = \frac{k_b\left(\bar{\rho}_A^0\right)^{\alpha_b+\beta_b}}{k_f\left(\bar{\rho}_A^0\right)^{\alpha_f+\beta_f}}$ to find the dimensionless equations

$$\frac{d\hat{\rho}_A}{dt} = -\frac{d\hat{\rho}_B}{dt} = \hat{\rho}_A^{\alpha_f}\hat{\rho}_B^{\beta_f}e^{-\frac{\hat{E}_f}{\hat{T}}} - \hat{k}\hat{\rho}_A^{\alpha_b}\hat{\rho}_B^{\beta_b}e^{-\frac{\hat{E}_b}{\hat{T}}}$$

$$\hat{T}(t) = 1 + \frac{1-\hat{\rho}_A}{1+\hat{\rho}_B^0}\frac{\Delta\hat{E}}{\bar{c}_v/\bar{R}}$$

4. Set $\hat{k} = 1$, $\bar{c}_v/\bar{R} = \frac{5}{2}$, $\hat{E}_b = 1$, $\hat{E}_f = 10$, and chose initial conditions $\hat{\rho}_A^0 = 1$, $\hat{\rho}_B^0 = 0.001$. Use a suitable software to numerically solve the initial value problem, and plot curves $\hat{\rho}_A(t)$, $\hat{\rho}_B(t)$, $\hat{T}(t)$ over a suitable range for the following cases:

   a. First order reactions: $\alpha_b = \beta_f = 1$, and $\alpha_f = \beta_b = 0$.
   b. First order and second order reactions: $\alpha_b = \beta_f = \beta_b = 1$, and $\alpha_f = 0$.
   c. Second order and first order reactions: $\alpha_b = \beta_f = \alpha_f = 1$, and $\beta_b = 0$.
   d. Second order reactions: $\alpha_b = \beta_f = \alpha_f = \beta_b = 1$.

5. For each of the above cases discuss the law of mass action, and the equilibrium states.

9.23. Reaction Towards Equilibrium

Consider the isobaric and adiabatic reaction of methane with oxygen to water and carbon dioxide according to the reaction $CH_4 + 2O_2 \rightleftharpoons CO_2 + 2H_2O$. All substances can be considered as ideal gases. You will need a computer to solve the following questions:

1. Set up the chemical potentials $\bar{\mu}_\alpha$ for the constituents on a molar basis as

$$\bar{\mu}_\alpha = \bar{g}_\alpha(T,p) + \bar{R}T\ln X_\alpha$$

where $\bar{g}_\alpha(T,p) = \bar{h}_\alpha(T) - T\bar{s}_\alpha(T,p)$ is the Gibbs free energy of constituent α alone at (T,p) and X_α is the mole fraction
For this, compute the enthalpy from specific heat data

$$\bar{h}_\alpha(T) - \bar{h}_{R,\alpha} = \int_{T_R}^{T}\bar{c}_{p,\alpha}(T')\,dT'$$

where $T_R = 298\,\mathrm{K}$. Use tabulated data to find $\bar{h}_{R,\alpha}$, $c_{p,\alpha}(T)$.
Determine the entropy by integrating the Gibbs equation

$$\bar{s}_\alpha - \bar{s}_{R,\alpha} = \int_{T_R}^{T} \frac{\bar{c}_{p,\alpha}(T')}{T'} dT' - \int_{p_R}^{p} \bar{\mathrm{v}} dp$$

where $p_R = 1\,\text{bar}$ is the reference pressure. $\bar{s}_{R,\alpha}$ can also be found in standard tables.

2. Set up the law of mass action for this reaction in the form

$$\prod_\alpha p_\alpha^{\gamma_\alpha} = K_p(T)$$

where $p_\alpha = X_\alpha p$ is the partial pressure of α. Plot the chemical constant $K_p(T)$ as a function of temperature. Chose a suitable temperature range.

3. Assume a well-stirred reactor, so that all quantities are homogeneous in space. Set up the equations for the conservation of mass, concentration, and energy for this case (= time dependent differential equations). For the reaction rate λ assume that

$$\lambda = \Lambda \left(\exp \sum_{\alpha^-} \frac{\bar{\mu}_\alpha |\gamma_a|}{\bar{R}T} - \exp \sum_{\alpha^+} \frac{\bar{\mu}_\alpha \gamma_a}{\bar{R}T} \right)$$

where α^+ indicates positive stoichiometric coefficients, and α^- negative stoichiometric coefficients. Introduce dimensionless quantities (e.g. $\hat{T} = T/T_R$ etc. Hint: introduce a dimensionless time scale to absorb the unknown constant Λ).

4. Simplify the equations as far as possible, then use a computer to solve them for the case that 1 litre of stoichiometric mixture of CH_4 and O_2 at a pressure of 10 bar, and 298 K starts to react. Plot partial masses, temperature and volume as functions of time.

5. Assume now that an activation energy needs to be overcome for the reaction to take place, so that Λ follows an Arrhenius law

$$\Lambda = \Lambda_0 \exp\left[-\frac{\bar{E}}{\bar{R}T}\right].$$

Introduce a dimensionless activation energy $\hat{E} = \frac{\bar{E}}{\bar{R}T_R}$ and solve the same problem for several values of $\hat{E}$.

Appendix A
Scalars, Vectors, Tensors

Throughout the main text, transport equations and constitutive relations are written either in direct tensor notation, or in tensor index notation, where the latter is preferred. This appendix offers a compact summary of symbolic and index notation for tensors and their operations.

A.1 Thermodynamic Properties as Fields

When studying *Transport Phenomena* thermodynamic properties are resolved in space and time, and we speak of time dependent fields, e.g., the temperature field $T(\mathbf{x},t)$ where $\mathbf{x}$ denotes the location vector and t denotes time.

Temperature $T(\mathbf{x},t)$, mass density $\rho(\mathbf{x},t)$, pressure $p(\mathbf{x},t)$, internal energy $u(\mathbf{x},t)$, entropy $s(\mathbf{x},t)$ are examples for scalar fields. Flow velocity $\mathbf{v}(\mathbf{x},t)$ and heat flux vector $\mathbf{q}(\mathbf{x},t)$ are examples for vector fields, and the stress tensor $\mathbf{t}(\mathbf{x},t)$ is an example for a tensor field.

The resolved thermodynamic equations contain scalars, vectors, tensors and their space and time derivatives. Proper and consistent notation is key for understanding and solving the equations. This appendix introduces the notation used for scalars, vectors, tensors, and operations between them.

A.2 Tensor Notation

We consider Cartesian three-dimensional frames with base vectors

$$\{\mathbf{e}_1,\ \mathbf{e}_2,\ \mathbf{e}_3\} \tag{A.1}$$

H. Struchtrup, *A Thermodynamic Introduction to Transport Phenomena*, https://doi.org/10.1007/978-3-031-61868-0

The base vectors have unit length and are orthogonal to each other, so that for their scalar products we have

$$\mathbf{e}_i \cdot \mathbf{e}_j = \delta_{ij} = \begin{cases} 1 \text{ if } i = j \\ 0 \text{ if } i \neq j \end{cases} \tag{A.2}$$

Scalar quantities are written in italics: a, b, etc.

Vectors are written in bold face when using direct notation: **a**, **b**, etc. A vector is expressed in terms of its components a_i $(i = 1, 2, 3)$ with respect to the base $\mathbf{e}_i$ as

$$\mathbf{a} = a_1\mathbf{e}_1 + a_2\mathbf{e}_2 + a_3\mathbf{e}_3 = \sum_{i=1}^{3} a_i\mathbf{e}_i = a_i\mathbf{e}_i \tag{A.3}$$

Tensors are written in bold face, typically in capitals: **A, B**, etc. A **tensor of second order** is expressed through its components A_{ij} $(i, j = 1, 2, 3)$ with respect to the base as

$$\mathbf{A} = \sum_{i,j=1}^{3} A_{ij}\mathbf{e}_i\mathbf{e}_j = A_{ij}\mathbf{e}_i\mathbf{e}_j \tag{A.4}$$

Also **tensors of higher order** are typically written as capital letters, $\mathbf{A}, \mathbf{B}$, etc. Also these are expressed through their components, A_{ij}, B_{ijk}, C_{ijkl}, etc. where $\mathbf{A} = A_{ij}\mathbf{e}_i\mathbf{e}_j$, $\mathbf{B} = B_{ijk}\mathbf{e}_i\mathbf{e}_j\mathbf{e}_k$, etc.

For analytical or numerical solutions, one will consider the transport equations written for vector and tensor components. This is particularly easy for Cartesian frames, for which the base is universal, independent of location.

For general development and discussion of equations, one can either use direct notation, which is independent of the frame (hence also called component free), or use index notation for Cartesian coordinates. However, for many engineering application Cartesian coordinates are not the best choice, and curvilinear coordinates are better suited, where the local frame depends on location, e.g., cylindrical or spherical coordinates. Conversion of equations from Cartesian to curvilinear coordinates proceeds by first converting from Cartesian coordinates to direct notation, in which the equations are independent of any frame, and then convert these to curvilinear coordinates.

A.3 Einstein Summation Convention

In the above equations, we have already used the **Einstein summation convention** which declares it unnecessary to write the summation symbol, and states that a sum has to be taken over an index that appears twice. The Einstein convention makes equations in index notation much easier to read and write, and we use it extensively. The convention relies on the following rules:

1. Any term in a tensor equation in index notation can contain the same index only once or twice.
2. If an index appears twice in a term, summation is implied (as above in $a_i\mathbf{e}_i$ and $A_{ij}\mathbf{e}_i\mathbf{e}_j$). A summation index (also called dummy index) can be renamed at will, as long as the index is not used already, i.e., $a_i\mathbf{e}_i = a_j\mathbf{e}_j = a_r\mathbf{e}_r$ etc.
3. Free indices are those indices that appear only once in a term; in a tensor equation all terms must have the same free indices.

A.4 Tensor Index Notation and Scalar Product

With Cartesian coordinates and the Einstein convention, a vector or tensor can be expressed through a place holder for its components, e.g., a_i $(i = 1, 2, 3)$, or A_{ij} $(i, j = 1, 2, 3)$ and the possible values for the indices i, j are not written, if the dimension of the vector (here $D = 3$) is obvious.

That is instead of, e.g., $\mathbf{a}$, $\mathbf{A}$, we write tensor equations in a_i, A_{ij}, while obeying the rules of the summation convention.

The **scalar product** of two vectors $\mathbf{a} = a_i\mathbf{e}_i$ and $\mathbf{b} = b_j\mathbf{e}_j$ gives a scalar c. In some detail, using orthogonality of the base and the summation convention (summation over i and j):

$$\mathbf{a} \cdot \mathbf{b} = (a_i\mathbf{e}_i) \cdot (b_j\mathbf{e}_j) = a_i b_j\,(\mathbf{e}_i \cdot \mathbf{e}_j) = a_i b_j \delta_{ij} = a_i b_i = c \tag{A.5}$$

In index notation the scalar product is expressed directly through the operation that one will perform, i.e., summation of components, $a_i b_i = \sum_i a_i b_i$.

The **absolute value** of a vector, i.e., its length, is

$$a = |\mathbf{a}| = \sqrt{\mathbf{a} \cdot \mathbf{a}} \quad , \quad a = |a_i| = \sqrt{a_i a_i} \tag{A.6}$$

and its **direction vector** of unit length is

$$\mathbf{n} = \frac{\mathbf{a}}{a} \quad , \quad n_i = \frac{a_i}{a} \quad , \quad \text{with} \quad n = \sqrt{\mathbf{n} \cdot \mathbf{n}} = \sqrt{n_i n_i} = 1 \tag{A.7}$$

A typical example for a well written tensor equation in the Einstein convention is the scalar product[1] of a tensor $\mathbf{A}$ and a vector $\mathbf{b}$, which gives the vector $\mathbf{d}$,

$$\mathbf{d} = \mathbf{A} \cdot \mathbf{b} \quad \Rightarrow \quad d_i = A_{ij} b_j \tag{A.8}$$

[1] Here the name *scalar product* is used in extension of the operation between two vectors. Other names found in the literature are *dot product, inner product*, or *contraction*.

where summation over j, and values $i = 1, 2, 3$ are implicitly assumed. Note that this vector equation stands for three separate equations for the indices:

$$\begin{aligned} i = 1: \quad d_1 &= A_{11}b_1 + A_{12}b_2 + A_{13}b_3 \\ i = 2: \quad d_2 &= A_{21}b_1 + A_{22}b_2 + A_{23}b_3 \\ i = 3: \quad d_3 &= A_{31}b_1 + A_{32}b_2 + A_{33}b_3 \end{aligned} \tag{A.9}$$

Renaming the summation index, say $j \to k$, we have the equivalent expression $d_i = A_{ik}b_k$.

When renaming the free index, say $i \to m$, the index must be changed in all terms to give the equivalent expression $d_m = A_{mj}b_j = A_{mk}b_k$.

Possible violations of the rules of the summation convention are as follows: in violation of Rule 1, in the equation $d_j = A_{jj}b_j$ the index j appears three times—i.e., more than twice—in the term on the right; in violation of Rule 3, in the equation $d_k = A_{ij}b_j$ the free index on the left (k) and the free index on the right (i) do not agree.

A.5 Tensors of Second Order

Tensors are a generalization of scalars and vectors, and can have any order. Tensors of n-th order are defined through their components in the base $\mathbf{e}_i$ as $\mathcal{A}^{(n)} = A_{i_1 i_2 \cdots i_n}\mathbf{e}_{i_1}\mathbf{e}_{i_2}\cdots\mathbf{e}_{i_n}$. In this sense, scalars (a) are tensors of 0th order, and vectors ($\mathbf{a} = a_i\mathbf{e}_i$) are tensors of 1st order.

Tensors of 2nd order are typically written as capital letters, $\mathbf{A}, \mathbf{B}$, etc. With respect to a Cartesian base they have the components A_{ij}:

$$\mathbf{A} = A_{ij}\mathbf{e}_i\mathbf{e}_j \quad , \quad A_{ij} = \begin{bmatrix} A_{11} & A_{12} & A_{13} \\ A_{21} & A_{22} & A_{23} \\ A_{31} & A_{32} & A_{33} \end{bmatrix}_{ij} \tag{A.10}$$

In index notation, the tensor is represented by a placeholder for its components, A_{ij}, B_{kl} etc.

The **unit tensor** is written in symbolic and index notation as

$$\mathbf{1} = \delta_{ij}\mathbf{e}_i\mathbf{e}_j \quad \text{or} \quad \delta_{ij} = \begin{bmatrix} 1 & 0 & 0 \\ 0 & 1 & 0 \\ 0 & 0 & 1 \end{bmatrix}_{ij} = \begin{cases} 1\,, \text{ if } i = j \\ 0\,, \text{ if } i \neq j \end{cases} \tag{A.11}$$

The scalar product of a vector $\mathbf{b}$ with a 2-tensor $\mathbf{A}$ rotates and stretches the vector, to give the new vector

$$\mathbf{d} = \mathbf{A} \cdot \mathbf{b} \quad , \quad d_i = A_{ij}b_j \tag{A.12}$$

Accordingly, the product of a vector with the unit tensor reproduces the vector. In symbolic notation this simply reads

$$\mathbf{a} \cdot \mathbf{1} = \mathbf{1} \cdot \mathbf{a} = \mathbf{a} \tag{A.13}$$

but in index notation the operation yields the change of the index at the vector,

$$a_i \delta_{ij} = a_j \quad , \quad \delta_{ij} a_j = a_i \tag{A.14}$$

The scalar products between vectors or between tensor and vector are examples the **contraction** of tensors. It is instructive to show some examples in direct and index notation

$$\begin{array}{ll} \mathbf{d} = \mathbf{A} \cdot \mathbf{b} & d_i = A_{ij} b_j \\ \mathbf{C} = \mathbf{A} \cdot \mathbf{B} & C_{ij} = A_{ik} B_{kj} \\ \mathbf{A} = \mathbf{A} \cdot \mathbf{1} = \mathbf{1} \cdot \mathbf{A} & A_{ik} = A_{ij} \delta_{jk} \quad A_{kj} = \delta_{ki} A_{ij} \end{array} \tag{A.15}$$

In the second equation, the contraction of two 2-tensors $\mathbf{A}$ and $\mathbf{B}$ yields a 2-tensor $\mathbf{C}$. The last equation is another example for the change of the index through contraction with the unit tensor.

In the **transposed tensor** rows and columns are exchanged

$$\mathbf{A}^T = \begin{bmatrix} A_{11} & A_{21} & A_{31} \\ A_{12} & A_{22} & A_{32} \\ A_{13} & A_{23} & A_{33} \end{bmatrix} \quad \text{or} \quad A^T_{ij} = A_{ji} \tag{A.16}$$

A tensor is **symmetric** if

$$\mathbf{A}^T = \mathbf{A} \quad , \quad A_{ij} = A_{ji} \tag{A.17}$$

A tensor is **antisymmetric** if

$$\mathbf{A}^T = -\mathbf{A} \quad , \quad A_{ij} = -A_{ji} \tag{A.18}$$

Any 2-tensor can be decomposed into a **symmetric** part $\mathbf{A}^{(s)} = A_{(ij)} \mathbf{e}_i \mathbf{e}_j$ and an **antisymmetric** part $\mathbf{A}^{[a]} = A_{[ij]} \mathbf{e}_i \mathbf{e}_j$, where round and square brackets are used to indicate symmetry and anti-symmetry, respectively. These are obtained from the tensor and its transpose as

$$\mathbf{A}^{(s)} = \frac{1}{2} \left(\mathbf{A} + \mathbf{A}^T \right) \quad , \quad A_{(ij)} = A_{(ji)} = \frac{1}{2} \left(A_{ij} + A_{ji} \right) \tag{A.19}$$

$$\mathbf{A}^{[a]} = \frac{1}{2} \left(\mathbf{A} - \mathbf{A}^T \right) \quad , \quad A_{[ij]} = -A_{[ji]} = \frac{1}{2} \left(A_{ij} - A_{ji} \right) \tag{A.20}$$

The sum of symmetric and antisymmetric parts yields the tensor itself,

$$\mathbf{A} = \mathbf{A}^{(s)} + \mathbf{A}^{[a]} \quad , \quad A_{ij} = A_{(ij)} + A_{[ij]} \tag{A.21}$$

The **trace** of a tensor is the sum of its diagonal components,

$$\mathrm{tr}\mathbf{A} = A_{jk}\delta_{jk} = A_{kk} \; ; \tag{A.22}$$

for instance

$$\mathrm{tr}\mathbf{1} = \delta_{ij}\delta_{ij} = \delta_{ii} = \delta_{11} + \delta_{22} + \delta_{33} = 3 \quad , \quad A_{(ii)} = A_{ii} \quad , \quad A_{[ii]} = 0 \tag{A.23}$$

The double scalar product (double dot product) of two 2-tensors is a scalar[2]

$$c = \mathbf{A} : \mathbf{B} = \mathrm{tr}\left(\mathbf{A} \cdot \mathbf{B}^T\right) \quad , \quad c = A_{ij} B^T_{ji} = A_{ij} B_{ij} \tag{A.24}$$

The **tracefree and symmetric part** of a tensor, also known as the **deviatoric part**, indicated by angle brackets, is defined as

$$\mathbf{A}^{\langle d \rangle} = \mathbf{A}^{(s)} - \frac{1}{3}\left(\mathrm{tr}\mathbf{A}\right)\mathbf{1} \quad , \quad A_{\langle ij \rangle} = A_{(ij)} - \frac{1}{3} A_{kk}\delta_{ij} \tag{A.25}$$

so that

$$\mathrm{tr}\mathbf{A}^{\langle d \rangle} = A_{\langle ii \rangle} = 0 \quad \text{and} \quad \mathbf{A}^{\langle d \rangle T} = \mathbf{A}^{\langle d \rangle} \; , \quad A_{\langle ij \rangle} = A_{\langle ji \rangle} \tag{A.26}$$

Combining the above definitions, a 2-tensor can be uniquely decomposed into its trace, deviatoric part, and antisymmetric part,

$$\mathbf{A} = \frac{1}{3}\left(\mathrm{tr}\mathbf{A}\right)\mathbf{1} + \mathbf{A}^{\langle d \rangle} + \mathbf{A}^{[a]} \quad , \quad A_{ij} = \frac{1}{3} A_{kk}\delta_{ij} + A_{\langle ij \rangle} + A_{[ij]} \tag{A.27}$$

In three dimensional space, a 2-tensor has 9 independent components A_{ij}. A symmetric tensor has 6 independent components $A_{ij} = A_{ji}$. An antisymmetric tensor has 3 independent components $A_{ij} = -A_{ji}$, its diagonal elements are zero. A deviatoric tensor has 5 independent components (symmetry, and vanishing trace).

For the **inverse tensor** we write $\mathbf{A}^{-1}$ or A^{-1}_{ij}. Scalar product of a tensor with its inverse from either side yields the unit tensor,

$$\mathbf{A} \cdot \mathbf{A}^{-1} = \mathbf{1} \quad , \quad \mathbf{A}^{-1} \cdot \mathbf{A} = \mathbf{1} \tag{A.28}$$

$$A_{ij} A^{-1}_{jk} = \delta_{ik} \quad , \quad A^{-1}_{ij} A_{jk} = \delta_{ik} \tag{A.29}$$

Problem A.3 proves some useful relations between 2-tensors.

[2] Other authors might use different definitions of the double dot product, as $c = \mathbf{A} : \mathbf{B} = \mathrm{tr}(\mathbf{A} \cdot \mathbf{B})$, $c = A_{ij} B_{ji}$. It is always a good idea to carefully clarify notation used in the literature, instead of assuming what might be meant.

A.6 Tensor Product

The **tensor product** of two vectors results in a tensor,

$$\mathbf{A} = \mathbf{ab} \quad , \quad A_{ij} = a_i b_j \tag{A.30}$$

Extension to higher order tensors is evident, e.g., the tensor product of a 3- and a 2-tensor yields a 5-tensor,

$$\mathbf{C} = \mathbf{AB} \quad , \quad C_{ijklm} = A_{ijk} B_{lm} \tag{A.31}$$

A.7 Cross Product

The **cross product** (or vector product) of two vectors is written as

$$\mathbf{a} \times \mathbf{b} = \mathbf{c} \tag{A.32}$$

where the resulting vector **c** is perpendicular to **a** and **b**. In index notation we write

$$\varepsilon_{ijk} a_j b_k = c_i \tag{A.33}$$

where the completely **antisymmetric tensor of third order** is given by the Levi-Civita symbol

$$\varepsilon_{ijk} = \begin{cases} 1 \text{ if} & ijk = 123 \ / \ 312 \ / \ 231 \\ -1 \text{ if} & ijk = 321 \ / \ 132 \ / \ 213 \\ 0 \text{ else} & \end{cases} \tag{A.34}$$

The ε-tensor is antisymmetric in all index pairs,

$$\varepsilon_{ijk} = -\varepsilon_{jik} = \varepsilon_{jki} = -\varepsilon_{kji} = \varepsilon_{kij} = -\varepsilon_{ikj} \tag{A.35}$$

An antisymmetric 2-tensor $A_{[ij]}$ has only three independent components, hence can be expressed through a vector and the ε-tensor as

$$A_{[ij]} = \varepsilon_{ijk} \bar{A}_k \quad , \quad \bar{A}_k = \frac{1}{2} \varepsilon_{ijm} A_{[ij]} \tag{A.36}$$

where $\bar{A}_k$ is the **vector associated with the antisymmetric tensor.**

A.8 Orthogonal Tensors, Coordinate Transformations

For an **orthogonal tensor** the transpose equals the inverse,

$$\mathbf{O}^T = \mathbf{O}^{-1} \quad \Rightarrow \quad \mathbf{O^T} \cdot \mathbf{O} = \mathbf{O} \cdot \mathbf{O^T} = \mathbf{1} \tag{A.37}$$

$$O_{ji} = O_{ij}^{-1} \quad \Rightarrow \quad O_{ji}O_{jk} = O_{ij}O_{kj} = \delta_{ik} \tag{A.38}$$

For the determinant of an orthogonal tensor we find with the rules for determinants

$$\begin{aligned}(\det \mathbf{O})^2 &= \det \mathbf{O} \det \mathbf{O} = \det \mathbf{O} \det \mathbf{O}^T \\ &= \det \mathbf{O} \det \mathbf{O}^{-1} = \det\left(\mathbf{O} \cdot \mathbf{O}^{-1}\right) = \det(\mathbf{1}) = 1\end{aligned} \tag{A.39}$$

hence $\det \mathbf{O} = \pm 1$.

An orthogonal tensor O_{ij} rotates a vector, but does not change its length,

$$\hat{v}_i = O_{ij}v_j \qquad \Rightarrow \hat{v}^2 = (O_{ij}v_j)(O_{ik}v_k) = (O_{ij}O_{ik})\, v_j v_k = \delta_{jk}v_j v_k = v^2 \tag{A.40}$$

A new Cartesian base $\hat{\mathbf{e}}_i$ is obtained by rotating the original base $\mathbf{e}_j$ by means of an orthogonal tensor,

$$\hat{\mathbf{e}}_i = O_{ij}\mathbf{e}_j \quad \Rightarrow \quad O_{ik}\hat{\mathbf{e}}_i = O_{ik}O_{ij}\mathbf{e}_j = \delta_{jk}\mathbf{e}_j = \mathbf{e}_k \tag{A.41}$$

A vector or tensor can be expressed with respect to either of the two bases, where their coordinates v_i, $\hat{v}_i$ and A_{ij}, $\hat{A}_{ij}$ are different for the bases

$$\mathbf{v} = v_i\mathbf{e}_i = \hat{v}_i\hat{\mathbf{e}}_i \quad , \quad \mathbf{A} = A_{ij}\mathbf{e}_i\mathbf{e}_j = \hat{A}_{ij}\hat{\mathbf{e}}_i\hat{\mathbf{e}}_j \tag{A.42}$$

With the relations between the bases the coordinates transform as

$$v_k = O_{ik}\hat{v}_i \quad , \quad \hat{v}_k = O_{ki}v_i \tag{A.43}$$

$$A_{kl} = O_{ik}O_{jl}\hat{A}_{ij} \quad , \quad \hat{A}_{ij} = O_{ik}O_{jl}A_{kl} \tag{A.44}$$

and so on for higher order tensors.

The transformation does not change tensor properties. Specifically the trace is invariant,

$$\hat{A}_{kk} = A_{kk} \tag{A.45}$$

and symmetry properties transfer, that is (Prob. A.6)

$$\hat{A}_{(ij)} = O_{ik}O_{jl}A_{(kl)} \quad , \quad \hat{A}_{\langle ij\rangle} = O_{ik}O_{jl}A_{\langle kl\rangle} \quad , \quad \hat{A}_{[ij]} = O_{ik}O_{jl}A_{[kl]} \,. \tag{A.46}$$

A.9 Isotropic Tensors

Tensors whose coordinates do not change in the transformation between different Cartesian bases are called isotropic tensors.

All scalars are isotropic.

There are no isotropic vectors.

The unit tensor δ_{ij} is isotropic, hence all tensors of the form $A_{ij} = a\delta_{ij}$ with arbitrary scalar a are isotropic; these are the only isotropic 2-tensors.

The ε-tensor ε_{ijk} is isotropic, and so are all combinations with a scalar, $A_{ijk} = a\varepsilon_{ijk}$. These are the only isotropic 3-tensors.

Isotropic tensors of higher orders are suitable combinations of scalars, δ-tensors, and ε-tensors.

A.10 Tensor Calculus

In index notation, the nabla (or Del) vector operator $\boldsymbol{\nabla}$ can be expressed through its coordinates,

$$\nabla_i = \frac{\partial}{\partial x_i} \tag{A.47}$$

The **gradients** of a scalar, vector, or tensor field are written in direct and index notation as

$$\operatorname{grad} a = \boldsymbol{\nabla} a \quad , \quad \nabla_i a = \frac{\partial a}{\partial x_i} \tag{A.48}$$

$$\operatorname{grad} \mathbf{a} = \boldsymbol{\nabla} \mathbf{a} \quad , \quad \nabla_j a_i = \frac{\partial a_i}{\partial x_j} \tag{A.49}$$

$$\operatorname{grad} \mathbf{A} = \boldsymbol{\nabla} \mathbf{A} \quad , \quad \nabla_k A_{ij} = \frac{\partial A_{ij}}{\partial x_k} \tag{A.50}$$

The gradient of a scalar, $\operatorname{grad} a$, is a vector that points into the direction of greatest change of the scalar. The gradient of a vector component, $\operatorname{grad} a_i$, is a vector that points in the direction of the greatest change of the vector component. The gradient of a vector, $\operatorname{grad} \mathbf{a}$ or $\frac{\partial a_i}{\partial x_j}$, is a tensor comprised of the gradient vectors of the components.

The **divergence** is defined for a vector as

$$\operatorname{div}\mathbf{a} = \operatorname{tr}(\operatorname{grad} \mathbf{a}) = \boldsymbol{\nabla} \cdot \mathbf{a} \quad , \quad \nabla_i a_i = \frac{\partial a_i}{\partial x_i} \tag{A.51}$$

and for a tensor as

$$\operatorname{div} \mathbf{A} = \nabla \cdot \mathbf{A}^T \quad , \quad \nabla_j A_{ij} = \frac{\partial A_{ij}}{\partial x_j} \tag{A.52}$$

Similar equations hold for gradient and divergence of higher order tensors, e.g. for $\mathbf{A} = \{A_{ijk}\}$; written in components

$$[\operatorname{grad} \mathbf{A}]_{ijkl} = \frac{\partial A_{ijk}}{\partial x_l} \quad , \quad [\operatorname{div} \mathbf{A}]_{ij} = [\nabla \cdot \mathbf{A}]_{ij} = \frac{\partial A_{ijk}}{\partial x_k} \tag{A.53}$$

This example shows that the index notation becomes particularly helpful when tensors of higher orders are considered. Indeed, in taking the divergence it is not always clear on which index the nabla operator should act. Index notation allows to distinguish easily between, e.g., $\frac{\partial A_{ijk}}{\partial x_k}$, $\frac{\partial A_{ijk}}{\partial x_j}$, $\frac{\partial A_{ijk}}{\partial x_i}$.

The Laplacian is defined, e.g., for a vector, as

$$\triangle \mathbf{a} = \operatorname{div} \operatorname{grad} \mathbf{a} = \nabla \cdot \nabla \mathbf{a} \quad , \quad \triangle a_i = \nabla_k \nabla_k a_i = \frac{\partial^2 a_i}{\partial x_k \partial x_k} \tag{A.54}$$

The curl of a vector is

$$\operatorname{curl} \mathbf{a} = \nabla \times \mathbf{a} \quad , \quad \varepsilon_{ijk} \nabla_j a_k = \varepsilon_{ijk} \frac{\partial a_k}{\partial x_j} \; . \tag{A.55}$$

A.11 Eigenvalues and Eigenvectors

Eigenvalues λ and eigenvectors $\mathbf{X}$ of a tensor are determined from the relation

$$\mathbf{A} \cdot \mathbf{X} = \lambda \mathbf{X} \quad , \quad A_{ij} X_j = \lambda X_i \tag{A.56}$$

That is the tensor A_{ij} stretches its eigenvectors $\mathbf{X}$ with factor λ, but leaves their direction unchanged. The relation can be rewritten as a homogeneous system for the eigenvectors,

$$\left(A_{ij} - \lambda \delta_{ij} \right) X_j = 0 \tag{A.57}$$

A homogeneous system only has solutions when the determinant vanishes, i.e., it must hold

$$\det \left(A_{ij} - \lambda \delta_{ij} \right) = 0 \tag{A.58}$$

This leads to the cubic equation

$$A - A' \lambda + A'' \lambda^2 - \lambda^3 = 0 \tag{A.59}$$

with the so-called invariants of the tensor

$$A = \det A_{ij} \; , \quad A' = \frac{1}{2} \left(\left(A_{ii} \right)^2 - A_{ii}^2 \right) \; , \quad A'' = A_{ii} \tag{A.60}$$

where $A_{ii}^2 = A_{ik} A_{ki}$ is the trace of the square of the tensor. Solution of the cubic equation gives three eigenvalues λ_i, and three corresponding eigenvectors $\mathbf{X}_i$.

A symmetric tensor has three real eigenvalues[3] λ_i with the corresponding real eigenvectors $\mathbf{X}^\alpha$, so that

---

[3] An antisymmetric tensor has purely imaginary eigenvalues, and a neither symmetric nor antisymmetric tensor has complex eigenvalues.

$$\mathbf{A} \cdot \mathbf{X}_i = \lambda_{\underline{i}} \mathbf{X}_i \quad \text{for} \quad i = 1, 2, 3 \tag{A.61}$$

where the underlined index indicates that no summation is implied.

Since the tensor $\mathbf{A}$ is symmetric, we have for any choice of i and j

$$\mathbf{X}_i \cdot \mathbf{A} \cdot \mathbf{X}_j = \mathbf{X}_j \cdot \mathbf{A} \cdot \mathbf{X}_i \quad \Rightarrow \quad \mathbf{X}_i \cdot \left(\lambda_{\underline{j}} \mathbf{X}_j \right) = \mathbf{X}_j \cdot \left(\lambda_{\underline{i}} \mathbf{X}_i \right) \tag{A.62}$$

and hence

$$\left(\lambda_{\underline{j}} - \lambda_{\underline{i}} \right) \mathbf{X}_i \cdot \mathbf{X}_j = 0 \tag{A.63}$$

where notation is such that no summation takes places over underlined indices. With this, in case that all eigenvalues are different, the three eigenvectors are orthogonal to each other. Note that the length of the $\mathbf{X}_i$ is not determined by the defining equation, hence they can (and should) be normalized. Normalized eigenvectors form a Cartesian base.

If two eigenvalues are equal, a complete orthonormal base can be constructed nevertheless, by choosing two perpendicular vectors that are perpendicular to the eigenvector of the single eigenvalue.

If all three eigenvalues agree, the tensor is isotropic and any orthonormal base is a set of eigenvectors.

The transformation between a Cartesian base $\mathbf{e}_i$ and the normalized eigensystem $\mathbf{X}_i$ is given through the transformation tensor X_{ij}, so that

$$\mathbf{X}_i = X_{ik} \mathbf{e}_k \ . \tag{A.64}$$

Scalar multiplication with $\mathbf{e}_j$ shows (with $\mathbf{e}_k \cdot \mathbf{e}_j = \delta_{kj}$) that

$$\mathbf{X}_i \cdot \mathbf{e}_j = X_{ij} \ . \tag{A.65}$$

Scalar multiplication of $\mathbf{X}_i$ and $\mathbf{X}_j$ shows that X_{ij} is orthogonal (assuming the $\mathbf{X}_i$ are normalized),

$$\delta_{ij} = \mathbf{X}_i \cdot \mathbf{X}_j = \left(X_{ik} \mathbf{e}_k \right) \cdot \mathbf{X}_j = X_{ik} \left(\mathbf{X}_j \cdot \mathbf{e}_k \right) = X_{ik} X_{jk} = X_{ik} X_{kj}^T \ . \tag{A.66}$$

Representing the tensor in the normalized eigensystem and the Cartesian frame we have

$$\mathbf{A} = \hat{A}_{ij} \mathbf{X}_i \mathbf{X}_j = A_{ij} \mathbf{e}_i \mathbf{e}_j \tag{A.67}$$

Scalar multiplication with one and two eigenvectors gives

$$\mathbf{A} \cdot \mathbf{X}_k = \lambda_{\underline{k}} \mathbf{X}_k = \hat{A}_{ik} \mathbf{X}_i = A_{ij} \mathbf{e}_i X_{kj} \tag{A.68}$$

$$\mathbf{X}_l \cdot \mathbf{A} \cdot \mathbf{X}_k = \lambda_{\underline{k}} \delta_{kl} = \hat{A}_{lk} = A_{ij} X_{li} X_{kj} = X_{li} A_{ij} X_{jk}^T \tag{A.69}$$

From these we recognize that in the eigensystem the (symmetric!) tensor is diagonal, with the eigenvalues as entries,

$$\hat{A}_{ij} = \lambda_i \delta_{ij} = \begin{bmatrix} \lambda_1 & 0 & 0 \\ 0 & \lambda_2 & 0 \\ 0 & 0 & \lambda_3 \end{bmatrix}_{ij} \quad . \tag{A.70}$$

The invariants of the tensor thus can be expressed through its eigenvalues,

$$\begin{aligned} A &= \det A_{ij} = \lambda_1 \lambda_2 \lambda_3 \ , \\ A' &= \frac{1}{2}\left((A_{ii})^2 - A_{kk}^2\right) = \lambda_1\lambda_2 + \lambda_1\lambda_3 + \lambda_2\lambda_3 \ , \\ A'' &= A_{ii} = \lambda_1 + \lambda_2 + \lambda_3 \ . \end{aligned} \tag{A.71}$$

A.12 Curvilinear Coordinates

Sometimes, due to the geometry of a problem, it is advantageous to use curvilinear coordinates. For the often used cases of cylindrical and spherical coordinates, some transformations and standard calculus operations are listed below.

A.12.1 Cylindrical Coordinates

Coordinates

$$\text{cylindrical: } \{r, \phi, z\} \quad , \quad \text{Cartesian: } \{x, y, z\}$$

with

$$x = r\cos\phi \quad , \quad y = r\sin\phi \quad , \quad z = z \tag{A.72}$$

$$r = \sqrt{x^2 + y^2} \quad , \quad \phi = \arctan\frac{y}{x} \quad , \quad z = z \tag{A.73}$$

Vector components

$$\text{cylindrical: } \{a_r, a_\phi, a_z\} \quad , \quad \text{Cartesian: } \{a_x, a_y, a_z\} \tag{A.74}$$

with

$$a_r = a_x\cos\phi + a_y\sin\phi \ , \quad a_\phi = -a_x\sin\phi + a_y\cos\phi \ , \quad a_z = a_z \tag{A.75}$$

$$a_x = a_r\cos\phi - a_\phi\sin\phi \ , \quad a_y = a_r\sin\phi + a_\phi\cos\phi \ , \quad a_z = a_z \tag{A.76}$$

Gradient of a scalar:

$$\operatorname{grad} a = \left\{\frac{\partial a}{\partial r}, \frac{1}{r}\frac{\partial a}{\partial \phi}, \frac{\partial a}{\partial z}\right\} \tag{A.77}$$

Divergence of a vector:

$$\operatorname{div}\mathbf{a} = \frac{\partial a_r}{\partial r} + \frac{a_r}{r} + \frac{1}{r}\frac{\partial a_\phi}{\partial \phi} + \frac{\partial a_z}{\partial z} \tag{A.78}$$

Curl of a vector:

$$\operatorname{curl}\mathbf{a} = \left\{ \frac{1}{r}\frac{\partial a_z}{\partial \phi} - \frac{\partial a_\phi}{\partial z}, \frac{\partial a_r}{\partial z} - \frac{\partial a_z}{\partial r}, \frac{\partial a_\phi}{\partial r} + \frac{a_\phi}{r} - \frac{1}{r}\frac{\partial a_r}{\partial \phi} \right\} \tag{A.79}$$

Gradient of a vector:

$$\operatorname{grad}\mathbf{a} = \left\{ \begin{array}{ccc} \frac{\partial a_r}{\partial r} & \frac{1}{r}\frac{\partial a_r}{\partial \phi} - \frac{a_\phi}{r} & \frac{\partial a_r}{\partial z} \\ \frac{\partial a_\phi}{\partial r} & \frac{1}{r}\frac{\partial a_\phi}{\partial \phi} + \frac{a_r}{r} & \frac{\partial a_\phi}{\partial z} \\ \frac{\partial a_z}{\partial r} & \frac{1}{r}\frac{\partial a_z}{\partial \phi} & \frac{\partial a_z}{\partial z} \end{array} \right\} \tag{A.80}$$

Divergence of a tensor:

$$\operatorname{div}\mathbf{A} = \left\{ \begin{array}{c} \frac{\partial A_{rr}}{\partial r} + \frac{1}{r}\frac{\partial A_{r\phi}}{\partial \phi} + \frac{\partial A_{rz}}{\partial z} + \frac{A_{rr} - A_{\phi\phi}}{r} \\ \frac{\partial A_{\phi r}}{\partial r} + \frac{1}{r}\frac{\partial A_{\phi\phi}}{\partial \phi} + \frac{\partial A_{\phi z}}{\partial z} + \frac{A_{\phi r} + A_{r\phi}}{r} \\ \frac{\partial A_{zr}}{\partial r} + \frac{1}{r}\frac{\partial A_{z\phi}}{\partial \phi} + \frac{\partial A_{zz}}{\partial z} + \frac{A_{zr}}{r} \end{array} \right\} \tag{A.81}$$

A.12.2 Spherical Coordinates

Coordinates:

$$\text{spherical: } \{r, \theta, \phi\} \quad , \quad \text{Cartesian: } \{x, y, z\}$$

with

$$x = r\sin\theta\cos\phi \quad , \quad y = r\sin\theta\sin\phi \quad , \quad z = r\cos\theta \tag{A.82}$$

$$r = \sqrt{x^2 + y^2 + z^2} \quad , \quad \theta = \arccos\frac{z}{\sqrt{x^2 + y^2 + z^2}} \quad , \quad \phi = \arctan\frac{y}{x} \tag{A.83}$$

Vector components

$$\{a_r, a_\theta, a_\phi\} \quad , \quad \{a_x, a_y, a_z\} \tag{A.84}$$

with

$$\begin{aligned} a_r &= a_x \sin\theta\cos\phi + a_y \sin\theta\sin\phi + a_z\cos\theta \\ a_\theta &= a_x\cos\theta\cos\phi + a_y\cos\theta\sin\phi - a_z\sin\theta \\ a_\phi &= -a_x\sin\phi + a_y\cos\phi \end{aligned} \tag{A.85}$$

$$\begin{aligned} a_x &= a_r \sin\theta \cos\phi + a_\theta \cos\theta \cos\phi - a_\phi \sin\phi \\ a_y &= a_r \sin\theta \sin\phi + a_\theta \cos\theta \sin\phi + a_\phi \cos\phi \\ a_z &= a_r \cos\theta - a_\theta \sin\theta \end{aligned} \tag{A.86}$$

Gradient of a scalar:

$$\operatorname{grad} a = \left\{ \frac{\partial a}{\partial r}, \frac{1}{r}\frac{\partial a}{\partial \theta}, \frac{1}{r\sin\theta}\frac{\partial a}{\partial \phi} \right\} \tag{A.87}$$

Divergence of a vector:

$$\operatorname{div} \mathbf{a} = \frac{\partial a_r}{\partial r} + \frac{2a_r}{r} + \frac{1}{r}\frac{\partial a_\theta}{\partial \theta} + \frac{a_\theta \cot\theta}{r} + \frac{1}{r\sin\theta}\frac{\partial a_\phi}{\partial \phi} \tag{A.88}$$

Curl of a vector:

$$\operatorname{curl} \mathbf{a} = \left\{ \begin{array}{c} \frac{1}{r}\frac{\partial a_\phi}{\partial \theta} + \frac{a_\phi \cot\theta}{r} - \frac{1}{r\sin\theta}\frac{\partial a_\theta}{\partial \phi} \\ \frac{1}{r\sin\theta}\frac{\partial a_r}{\partial \phi} - \frac{\partial a_\phi}{\partial r} - \frac{a_\phi}{r} \\ \frac{\partial a_\theta}{\partial r} + \frac{a_\theta}{r} - \frac{1}{r}\frac{\partial a_r}{\partial \theta} \end{array} \right\} \tag{A.89}$$

Gradient of a vector:

$$\operatorname{grad} \mathbf{a} = \left\{ \begin{array}{ccc} \frac{\partial a_r}{\partial r} & \frac{1}{r}\frac{\partial a_r}{\partial \theta} - \frac{a_\theta}{r} & \frac{1}{r\sin\theta}\frac{\partial a_r}{\partial \phi} - \frac{a_\phi}{r} \\ \frac{\partial a_\theta}{\partial r} & \frac{1}{r}\frac{\partial a_\theta}{\partial \theta} + \frac{a_r}{r} & \frac{1}{r\sin\theta}\frac{\partial a_\theta}{\partial \phi} - \frac{a_\phi \cot\theta}{r} \\ \frac{\partial a_\phi}{\partial r} & \frac{1}{r}\frac{\partial a_\phi}{\partial \theta} & \frac{1}{r\sin\theta}\frac{\partial a_\phi}{\partial \phi} + \frac{a_r}{r} + \frac{a_\theta \cot\theta}{r} \end{array} \right\} \tag{A.90}$$

Divergence of a tensor:

$$\operatorname{div} \mathbf{A} = \left\{ \begin{array}{c} \frac{\partial A_{rr}}{\partial r} + \frac{2A_{rr}}{r} + \frac{1}{r}\frac{\partial A_{r\theta}}{\partial \theta} + \frac{A_{r\theta}\cot\theta}{r} + \frac{1}{r\sin\theta}\frac{\partial A_{r\phi}}{\partial \theta} - \frac{A_{\theta\theta}+A_{\phi\phi}}{r} \\ \frac{\partial A_{\theta r}}{\partial r} + \frac{2A_{\theta r}+A_{r\theta}}{r} + \frac{1}{r}\frac{\partial A_{\theta\theta}}{\partial \theta} + \frac{1}{r\sin\theta}\frac{\partial A_{\theta\phi}}{\partial \phi} + \frac{\cot\theta}{r}\left(A_{\theta\theta} - A_{\phi\phi}\right) \\ \frac{\partial A_{\phi r}}{\partial r} + \frac{2A_{\phi r}+A_{r\phi}}{r} + \frac{1}{r}\frac{\partial A_{\phi\theta}}{\partial \theta} + \frac{1}{r\sin\theta}\frac{\partial A_{\phi\phi}}{\partial z} + \frac{\cot\theta}{r}\left(A_{\phi\theta} - A_{\theta\phi}\right) \end{array} \right\} \tag{A.91}$$

Problems

A.1. Tensor Index Notation

Write explicitly each component of

$$\text{a) } B_{ij}v_j \quad \text{b) } v_i w_j + \sigma\delta_{ij} \quad \text{c) } \frac{\partial f_k}{\partial x_l} \quad \text{d) } \frac{\partial A_{ij}}{\partial x_j}$$

A.2. Tensor Index Notation

Write the equations below in direct and index notation. Simplify as much as possible.

a) $\alpha = a_i b_k a_i c_k$	b) $\mathbf{C} = \mathbf{A}^T \cdot \mathbf{B}$	c) $C_{ij} = A_{ki} B_{jk}$
d) $u_j = A_{ijk} v_i w_k$	e) $\mathbf{w} = \mathbf{v} \cdot \mathbf{B}$	f) $a = \varepsilon_{ijk} u_i v_j w_k$
g) $v_k \frac{\partial v_i}{\partial x_k}$	h) $v_k \frac{\partial v_k}{\partial x_i}$	i) $h_i = \varepsilon_{ijk} \varepsilon_{klm} g_m \frac{\partial f_l}{\partial x_j}$
j) $\gamma = \text{curl}\,(\text{grad}\,\varphi)$	k) $\mathbf{b} = \text{curl}\,\text{curl}\,\mathbf{v}$	

A.3. Symmetric and Antisymmetric Tensors

Recall the definitions for symmetric part, anti-symmetric part, trace-free and symmetric part, and transposed tensors:

$$A_{(ij)} = \frac{1}{2}(A_{ij} + A_{ji}) \quad , \; A_{[ij]} = \frac{1}{2}(A_{ij} - A_{ji}) \; ,$$
$$A_{\langle ij \rangle} = A_{(ij)} - \frac{1}{3} A_{kk} \delta_{ij}, \; A^T_{ij} = A_{ji}$$

Show that

1. $\delta_{\langle ij \rangle} = \delta_{[ij]} = 0$
2. $A_{(ij)} B_{[ij]} = 0$
3. $A_{\langle ij \rangle} B_{[ij]} = 0$
4. $A_{\langle ij \rangle} B_{ij} = A_{\langle ij \rangle} B_{\langle ij \rangle} = A_{ij} B_{\langle ij \rangle}$
5. $(\mathbf{A} \cdot \mathbf{B})^T = \mathbf{B}^T \cdot \mathbf{A}^T$
6. $\mathbf{A} \cdot \mathbf{v} = \mathbf{v} \cdot \mathbf{A}^T$
7. $\left(\mathbf{A}^{-1}\right)^T = \left(\mathbf{A}^T\right)^{-1}$

A.4. Cross Product

The completely antisymmetric symbol of third order is given by

$$\varepsilon_{ijk} = \begin{cases} 1 \;\text{ if } & ijk = 123/312/231 \\ -1 \text{ if } & ijk = 321/132/213 \\ 0 \;\text{ else} & \end{cases}$$

where exchange of two indices leads changes the sign, e.g., $\varepsilon_{ijk} = -\varepsilon_{jik} = \varepsilon_{jki}$

The cross product (or vector product) of two vectors a_i and b_i is

$$\mathbf{a} \times \mathbf{b} = \mathbf{c} \qquad \Leftrightarrow \qquad \varepsilon_{ijk} a_j b_k = c_i$$

1. Express the three components of c_i explicitly through the components of a_i and b_i.
2. Use properties of ε_{ijk} to show that $\mathbf{b} \times \mathbf{a} = -(\mathbf{a} \times \mathbf{b})$.
3. Use antisymmetry of ε_{ijk} to show that the resulting vector c_i is perpendicular to a_i and b_i, that is the scalar products $a_i c_i$ and $b_i c_i$ both vanish.

A.5. Properties of Epsilon-Tensor

1. Show that $\varepsilon_{ijk} \varepsilon_{lmk} = \delta_{il} \delta_{jm} - \delta_{im} \delta_{jl}$.
2. Determine $\varepsilon_{ijk} \varepsilon_{ljk}$.
3. Determine $\varepsilon_{ijk} \varepsilon_{ijk}$.
4. Verify that $\mathbf{a} \times (\mathbf{b} \times \mathbf{c}) = \mathbf{b}(\mathbf{a} \cdot \mathbf{c}) - \mathbf{c}(\mathbf{a} \cdot \mathbf{b})$

5. Show that the vector $\bar{A}_k$ associated with the antisymmetric tensor as $A_{[ij]} = \varepsilon_{ijk}\bar{A}_k$ is given by $\bar{A}_k = \frac{1}{2}\varepsilon_{ijm}A_{[ij]}$.

A.6. Coordinate Transforms

Under a coordinate transform from base $\mathbf{e}_i$ to base $\hat{\mathbf{e}}_i = O_{ij}\mathbf{e}_j$ with the orthogonal tensor O_{ik} the components of a tensor transform as $\hat{A}_{ij} = O_{ik}O_{jl}A_{kl}$.

Show that:

1. The trace of the tensor and the trace of the square of the tensor are invariant against transformation, that is $\hat{A}_{kk} = A_{kk}$, $\hat{A}_{ij}\hat{A}_{ji} = A_{ij}A_{ji}$.
2. The transform of the symmetric part of a tensor is symmetric, that is if $\hat{A}_{ij}^{(s)} = O_{ik}O_{jl}A_{(kl)}$ then $\hat{A}_{ij}^{(s)} = \hat{A}_{ji}^{(s)}$.
3. The transform of the deviatoric part of a tensor is symmetric and tracefree, that is if $\hat{A}_{ij}^{(d)} = O_{ik}O_{jl}A_{\langle kl\rangle}$ then $\hat{A}_{ij}^{(d)} = \hat{A}_{ji}^{(d)}$ and $\hat{A}_{kk}^{(d)} = 0$.
4. The transform of the anti-symmetric part of a tensor is anti-symmetric, that is if $\hat{A}_{ij}^{(a)} = O_{ik}O_{jl}A_{[kl]}$ then $\hat{A}_{ij}^{(a)} = -\hat{A}_{ji}^{(a)}$.

A.7. Velocity Gradient

The velocity field of a flow is, with $x_1 = x$, $x_2 = y$, $x_3 = z$,

$$v_i\left(x,y,z\right) = \left\{x^2y,\ xy^2,\ 0\right\}_i$$

Determine in matrix form:

1. Velocity gradient $\frac{\partial v_i}{\partial x_j}$ and its symmetric and antisymmetric parts, $\frac{\partial v_{(i}}{\partial x_{j)}}$, $\frac{\partial v_{[i}}{\partial x_{j]}}$
2. Divergence of velocity, $\frac{\partial v_k}{\partial x_k}$, and the trace-free and symmetric part of the gradient, $\frac{\partial v_{\langle i}}{\partial x_{j\rangle}}$
3. Curl of velocity $\varepsilon_{ijk}\frac{\partial v_i}{\partial x_k}$
4. Internal energy generation rate $\pi_u = \lambda\left(\frac{\partial v_k}{\partial x_k}\right)^2 + 2\eta\frac{\partial v_{\langle i}}{\partial x_{j\rangle}}\frac{\partial v_{\langle i}}{\partial x_{j\rangle}}$.

A.8. Temperature and Velocity Derivatives

The temperature and velocity fields of a flow are, with $x_1 = x$, $x_2 = y$, $x_3 = z$,

$$T\left(t,x,y,z\right) = T_0\left(1 - \frac{x^2yz}{L^4}\right)e^{-t/\tau} \quad , \quad \mathbf{v}\left(x,y,z,t\right) = \frac{v_0}{L^2}\left\{xy, -y^2,\ 0\right\}$$

where T_0, v_0, τ, L are constants.

Write explicitly in all components:

1. Temperature gradient $\frac{\partial T}{\partial x_i}$
2. Material derivative of temperature $\frac{DT}{Dt}$
3. Divergence of velocity $\frac{\partial v_k}{\partial x_k}$

4. Trace-free and symmetric part of the velocity gradient $\frac{\partial v_{\langle i}}{\partial x_{j\rangle}}$

A.9. Tensor Identities

Use index notation to show that

1. For any f:

$$\operatorname{curl}(\operatorname{grad} f) = 0 \;, \qquad \operatorname{div}(\operatorname{curl} f) = 0$$

2. For any scalar a

$$\operatorname{curl}\operatorname{curl} a = 0 \;,$$

3. For any vector $\mathbf{a}$

$$\mathbf{a}\cdot\operatorname{grad}\mathbf{a} = \frac{1}{2}\operatorname{grad} a^2 - \mathbf{a}\times\operatorname{curl}\mathbf{a} \;.$$

A.10. Vorticity

Show that the vorticity vector $\boldsymbol{\omega} = \operatorname{curl}\mathbf{v}$, in index notation $\omega_i = \varepsilon_{ijk}\frac{\partial v_k}{\partial x_j}$, obeys

$$\frac{\partial \mathbf{v}}{\partial t} + \boldsymbol{\omega}\times\mathbf{v} + \frac{1}{2}\boldsymbol{\nabla} v^2 = \frac{D\mathbf{v}}{Dt} \qquad \text{or} \qquad \frac{\partial v_i}{\partial t} + \varepsilon_{ijk}\omega_j v_k + \frac{1}{2}\frac{\partial v^2}{\partial x_i} = \frac{Dv_i}{Dt}$$

where $v^2 = \mathbf{v}\cdot\mathbf{v} = v_i v_i$ and $\frac{Dv_i}{Dt} = \frac{\partial \mathbf{v}}{\partial t} + v_k\frac{\partial v_i}{\partial x_k}$ is the convective derivative.

A.11. Transforms of Derivatives

The orthogonal tensor O_{ik} transforms from base $\mathbf{e}_i$ to base $\hat{\mathbf{e}}_i = O_{ij}\mathbf{e}_j$. Show that:

1. The coordinates of the location vector $\mathbf{x}$ transform as $\hat{x}_k = O_{kj}x_j$ and $x_j = O_{ij}\hat{x}_i$.
2. The coordinates of the gradient of a scalar a transform as $\frac{\partial a}{\partial \hat{x}_i} = O_{ij}\frac{\partial a}{\partial x_j}$ and $\frac{\partial a}{\partial x_i} = O_{ji}\frac{\partial a}{\partial \hat{x}_j}$
3. The coordinates of the gradient of a vector a_i transform as $\frac{\partial \hat{a}_i}{\partial \hat{x}_j} = O_{ik}O_{jl}\frac{\partial a_k}{\partial x_l}$ and $\frac{\partial a_i}{\partial x_j} = O_{ki}O_{lj}\frac{\partial a_k}{\partial x_l}$.
4. The divergence of the vector a_i is invariant, $\frac{\partial \hat{a}_k}{\partial \hat{x}_k} = \frac{\partial a_k}{\partial x_k}$.

A.12. Invariants

Prove that for a symmetric 3×3 tensor $\mathbf{A}$ the invariants—i.e. the factors in the characteristic equation $\det(\mathbf{A} - \lambda\mathbf{1}) = A - A'\lambda + A''\lambda^2 - \lambda^3 = 0$—are

$$\begin{aligned} A &= \det\mathbf{A} = \lambda_1\lambda_2\lambda_3 \\ A' &= \frac{1}{2}\left((\operatorname{tr}\mathbf{A})^2 - \operatorname{tr}\left(\mathbf{A}^2\right)\right) = \lambda_1\lambda_2 + \lambda_1\lambda_3 + \lambda_2\lambda_3 \\ A'' &= \operatorname{tr}\mathbf{A} = \lambda_1 + \lambda_2 + \lambda_3 \end{aligned}$$

Also, show that

$$\det \mathbf{A} = \frac{1}{6}(\mathrm{tr}\mathbf{A})^3 - \frac{1}{2}(\mathrm{tr}\mathbf{A})\,\mathrm{tr}\left(\mathbf{A}^2\right) + \frac{1}{3}\mathrm{tr}\mathbf{A}^3$$

Hint: Consider the relations in the eigensystem of $\mathbf{A}$, with the eigenvalues λ_1, λ_2, λ_3.

A.13. Cayley-Hamilton Theorem

The Cayley-Hamilton Theorem states that a tensor $\mathbf{A}$ obeys its own characteristic equation (see previous problem). Show that for a (symmetric) tensor in $\mathbb{R}^3$

$$A\delta_{ij} - A' A_{ij} + A'' \left(\mathbf{A}^2\right)_{ij} - \left(\mathbf{A}^3\right)_{ij} = 0$$

and

$$A\left(\mathbf{A}^n\right)_{ij} - A'\left(\mathbf{A}^{n+1}\right)_{ij} + A''\left(\mathbf{A}^{n+2}\right)_{ij} - \left(\mathbf{A}^{n+3}\right)_{ij} = 0$$

where $\mathbf{A}^n$ is the n-th power of the tensor $\mathbf{A}$. Note: n can be positive or negative, as long as the inverse exists.

What is the corresponding equation in $\mathbb{R}^2$?

A.14. Euclidian Transformation

The coordinates in two frames are related as

$$x_i^* = O_{ij}(t)\, x_j + b_i(t)$$

where $O_{ij}(t)$ is a time dependent orthogonal matrix ($O_{ik}O_{jk} = \delta_{ij}$) which describes the relative rotation of the axes of the two frames, and $b_i(t)$ is a time-dependent vector that determines the relative motion of the origins of the two frames. Below, time derivatives are written as $\frac{d}{dt}O_{ij} = \dot{O}_{ij}$, $\frac{d}{dt}b_i = \dot{b}_i$.

1. Show that the matrix of the relative angular velocity

$$W_{ij} = \dot{O}_{ik}O_{jk}\,,$$

is antisymmetric.

2. Show that velocity transforms as

$$v_i^* = O_{ij}v_j + \dot{O}_{ij}x_j + \dot{b}_i$$

and

$$v_k = O_{ik}\left(v_i^* - W_{il}\left(x_l^* - b_l\right) - \dot{b}_i\right)$$

3. Show that the velocity gradient transforms as

$$\frac{\partial v_i^*}{\partial v_j^*} = O_{ik}O_{jl}\frac{\partial v_k}{\partial x_l} + W_{ij}$$

and that the symmetric and antisymmetric parts of the velocity gradient transform as

$$\frac{\partial v^*_{(i}}{\partial v^*_{j)}} = O_{ik}O_{jl}\frac{\partial v_{(k}}{\partial x_{l)}} \quad , \quad \frac{\partial v^*_{[i}}{\partial v^*_{j]}} = O_{ik}O_{jl}\frac{\partial v_{[k}}{\partial x_{l]}} + W_{ij}$$

4. Show that acceleration transforms as

$$\dot{v}^*_i = O_{ij}\dot{v}_j + 2\dot{O}_{ij}v_j + \ddot{O}_{ij}x_j + \ddot{b}_i$$

or

$$\dot{v}^*_i = O_{ij}\dot{v}_j + 2W_{ik}\left(v^*_k - \dot{b}_k\right) - W^2_{ik}\left(x^*_k - b_k\right) + \dot{W}_{ik}\left(x^*_k - b_k\right) + \ddot{b}$$

and

$$\dot{v}_j = O_{ij}\left[\dot{v}^*_i - 2W_{ik}\left(v^*_k - \dot{b}_k\right) + W^2_{ik}\left(x^*_k - b_k\right) - \dot{W}_{ik}\left(x^*_k - b_k\right) - \ddot{b}_i\right]$$

Identify:
Coriolis acceleration $2W_{ik}\left(v^*_k - \dot{b}_k\right)$,
centrifugal acceleration $-W^2_{ik}\left(x^*_k - b_k\right)$,
Euler acceleration $\dot{W}_{ik}\left(x^*_k - b_k\right)$,
acceleration of relative translation $\ddot{b}$.

5. Introduce the vector of angular acceleration $\mathbf{w}$ as $W_{ij} = \varepsilon_{ijs}w_s$and show that the above can be written as

$$\dot{\mathbf{v}}^* = \mathbf{O}\cdot\dot{\mathbf{v}} - 2\mathbf{w}\times\left(\mathbf{v}^* - \dot{\mathbf{b}}\right) - \mathbf{w}\times\left(\mathbf{w}\times\left(\mathbf{x}^* - \mathbf{b}\right)\right) - \dot{\mathbf{w}}\times\left(\mathbf{x}^* - \mathbf{b}\right) + \ddot{\mathbf{b}}_i$$

A.15. Transformation between two NIF

Consider two non-inertial frames, with the Euclidian transformations against an inertial frame (x_i) given by

$$x^*_i = O^*_{ij}x_j + b^*_i \quad , \quad x^{**}_i = O^{**}_{ij}x_j + b^{**}_i \; .$$

Show that the transformation between the two NIFs is an Euclidian transformation as well, i.e. it can be written as

$$x^{**}_i = Q^*_{ij}x^*_j + B_i$$

Identify Q^*_{ij}, B_i and show that Q^*_{ij} is an orthogonal matrix.

A.16. Non-inertial Forces

Consider the relations of Prob. A.14, and assume the starred system is a non-inertial frame (NIF), while the un-starred system is an inertial frame (IF).

1. Determine the actual force acting on a mass point that rests in NIF. Relate to your own experience as passenger in a car on the highway or a curvy mountain road, and a carousel.

2. For an observer resting in NIF, determine the apparent force acting on a mass point resting in IF.
3. When travelling in, e.g., a train, did you ever experience the feeling at a station that your train was accelerating, only to then notice that it was the train on the other platform that began to move? Relate this experience to the transformation between IF and NIF.
4. Consider that NIF is a disk rotating around the origin of the origin of IF. Determine the force on a mass point that in NIF moves on a straight line outward from the origin.
5. Discuss centrifugal and centripetal forces.

A.17. Balloon in Accelerating Car

A helium filled balloon is attached to the bottom of an accelerating car by a string. The mass of the empty balloon is m_B, the mass of helium in the balloon is m_{He}, and the filled balloon has the volume V_B. Set up equations for the force in the string, and the angle α between string and bottom, starting from the balance of momentum in integral form.

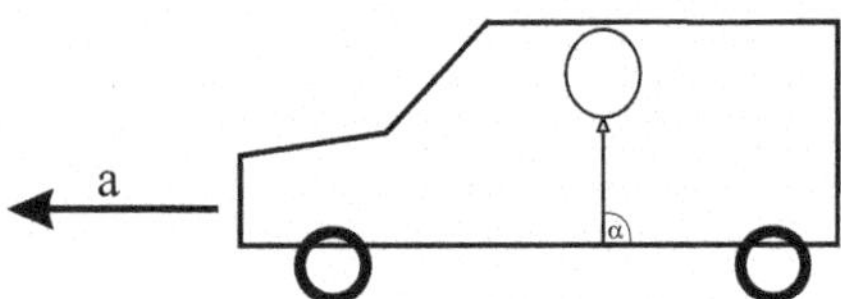

A.18. Cylindrical Coordinates

Use the listed relations for cylindrical coordinates to determine general expressions for

1. $\triangle a = \operatorname{div} \operatorname{grad} a$
2. $\operatorname{curl} \operatorname{grad} a$
3. $\operatorname{div} \operatorname{grad} \mathbf{a}$
4. Simplify the above for the case that scalar a and vector $\mathbf{a}$ depend only on radius r.
5. Write the Navier-Stokes Fourier equations in cylindrical coordinates.

A.19. Spherical Coordinates

Use the listed relations for spherical coordinates to determine general expressions for

1. $\triangle a = \operatorname{div} \operatorname{grad} a$
2. $\operatorname{curl} \operatorname{grad} a$
3. $\operatorname{div} \operatorname{grad} \mathbf{a}$
4. Simplify the above for the case that scalar a and vector $\mathbf{a}$ depend only on radius r.
5. Write the Navier-Stokes Fourier equations in spherical coordinates.

A.20. Gas in a Centrifuge

Consider an ideal gas in a cylinder rotating around its axis with frequency ω, as in Fig. 9.1, where the gas is at rest relative to the cylinder walls.

1. Show that for rotation around the z-axis the Euclidian transformation is given by $O_{ik} = \begin{Bmatrix} \cos\omega t & -\sin\omega t & 0 \\ \sin\omega t & \cos\omega t & 0 \\ 0 & 0 & 1 \end{Bmatrix}$
2. Show that O_{ik} is orthogonal, that is $O_{ik}^{-1} = O_{ki}$.
3. Determine the matrix of relative angular velocity $W_{ij} = \dot{O}_{ik}O_{jk}$, show that it is antisymmetric, and determine the vector of angular acceleration Ω_k, defined through $W_{ij} = \varepsilon_{ijk}\Omega_k$.
4. Use the transformations of Prob. A.14, to formulate the balance of momentum for an observer in the rotating cylinder, with the apparent body force
$$f_i^{\omega} = 2W_{ik}\left(v_k^* - \dot{b}_k\right) - W_{ik}^2\left(x_k^* - b_k\right) + \dot{W}_{ik}\left(x_k^* - b_k\right) + \ddot{b}$$
simplify for $b_i = 0$, $\omega = const.$
5. Introduce cylindrical coordinates and determine the pressure $p(r)$ as function of radius in the ideal gas in dependence of ω.

A.21. Non-inertial Forces in Fluid Flow

Consider the following system: water enters a rotating pipe of length L at the origin, flows to the end and back, and leaves again at the origin. This is the basic setup for the Centrifugal Membrane and Density Separation method [J.G. Pharoah, N. Djilali, G.W. Vickers, J. Membrane Sci. **176**,277–289 (2000)].

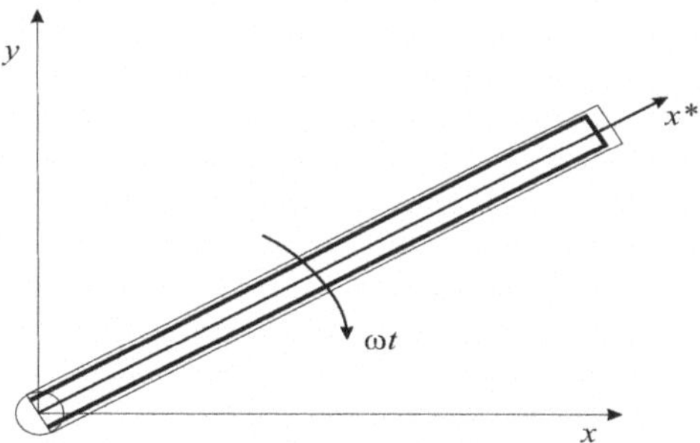

The diameter of the pipe is constant, and the water can be considered as being incompressible; viscous effects can be ignored, and temperature is constant.

1. Find the transformation matrix O_{ij}, the matrix of angular velocities W_{ij} and the vector of angular velocities ω_i. Also show that O_{ij} is orthogonal.
2. Formulate the balances of mass and momentum for an observer in the NIF, i.e., for an observer resting in the frame of the pipe.

3. Determine the pressure as a function of location x^* and flow velocity $v_i^* = \{v_{x^*}, v_{y^*}, 0\}$. Where is the pressure largest? Can the Coriolis force be used to increase the pressure?
4. Assume that the same amount of water leaves and enters at the origin: how much work does it take to run the process. What if some or all of the fluid leaves through a membrane at the end of the pipe? Discuss!

Index

H. Struchtrup, *A Thermodynamic Introduction to Transport Phenomena*,
https://doi.org/10.1007/978-3-031-61868-0

The manufacturer's authorised representative in the EU is Springer Nature Customer Service Centre GmbH, Europaplatz 3, 69115 Heidelberg, Germany. If you have any concerns regarding our products, please contact ProductSafety@springernature.com

Printed and bound by CPI Group (UK) Ltd, Croydon, CR0 4YY
15/07/2026
02167633-0002